£54.00

Principles of Modern Radar Systems

The Artech House Radar Library

Arams, F.R., ed.	Infrared-to-Millimeter Wavelength Detectors
Banakh, V.A. and V.L. Mironov	Lidar in a Turbulent Atmosphere
Barton, D.K.	Modern Radar System Analysis
Barton, D.K. and H.R. Ward	Handbook of Radar Measurement
Beckmann, P. and A. Spizzichino	The Scattering of Electromagnetic Waves from Rough Surfaces
Blackman, S.S.	Multiple-Target Tracking with Radar Applications
Blake, L.V.	Radar Range-Performance Analysis
Brookner, E., ed.	Radar Technology
Currie, N.C. and C.E. Brown, eds	Principles and Applications of Millimeter-Wave Radar
DiFranco, J.V. and W.L. Rubin	Radar Detection
Erst, S.J.	Receiving System Design
Fielding, J.E. and G.D. Reynolds	RGCALC: Radar Range Detection Software and User's Manual
Hovanessian, S.A.	Radar System Design and Analysis
Hughes, R.S.	Logarithmic Amplification
Knott, E.F., J.F. Shaeffer and M.T. Tuley	Radar Cross Section
Kolosov, A.A. et al	Over the Horizon Radar
Leonov, A.I. and K.I. Fomichev	Monopulse Radar
Lewis, B., F. Kretschmer and W. Shelton	Aspects of Radar Signal Processing
Maksimov, M.V. et al	Radar Anti-Jamming Techniques
Meeks, M.L.	Radar Propagation at Low Altitudes
Mensa, D.L.	High Resolution Radar Imaging
Ostroff, E.D. et al	Solid-State Radar Transmitters
Ostrovityanov, R.V. and F.A. Basalov	Statistical Theory of Extended Radar Targets
Schleher, D.C.	Introduction to Electronic Warfare
Sherman, S.M.	Monopulse Principles and Techniques
Skillman, W.A.	SIGCLUT: Surface and Volumetric Clutter-to-Noise, Jammer and Target Signal-to-Noise Radar Calculation Software and User's Manual
Stevens, M.C.	Secondary Surveillance Radar
Torrieri, D.J.	Principles of Secure Communication Systems
Wehner, D.R.	High Resolution Radar
Wiley, R.G.	Electronic Intelligence: The Analysis of Radar Signals
Wiley, R.G.	Electronic Intelligence: The Interception of Radar Signals
Wiley, R.G. and M.B. Szymanski	Pulse Train Analysis Using Personal Computers

Principles of Modern Radar Systems

Michel H. Carpentier

Artech House
London and Boston

British Library Cataloguing in Publication Data

Carpentier, Michel H.
Principles of modern radar systems.

1. Radar systems
I. Title II. Radars bases modernes.
621.3848′5

ISBN 0-89006-285-4

Library of Congress Cataloging-in-Publication Data

Carpentier, Michel H.

Principles of modern radar systems.
Translation of: Radar bases modernes.
Bibliography: p.
Includes index.
1. Radar. I. Title.

TK6575.C3613 1988 621.3848 88-10563

ARTECH HOUSE, INC.
685 Canton Street
Norwood, MA 02062, USA

International Standard Book Number: 0-89006-285-4
Library of Congress Catalog Card Number: 88-10563

Translation from the French of *Radars bases modernes*, copyright 1984 by Masson, Paris.

Contents

Preface

Principles of Modern Radar Systems is the latest version of a book originally titled *Radars—Philosophie et Principes* which was first published in 1961 and was available on a limited scale, primarily for students of the French Ecole Nationale Supérieure de l'Aéronautique.

The book presented the principles of modern radar theory and their first applications at a time when practical implementation was in its early stages. The first pulse compression radar systems were being designed in the USA in the late 1950s and the first experiments on pulse compression and randomly phase-coded radar systems were being carried out in Europe in 1959 and 1960.

The success of *Philosophie et Principes* led to the publication of a library edition at the end of 1962 and a Russian translation in 1965. A revised version, *Radars—Concepts Nouveaux* published in 1966, included a number of results from experimentation with modern radar prototypes.

Subsequent editions have been extensively revised and updated in the light of rapid advances in this field and the current English language version is a translation of the fifth French edition, *Radars—Bases Modernes*.

It is not the purpose of this book to give detailed descriptions of all types of existing radar systems, rather it aims to explain, in a tutorial manner, the principles of several radar architectures, the methods of and reasons for their introduction and the theoretical and technological limitations in radar performance.

The first two chapters cover the minimum amount of mathematics that is required for an understanding of the ideal receiver, with exercises to assist the reader. Chapter 3 defines the theoretical limitations of radar performance in target detection and in range and velocity measurement. Applications to existing radar systems are covered, with practical examples, in chapter 4. Chapter 5 describes target reflection and analyzes echo fluctuation, providing solutions to fluctuation problems. Echo fluctuation is also explored in chapter 6, within the context of angle measurement and its performance limitations.

Chapter 7 covers the technique of integrating several consecutive radar measurements, a method used in most radar systems, both coherent and non-coherent. Finally, chapter 8 discusses the advantages and limitations of electronically scanned phased-arrays.

I welcome the publication of Artech's English edition of my book and I am grateful to my good friend Dave Barton for his efficient help in its preparation. I send my best regards to all English-speaking friends who use this book, and to all readers who belong to the congenial and influential radar community.

MICHEL H. CARPENTIER
Paris, March 1988

Chapter 1

Introduction to Random Functions

The basic properties of random functions which are required to follow the arguments in the following chapters are presented in this chapter. There is no intention whatsoever of presenting the mathematics for the sake of its intrinsic beauty. Therefore the exposition is seldom rigorous; we are aware of this and do not regret it in the slightest provided that this approach makes random functions less forbidding. The fact still remains that this chapter (and therefore the entire book) would not be possible if there had not been "pure" mathematicians who were totally dedicated to the rigor of definitions and proofs and who validated the "mathematical massacre" which follows.

1.1 General considerations. Steady state conditions

In physics we often arrive at the concept of a random function via the concept of noise. How can we define noise? One possible definition is as follows: noise is a random and disturbing phenomenon which accompanies useful information and degrades it. This definition implies that noise is a subjective concept: what is noise for one person may not be noise for another.

This is obvious. A lecture may be recorded by a microphone in the form of voltage as a function of time:

$$V = f(t)$$

Now, if the curve $V = f(t)$ corresponding to this lecture is considered, it seems perfectly random. However, when it is transformed by an amplifier and loudspeaker and heard by the lecturer who knows what he is saying, the curve becomes a much less random phenomenon. The lecture is less random for the student who knows English than for someone who does not know English. The same phenomenon may not be random for everybody. Similarly, jamming may be random for the person jammed or the jammed system but it may not be random for the person or system which is carrying out the jamming.

In a similar manner (and often in a corollary manner) a phenomenon may be disturbing for some and not for others: an exposition delivered in a lecture theatre is not disturbing for interested students but is disturbing for students who are gossiping. Similarly, the gossip which some students are exchanging during the lecture is not random for those who are passing it on, is random for those who are listening (otherwise it would not be gossip), is disturbing for the lecturer and is not disturbing for the students who are listening to the gossip and not to the lecture.

A phenomenon which exhibits random characteristics does not follow the "laws" of chance exactly, in the same way as a number appears in roulette, but it is not exactly predictable. This is logical since we have seen that phenomena can have different degrees of randomness. Let us take the distance D between an individual and his house as an example. This distance is a random phenomenon as a function of time. However, it has a certain number of properties.

(a) First, D is always positive. The probability that D is negative is zero by definition. Furthermore, the probability that D lies between 1000 and 1001 km at any time in the future is generally smaller than the probability that D lies between 0 and 1 km. The probability that D lies between 50 000 and 50 001 km is generally extremely low if not zero. The mathematical expression of these characteristics, which indicate the extent of the variations of D, gives the amplitude distribution of D.

(b) If it is known that $D = 15$ km at time $t = 0$, the probability that D lies between 14.9 and 15.1 km at $t = 1$ s is generally higher than the probability that D lies between 15.5 and 15.7 km (still at $t = 1$ s), especially if the person is not a pilot of a modern airplane, and the probability that D lies between 30.0 and 30.2 km (at $t = 1$ s) is almost zero. Furthermore, the value of D generally becomes zero at a rather regular rate, once or twice a day. The mathematical expression of these characteristics, which indicate the rate or the speed at which D varies, gives the frequency distribution of D or, better, the spectral density of D, which is related in a reciprocal manner to the autocorrelation function of D.

It can therefore be seen that if the variation in D as a function of time is random (it is not possible to predict the value of D exactly a long time in advance), it will be governed by "statistical" laws. In this case these laws depend on the profession of the person, his character and his age. They are quite well defined for a short interval of time (e.g. a month) during which the phenomenon is known as steady state, but they are often much less well defined over a long interval (e.g. 10 years) because the person may change house, change profession etc.

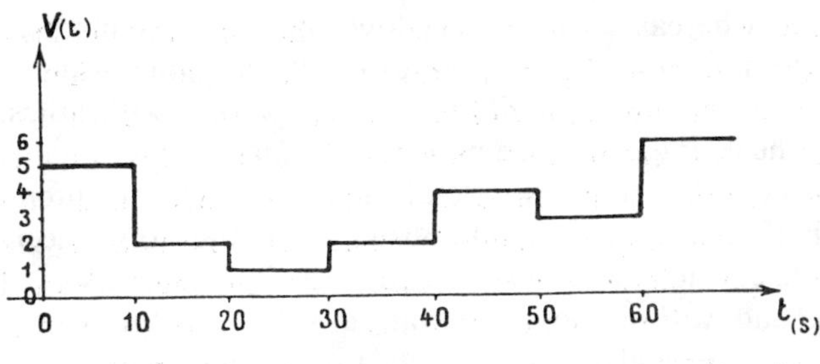

Fig. 1.1

The concept of steady state conditions of a random phenomenon can be better understood using examples.

(1) Let us consider a random function obtained in the following manner: a perfect die is thrown every 10 s and the number which turns up gives the value of V. In this way a discontinuous function of time is obtained (Fig. 1.1). The amplitude distribution here is simple: it is expressed by saying that all integral values between 1 and 6 have a probability of existence of 1/6 and that all other values of V have a zero probability of existence. The spectral density of $V(t)$ is less simple to calculate. However, if the interval of 10 s were replaced by an interval of 1 s, the amplitude distribution would be unchanged whereas the spectral density would indicate a higher rate in the ratio 1 : 10. As long as the die remains perfect and the sampling impartial, and as long as the clock giving the time sequence keeps the same time, $V(t)$ remains a steady state function. However, if over time the center of gravity of the die is displaced towards the face bearing the number one, for example, the number six will gradually become less probable and we can no longer assume that $V(t)$ remains a steady state function. Similarly, if the clock regularly slows down, the phenomenon can no longer be assumed to be steady state. If the displacement of the center of gravity is very slow with respect to 10 s, so that it takes a day, for example, for the probability that a six is thrown to go from 1/6 to 1.05/6 or it takes a day for the time interval to increase from 10 s to 10.5 s, it can generally be assumed that the phenomenon is sufficiently steady state over an interval of 1 h.

(2) Let us consider a directional antenna (e.g. a search radar antenna) rotating regularly around a vertical axis at a speed of 6 rev min^{-1} in the presence of a transmitter at some distance from it which is transmitting a perfectly steady state noise. It is clear that the antenna gain in the direction of the transmitter will vary as a function of time. In particular, the average power received in the antenna will be high when the radar beam passes over

the transmitter, whereas it will be very low when the antenna beam "looks" in the opposite direction. To be more exact, if the radar beam has a width of 1° in azimuth, the antenna will have a leaf pattern with lobes of similar width so that the average received power will remain approximately constant for a time of the order of magnitude of the time which the antenna takes to turn through 1°, i.e. approximately 30 ms. Thus the noise received by the antenna can be considered as steady state over time intervals of the order of 1 ms or less, and within these intervals it will have the same amplitude distribution and spectral density as the transmitted noise (except for the coefficient to allow for the attenuation). For time intervals of the order of 0.1 s, the noise received by the antenna is not steady state.

However, the problem is not so simple. If we consider the few seconds during which the antenna is "looking" in the opposite direction to the transmitter, in certain applications the pattern of the corresponding antenna lobes (backlobes) in the direction of the transmitter is random. It is difficult to make appropriate measurements from which this pattern can be obtained, and the pattern is not known *a priori* because it depends on the antenna and the landscape surrounding it. However, the statistical characteristics (amplitude distribution and spectral density) are known for such antennas in such landscapes. On the time scale of a few seconds and in the absence of anything better, the noise is assumed to be steady state in the calculations but with an amplitude distribution and a spectral density which take into account the transmitted noise and the rear lobes. If we now take the mean power of the received noise over a period of 1 s (by envelope detection and filtering the signal in a receiver with very narrow bandwidth of the order of 1 Hz), we obtain an almost periodic signal with a period of 10 s (the antenna rotation period) which is clearly not random.

In the light of the above considerations in the following section we consider only steady state random functions, i.e. functions where all the statistical properties are invariant for any change in the origin of time. We also accept the ergodic hypothesis in which it is assumed that the averages over time measured on one sample of the random function are the same as the averages of all possible different realizations of the random function.

1.2 Amplitude distribution

Let us consider a random function $D(t)$. Then $p(D_0)\,\mathrm{d}D$ defines the probability that, at any given time, the value of $D(t)$ lies between D_0 and $D_0 + \mathrm{d}D$, and $p(D)$, which is a probability density, characterizes the amplitude distribution of D.

At this point it is worth recalling some mathematical properties.

(1) For detecting errors in computing we have, by definition,

$$\int_{-\infty}^{+\infty} p(D)\,\mathrm{d}D = 1$$

(2) It is convenient for calculations to associate a "characteristic" function $\varphi(u)$ with $p(D)$ such that $\varphi(-2\pi f)$ is the Fourier transform of $p(D)$. It is known that the characteristic function of the sum of two independent variables is the product of the characteristic functions of each of these variables. It should be noted that $\varphi(u)$ does not generally have a simple physical interpretation.

(3) We define the moments of order 1, 2 etc. as the average values of D, D^2 etc. as follows:

$$\text{first-order moment} = \int_{-\infty}^{+\infty} Dp(D)\,\mathrm{d}D$$

$$\text{second-order moment} = \int_{-\infty}^{+\infty} D^2p(D)\,\mathrm{d}D \quad \text{etc.}$$

In the particular case when the average value of the first-order moment is zero, the second-order moment is called the "mean square value" and its square root is called the root mean square (r.m.s.) or the effective value. Electronic engineers generally apply the term "power" to the second-order moment (i.e. the square of the effective value) of a random function with a mean value of zero. Finally, the mean square value (power) of a gaussian variable with zero mean is called the "variance" and the effective value is called the "standard deviation".

(4) It can easily be proved that the characteristic function $\varphi(u)$ can be written

$$\varphi(u) = 1 + \mathrm{j}M_1u + \mathrm{j}^2M_2\frac{u^2}{2!} + \mathrm{j}^3M_3\frac{u^3}{3!} + \cdots$$

where M_1, M_2, M_3, . . . are the moments of order 1, 2, 3, ...

(5) As a result of (4), when two independent variables are summed, the first-order moment of the sum is the sum of the first-order moments of each of the variables, which is obvious *a priori*. Furthermore, if the first-order moment of at least one of the variables is zero, the second-order moment of the sum is equal to the sum of the second-order moments.

(6) If we consider the variable $Y = kD$, it is obvious that

$$p_1(Y)\,\mathrm{d}Y = p(D)\,\mathrm{d}D$$

where $p_1(Y)$ is the probability density of Y, and therefore

$$p_1(Y) = \frac{1}{k}\,p(D)$$

$$p_1(Y) = \frac{1}{k} p\left(\frac{Y}{k}\right)$$

Thus the characteristic function of Y is equal to $\varphi(ku)$.

(7) Finally, and particularly with respect to calculations, it should be noted that if $p(D)$ is symmetric with respect to $D = 0$ and because $p(D)$ tends to zero as D tends to infinity, $\varphi(u)$ can be obtained by taking the Laplace transform of $p(D)\,u(D)$, where $u(D)$ is the unit level. This is done by replacing the Laplace operator by $-ju$ and multiplying the real part of the resulting expression by 2, thus obtaining a function $\varphi(u)$ which is symmetric with respect to $u = 0$.

We now apply these useful properties in the following examples which illustrate the concept and calculation of amplitude distribution.

Example 1.1

Let us consider the random function $V(t)$ defined as follows: (a) $V(t)$ is constant between $t = 0$ and $t = 1$, between $t = 1$ and $t = 2$ etc.; (b) in each of these intervals, $V(t)$ has one chance in two of being equal to 3, and one chance in two of being equal to 1; (c) the sampling which gives values of V in each of these intervals is independent (Fig. 1.2).

The amplitude distribution can be simply represented by saying that V has a probability of 0.5 of being equal to 3, a probability of 0.5 of being equal to 1 and a probability of zero of being different from 3 or 1, and we can write

$$p(V) = \tfrac{1}{2}[\delta(V - 1) + \delta(V - 3)]$$

where $\delta(t)$ is the Dirac delta function. Since the mean value of $V(t)$ is obviously equal to 2, it is convenient (and often makes sense physically) to write $V(t)$ in the form

$$V(t) = 2 + v(t)$$

where $v(t)$ is a random variable with a mean value of zero and is symmetric with respect to zero (odd-order moments are all zero). Its power (mean

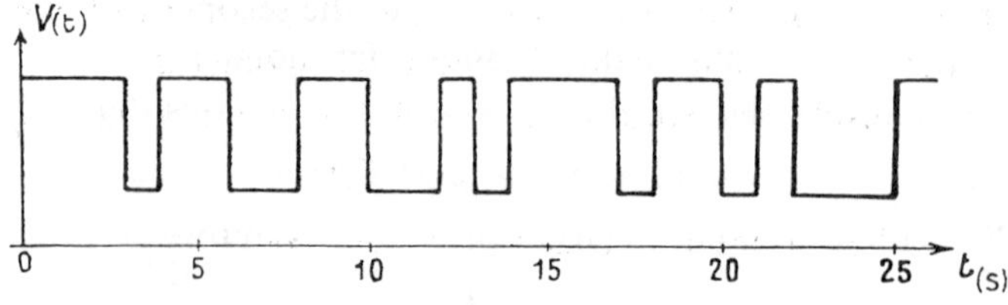

Fig. 1.2

square value) is given by

$$\tfrac{1}{2}(1 + 1) = 1$$

and its fourth-order moment is also equal to unity, i.e. the square of the power and all the even-order moments are equal to unity. We can also show directly that the characteristic function of $v(t)$ can be written

$$\varphi(u) = \tfrac{1}{2}[\exp(-ju) + \exp(+ju)]$$

since the probability density of v is given by

$$\tfrac{1}{2}[\delta(v + 1) + \delta(v - 1)]$$

Therefore

$$\varphi(u) = \cos u = 1 - \frac{u^2}{2!} + \frac{u^4}{4!} + \cdots$$

which confirms that all the moments of even order are equal to unity.

We now consider the random function $W_1(t)$ defined by

$$2W_1(t) = V(t) + V(t - 1)$$

which can be rewritten in the form

$$2W_1(t) = 4 + v(t) + v(t - 1)$$

where $v(t)$ and $v(t - 1)$ are two independent variables with the same probability density, or

$$W_1(t) = 2 + w_1(t)$$

where $w_1(t)$ is a random function with zero mean whose characteristic function is given by

$$\cos\left(\frac{u}{2}\right)\cos\left(\frac{u}{2}\right) = \left[\cos\left(\frac{u}{2}\right)\right]^2$$

according to (2) and (6), i.e.

$$\frac{1 + \cos u}{2} = 1 - \frac{u^2}{4} + \frac{u^4}{2 \times 4!} - \frac{u^6}{2 \times 6!} \cdots$$

The power (mean square value) of $w_1(t)$ is therefore equal to 0.5. This is obtained directly because it is twice the power of $v(t)/2$. The fourth-order moment of $w_1(t)$ is therefore also equal to 0.5, i.e. twice the square of the power, and all the even-order moments are equal to 0.5.

Finally, the random function $W_n(t)$ defined by

$$nW_n(t) = V(t) + V(t - 1) + \cdots + V(t - n)$$

where n is very large, involves the sum of a very large number of independent values of a random variable for each value of t. $W_n(t)$ therefore has a gaussian amplitude distribution which can be written in the form

$$W_n(t) = 2 + w_n(t)$$

where $w_n(t)$ is a random function with zero mean, a gaussian distribution and a power (variance) equal to n times the power of $v(t)/n$, i.e.

$$n\frac{1}{n^2} = \frac{1}{n}$$

The amplitude distribution of $w_n(t)$ is represented by

$$p(w_n) = \frac{n^{1/2}}{(2\pi)^{1/2}}\exp\left(-\frac{nw_n{}^2}{2}\right)$$

In other words, $W_n(t)$ lies between $2 - 1/n^{1/2}$ and $2 + 1/n^{1/2}$ for 70% of the time. The fourth-order moment of $w_n(t)$ is therefore three times the square of the power of $w_n(t)$.

Example 1.2

Let us consider the random function $x(t)$ defined by

$$x(t) = \sin[\varphi(t)]$$

where $\varphi(t)$ is constant between 0 and 1 s, between 1 and 2 s etc., the values of $\varphi(t)$ in each of these intervals are independent and $\varphi(t)$ can take all values between zero and 2π with the same probability. One value of x has two corresponding values of $\varphi(t)$ lying between zero and 2π so that the probability that x lies between x_0 and $x_0 + \mathrm{d}x$ is equal to the probability that φ lies between $\varphi_0 = \mathrm{Arcsin}\, x_0$ and $\varphi_0 + \mathrm{d}\varphi$ plus the probability that φ lies between $\pi - \varphi_0$ and $\pi - \varphi_0 - \mathrm{d}\varphi$ (Arcsin x is the principal value of arcsin x). We can therefore write

$$p_2(x_0)\,\mathrm{d}x = 2p_1(\varphi_0)\,\mathrm{d}\varphi$$

where φ_0 lies between $-\pi/2$ and $+\pi/2$, with

$$\int_0^{2\pi} p_1(\varphi)\,\mathrm{d}\varphi = 1$$

$$p_1(\varphi) = \frac{1}{2\pi}$$

and

$$\mathrm{d}x = \cos\varphi\,\mathrm{d}\varphi = \mathrm{d}\varphi(1 - x^2)^{1/2}$$

$p_2(x)$ is the amplitude distribution of x and $p_1(\varphi)$ is the amplitude distribution of φ. We thus have

$$p_2(x) = \frac{1}{\pi(1 - x^2)^{1/2}}$$

It is obvious that $|x|$ cannot exceed unity. This is verified by the following calculation:

$$\int_{-1}^{+1} p_2(x)\,\mathrm{d}x = \frac{2}{\pi}\int_0^1 \frac{1}{(1-x^2)^{1/2}}\,\mathrm{d}x = 1$$

Example 1.3

Let us consider the random function

$$Y(t) = y^2(t)$$

where $y(t)$ is a gaussian random function with zero mean and power (variance) equal to unity ($Y(t)$ is the output of a square-law detector whose input is gaussian noise). The mean value of $Y(t)$ is obviously unity, so that we can write

$$Y(t) = 1 + \mathscr{Y}(t)$$

where $\mathscr{Y}(t)$ is a random variable with zero mean. The amplitude distribution of $\mathscr{Y}(t)$ can be represented by its moments of order 2, 3, 4 etc. The second-order moment can be written

$$\text{mean value of } [y^2(t) - 1]^2 = \overline{(y^2-1)^2}$$
$$= \overline{y^4} - \overline{2y^2} + 1$$

The $2r$th-order moment of y (gaussian variable with zero mean) is equal to $1, 3, \ldots, 2r-1$ times the rth power of the second-order moment. Therefore the second-order moment of $\mathscr{Y}$ is equal to 2. The third-order moment is given by

$$\overline{(y^2-1)^3} = \overline{y^6} - \overline{3y^4} + \overline{3y^2} - 1$$
$$= 15 - 9 + 3 - 1$$
$$= 8$$

The fourth-order moment is given by

$$\overline{(y^2-1)^4} = \overline{y^8} - \overline{4y^6} + \overline{6y^4} - \overline{4y^2} + 1$$
$$= 105 - 60 + 18 - 4 + 1$$
$$= 60$$

Therefore the characteristic function associated with $\mathscr{Y}(t)$ can be written in the form

$$\varphi(u) = 1 - u^2 - \mathrm{j}\frac{4u^3}{3} + \frac{5u^4}{2} + \cdots$$

Remark 1.1

A random function can often be represented as the sum of a deterministic function and a random function with zero mean. This decomposition can frequently be used to simplify calculations. In particular, it is always possible (and this is often useful) to write a random function as the sum of a constant term equal to its mean value and a random function with zero mean. Therefore in what follows (Sections 1.3–1.7) we will only consider random functions with zero mean.

1.3 Spectral distribution $A^2(f)$

Consider a random function with infinite duration t which has constant statistical properties since it is a steady state function. For simplicity it is assumed to have zero mean (in view of Remark 1.1 this does not affect the generality of the analysis). To be more precise, let us take the function $v(t)$ as defined in Example 1.1 and shown in Fig. 1.2 (which in fact represents $v(t) + 2$). We consider a function equal to $v(t)$ for $0 < t < 3$ and zero for all other values of t. This can be written as $u(t) - u(t - 3)$, where $u(t)$ is the unit level, and we can calculate

$$\int_0^3 v(t)\,\mathrm{d}t = 3$$

Then, for $T = 3$,

$$\frac{1}{T}\int_0^T v(t)\,\mathrm{d}t = 1$$

This is the mean value of $v(t)$ in the interval $[0, T]$. We can also calculate

$$\frac{1}{T}\int_0^T v^2(t)\,\mathrm{d}t$$

for $T = 3$, and this expression is also equal to unity. This is the mean value of the power of $v(t)$ in the interval $[0, T]$. We can calculate the Fourier transform of a sample of $v(t)$ of duration T as follows:

$$A_{T=3}(f) = \int_0^{T=3} v(t)\exp(-2\pi\mathrm{j}ft)\,\mathrm{d}t = \frac{1}{2\pi\mathrm{j}f}[1 - \exp(-3 \times 2\pi\mathrm{j}f)]$$

$$A_T(f) = \frac{[1 - \cos(6\pi f)] + \mathrm{j}\sin(6\pi f)}{2\pi\mathrm{j}f}$$

It is found that $A_T(f) = A_T{}^*(-f)$, where the asterisk indicates that the function $A_T{}^*$ is the imaginary conjugate of A_T, and that

$$A_T(f)\,A_T(-f) = |A_T{}^2(f)|$$

which is an even function of f, and

$$A_3^2(f) = \frac{\sin^2(3\pi f)}{\pi^2 f^2}$$

Finally, we can define

$$\frac{A_T^2(f)}{T} = \frac{\sin^2(3\pi f)}{3\pi^2 f^2} \quad \text{for } T = 3$$

For illustration, let us assume that $v(t)$ is in volts. Then $A_T(f)$ is in volt seconds, $|A_T(f)|^2$ is in (volt seconds)^2 and $|A_T^2(f)|/T$ is in volt^2 second or volt^2 hertz^{-1}.

It is clear that

$$A_T(0) = \int_0^T v(t)\,\mathrm{d}t$$

and that

$$\frac{A_T(0)}{T} = \frac{1}{T}\int_0^T v(t)\,\mathrm{d}t$$

which is verified here since

$$A_3(0) = 3 \qquad \frac{A_3(0)}{3} = 1$$

In addition, Parseval's theorem states that

$$\int_0^T v^2(t)\,\mathrm{d}t = \int_{-\infty}^{+\infty} |A_T(f)|^2\,\mathrm{d}f$$

In other words

$$\frac{1}{T}\int_0^T v^2(t)\,\mathrm{d}t = \int_{-\infty}^{+\infty} \frac{|A_T(f)|^2}{T}\,\mathrm{d}f = 1$$

which is also verified for $T = 3$.

If we consider a longer sample of the same function $v(t)$ with $T = 6$, we have

$$u(t) - 2u(t-3) + 2u(t-4) - u(t-6)$$

and

$$\frac{1}{6}\int_0^6 v(t)\,\mathrm{d}t = \frac{2}{3} = \frac{A_6(0)}{6}$$

$$A_6(0) = |A_6(0)| = 4$$

$$\frac{|A_6^2(0)|}{6} = \frac{16}{6} = 2.7$$

$$\frac{1}{6}\int_0^6 v^2(t)\,\mathrm{d}t = 1 = \int_{-\infty}^{+\infty} \frac{|A_6(f)|^2}{6}\,\mathrm{d}f$$

The sample of $v(t)$ lying between $T = 0$ and $T = 12$ (see Fig. 1.2) gives

$$\frac{1}{12}\int_0^{12} v(t)\,dt = \frac{1}{6} = \frac{A_{12}(0)}{12}$$

$$A_{12}(0) = 2$$

$$\frac{|A_{12}(0)|^2}{12} = \frac{4}{12} = 0.33$$

$$\frac{1}{12}\int_0^{12} v^2(t)\,dt = 1 = \int_{-\infty}^{+\infty} \frac{|A_{12}(f)|^2}{12}\,df$$

The sample of $v(t)$ lying between $T = 0$ and $T = 24$ (see Fig. 1.2) gives

$$\frac{1}{24}\int_0^{24} v(t)\,dt = \frac{4}{24} = \frac{1}{6} = \frac{A_2(0)}{24}$$

$$A_{24}(0) = 4$$

$$\frac{|A_{24}(0)|^2}{24} = \frac{16}{24} = 0.67$$

$$\frac{1}{24}\int_0^{24} v^2(t)\,dt = 1 = \int_{-\infty}^{+\infty} \frac{|A_{24}(f)|^2}{24}\,df$$

Figure 1.3 shows the variations in $|A_T(f)|^2/T$ for various values of T.

By generalizing the process for infinite T we obtain the following definition: the limit for infinite T of $(1/T)\int_0^T v(t)\,dt$ is the mean value of $v(t)$, which in the present case is zero. Thus as T tends to infinity, $A_T(0)/T$ tends to zero because the random function considered has a zero mean. However, it can easily be shown that $A_T^2(0)/T$ tends to a nonzero limit which in this case is equal to unity.

In the same manner the limit for infinite T of $(1/T)\int_0^T v^2(t)\,dt$ is defined as the mean value of $v^2(t)$, which is the power of $v(t)$ and in this case is equal to unity. Under these conditions it can be seen that $|A_T(f)|^2/T$ tends to a limit which is a function of f, i.e. $A^2(f)$, such that

$$\text{power of } v(t) = \int_{-\infty}^{+\infty} A^2(f)\,df$$

In the example used here the function $A^2(f)$ is represented by the curve in Fig. 1.4 (the procedure for obtaining this curve is explained in Section 1.6). The above relationship explains why $A^2(f)$ is called the power spectral density of the random function $v(t)$. If $v(t)$ is expressed in volts, $A^2(f)$ is expressed in volt^2 hertz^{-1}. Since $|A_T(f)|^2/T$ is an even function of f, its limit $A^2(f)$ is also an even function of f so that we can write

$$\text{power of } v = \int_0^{\infty} 2A^2(f)\,df$$

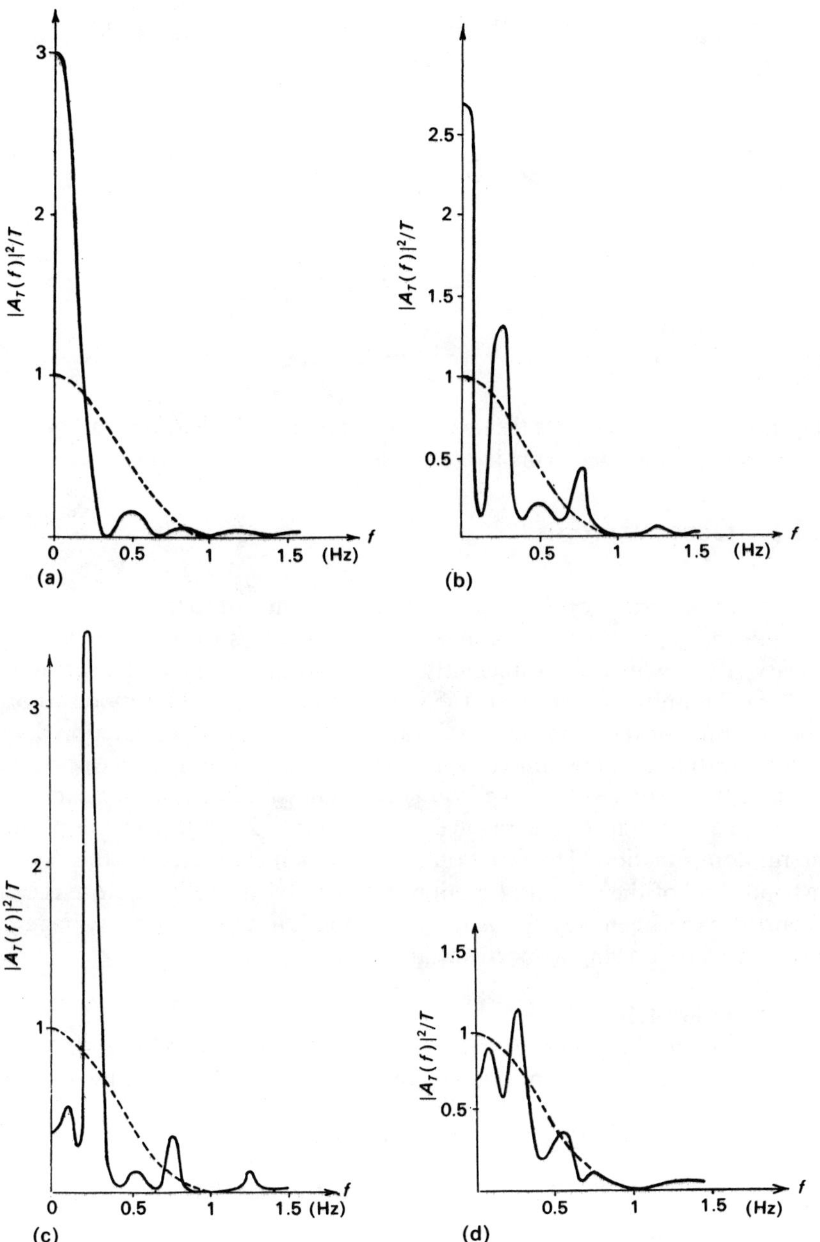

Fig. 1.3 Variations in $|A_T(f)|^2/T$ with f for (a) $T = 3$, (b) $T = 6$, (c) $T = 12$ and (d) $T = 24$.

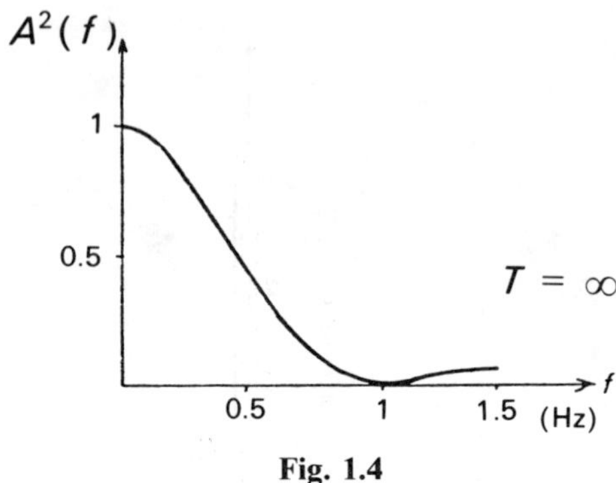

Fig. 1.4

and thus $2A^2(f)$ represents the spectral density of $v(t)$ when we limit ourselves to positive frequencies. This is often useful for calculations.

1.3.1 Generalization

By considering only random functions with zero mean in order to simplify the calculations, we can see that it is necessary to define a spectral density $A^2(f)$ which, for sufficiently large T (in the sense of the distribution of "large" numbers), represents the value of the square of the modulus of the Fourier transform of a sample of duration T of the random function divided by the duration T of the sample. The function $A^2(f)$, which has the dimensions of a power density (with respect to the frequency f), is a real even function of the frequency f whose integral over all frequencies is equal to the power of the random function. The function $A^2(f)$, which characterizes the "spectral distribution" of the random function, has a physical meaning since it can be obtained experimentally for a random function representing voltage as a function of time using a spectrum analyzer.

Definition 1.1

When $A^2(f)$ is constant over a certain frequency band (between f_1 and f_2 for example) the random phenomenon is called "white" in this frequency band.

1.4 Response of a linear filter

Theorem 1.1

Let us consider a random function $x(t)$ entering a linear filter characterized by a transfer function $F(p)$. If the spectral density of $x(t)$ is defined

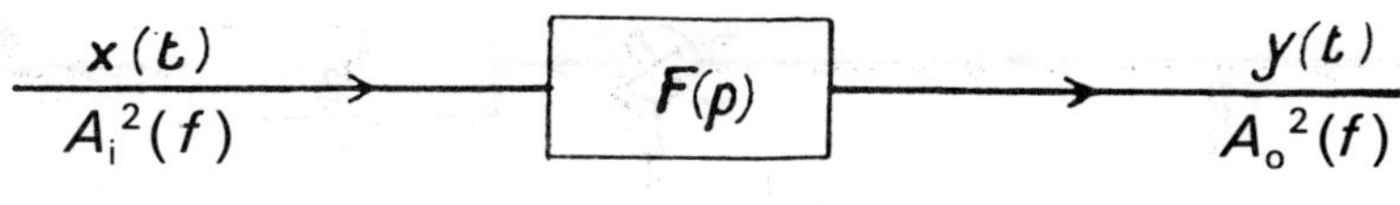

Fig. 1.5

as $A_i^2(f)$, what is the spectral density $A_o^2(f)$ of the output $y(t)$ of the filter (Fig. 1.5)?

We consider a sample of duration T of the function $x(t)$ whose Fourier transform is $A_T(f)$; it is known that the corresponding sample of $y(t)$ will have the following Fourier transform:

$$A_{To}(f) = A_{Ti}(f)\,F(2\pi jf)$$

Thus we have

$$\frac{|A_{To}{}^2(f)|}{T} = \frac{|A_{Ti}{}^2(f)|}{T}\,|F(2\pi jf)|^2$$

and, in the limit of infinite T,

$$A_o^2(f) = A_i^2(f)|F(2\pi jf)|^2$$

Remark 1.2

It should be noted that, just as the definition of $A^2(f)$ is not justified rigorously in Section 1.3, the preceding argument can be taken at most as a demonstration and not as a proof. A mathematically oriented mind will easily find that it does not prove anything. However, it is very satisfying physically (especially for an electronic engineer), and the formula has been rigorously proved by Blanc-Lapierre in ref. 1, Section 9.2. In what follows, and in accordance with the comments at the beginning of this chapter, we will not comment on inadequate proofs in this way. It is recommended in all cases that the reader refers to more rigorous treatments and in particular to ref. 1.

Example 1.4

Let us look again at Example 1.1. It was stated that $v(t)$ has the spectral density $A^2(f)$ shown in Fig. 1.4 which can be written as (see Section 1.6)

$$A^2(f) = \left[\frac{\sin(\pi f)}{\pi f}\right]^2$$

We will now determine the spectral density of the random function

$$w_1(t) = \tfrac{1}{2}[v(t) + v(t-1)]$$

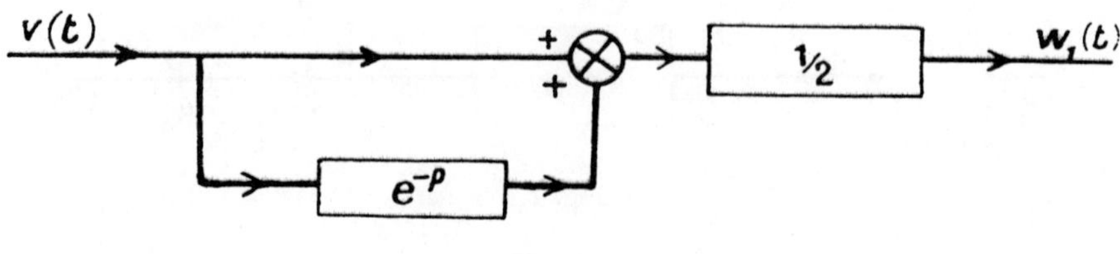

Fig. 1.6

This is easily obtained by noting that $w_1(t)$ is obtained from $v(t)$ as shown in Fig. 1.6, since the transfer function w_1/v is equal to

$$F(p) = \frac{1 + \exp(-p)}{2}$$

Thus the spectral density of $w_1(t)$ can be written as

$$\left[\frac{\sin(\pi f)}{\pi f}\right]^2 \times \tfrac{1}{4}|1 + \exp(-2\pi j f)|^2$$

$$= \left[\frac{\sin(\pi f)}{\pi f}\right]^2 \cos^2(\pi f)$$

$$= \left[\frac{\sin(2\pi f)}{2\pi f}\right]^2$$

1.5 Applications of the convolution theorem

1.5.1 Convolution theorem

Consider two functions $X(t)$ and $Y(t)$ with Fourier transforms $\Phi_x(f)$ and $\Phi_y(f)$ respectively. We can prove the following relationship, which is very useful in a large number of areas:

$$\int_{-\infty}^{+\infty} X(t)\, Y(t - t_0)\, \mathrm{d}t = \int_{-\infty}^{+\infty} \Phi_X(f)\, \Phi_Y(-f) \exp(2\pi j f t_0)\, \mathrm{d}f$$

By setting $Y(t) = X(t)$ and $t_0 = 0$, we return to Parseval's theorem, which has already been used in Section 1.3, since the relationship becomes

$$\int_{-\infty}^{+\infty} X^2(t)\, \mathrm{d}t = \int_{-\infty}^{+\infty} |\Phi_X(f)|^2\, \mathrm{d}f$$

1.5.2 The cross-correlation function

Consider two random functions $x(t)$ and $y(t)$. Further, consider two samples of duration T (between $t = 0$ and $t = T$) of each of these functions with Fourier transforms $A_{Tx}(f)$ and $A_{Ty}(f)$ respectively. We can then write

$$\frac{1}{T}\int_0^T x(t)\,y(t-\tau)\,\mathrm{d}t = \frac{1}{T}\int_{-\infty}^{+\infty} A_{Tx}(f)\,A_{Ty}(-f)\exp(2\pi \mathrm{j} f\tau)\,\mathrm{d}f$$

If we now let T tend to infinity, we see that the mean value of $[x(t)\,y(t-\tau)]$, which is known as the cross-correlation function of x with y and is given by

$$\overline{x(t)\,y(t-\tau)} = \varrho_{xy}(\tau)$$

has a Fourier transform in the limit of infinite T of

$$\frac{A_{Tx}(f)\,A_{Ty}(-f)}{T}$$

Permuting x and y, we can define the cross-correlation function of y with x in a similar manner:

$$\begin{aligned}\varrho_{yx}(\tau) &= \lim_{T\to\infty}\frac{1}{T}\int_{-\infty}^{+\infty} A_{Ty}(f)\,A_{Tx}(-f)\exp(2\pi \mathrm{j} f\tau)\,\mathrm{d}f \\ &= \varrho_{xy}{}^{*}(-\tau)\end{aligned}$$

In the particular case described in Section 1.4, where $y(t)$ is the output of a linear filter with transfer function $F(p)$ whose input is $x(t)$,

$$A_{Ty}(f) = A_{Tx}(f)\,F(2\pi \mathrm{j} f)$$

so that

$$\varrho_{xy}(\tau) = \lim_{T\to\infty}\frac{1}{T}\int_{-\infty}^{+\infty} |A_{Tx}{}^2(f)|\,F(-2\pi \mathrm{j} f)\exp(2\pi \mathrm{j} f\tau)\,\mathrm{d}f$$

The Fourier transform of $\varrho_{xy}(\tau)$ is

$$A_x{}^2(f)\,F^*(2\pi \mathrm{j} f)$$

If x and y are real, so are $\varrho_{xy}(\tau)$ and $\varrho_{yx}(\tau)$, so that we have

$$\varrho_{yx}(\tau) = \varrho_{xy}(-\tau)$$

1.5.3 Autocorrelation function: Wiener's theorem

If $y(t) \equiv x(t)$, we can write

$$\frac{1}{T}\int_0^T x(t)\,x(t-\tau)\,\mathrm{d}t = \frac{1}{T}\int_{-\infty}^{+\infty} A_{Tx}(f)\,A_{Tx}(-f)\exp(2\pi \mathrm{j} f\tau)\,\mathrm{d}f$$

In the limit of infinite T we have

$$\varrho_{xx}(\tau) = \overline{x(t)\,x(t-\tau)} = \int_{-\infty}^{+\infty} A_x{}^2(f)\exp(2\pi \mathrm{j} f\tau)\,\mathrm{d}f$$

This is known as the autocorrelation function of the random function $x(t)$ and is designated by $\varrho_{xx}(\tau)$ or, if there is no ambiguity, by $\varrho_x(\tau)$ or $\varrho(\tau)$. It is defined as the mean value of the product of the function $x(t)$ with the same function delayed by τ, i.e. $x(t - \tau)$. It can be proved that the autocorrelation function of the function $x(t)$ has a Fourier transform which is exactly the spectral density of this function. Therefore the function $\varrho(\tau)$ can be used instead of the spectral density to characterize the spectral distribution of the random function.

Mathematically, the function $\varrho(\tau)$ is an even function of τ since $A^2(f)$ is an even function of f, so that

$$\varrho(\tau) = \varrho(-\tau)$$

$$\varrho'(0) = 0$$

($\varrho(\tau)$ is a maximum for $\tau = 0$). The value of $\varrho(\tau)$ for $\tau = 0$ is the mean value of the square of the function $x(t)$. If we only consider random functions with zero mean, as has been justified above, $\varrho(0)$ represents the power of the random function. If we also assume sufficiently large values for τ, it is physically obvious that $x(t)$ and $x(t - \tau)$ will become independent variables and thus the mean of $x(t)x(t - \tau)$ will be equal to the square of the mean value of $x(t)$. If this is assumed to be zero, which we have accepted, we can deduce that $\varrho(\tau)$ tends to zero as τ tends to infinity.

Physically, $x(t)$ and $x(t - \tau)$ are very similar when $\varrho(\tau)$ is large. If $\varrho(\tau)$ is large and positive, $x(t)$ and $x(t - \tau)$ are more likely to have the same sign than to have opposite signs. If $\varrho(\tau)$ is large and negative, $x(t)$ and $-x(t - \tau)$ are more likely to have the same sign than to have opposite signs. In other words, if $\varrho(\tau)$ tends to zero very slowly, $x(t)$ changes slowly, but if $\varrho(\tau)$ tends to zero very rapidly, $x(t)$ changes very rapidly.

It should be noted that we can have $\varrho(\tau) = 0$ even if $x(t)$ and $x(t - \tau)$ are not independent; this is the case for $x(t) = \sin t$ and $\tau = \pi/2$.

1.5.4 Differentiation of $\varrho(\tau)$ when $x(t)$ is real

We can write

$$\varrho(\tau) = \lim_{T\to\infty} \frac{1}{T}\int_0^T x(t)\,x(t - \tau)\,\mathrm{d}t$$

$$\varrho'(\tau) = \lim_{T\to\infty} \left[-\frac{1}{T}\int_0^T x(t)\,x'(t - \tau)\,\mathrm{d}t\right]$$

$$= -\varrho_{xx'}(\tau)$$

$$= -\varrho_{x'x}(-\tau)$$

$$= \lim_{T\to\infty}\left[-\frac{1}{T}\int_0^T x'(t)\,x(t+\tau)\,\mathrm{d}t\right]$$

$$\varrho''(\tau) = \lim_{T\to\infty}\left[-\frac{1}{T}\int_0^T x'(t)\,x'(t+\tau)\,\mathrm{d}t\right]$$

The second derivative of the autocorrelation function of $x(t)$ is equal to the autocorrelation function of the first derivative of $x(t)$ but with opposite sign.

1.6 Examples of calculations of $\varrho(\tau)$ and $A^2(f)$

1.6.1 Autocorrelation function associated with a rectangular spectrum

If we consider only positive frequencies (cf. Section 1.3), the spectral density is zero outside the interval $(f_0 - \Delta f/2, f_0 + \Delta f/2)$ and is equal to N_0 inside this interval (the power of the signal is $N_0\Delta f$). This can be expressed in the following alternative way: for f positive, we have

$$2A^2(f) = N_0\left[u\left(f - f_0 + \frac{\Delta f}{2}\right) - u\left(f - f_0 - \frac{\Delta f}{2}\right)\right]$$

where $u(t)$ is the unit level. Then, from Section 1.2, point (7), we can deduce that

$$\varrho(\tau) = 2\,\mathrm{Re}\left(\frac{N_0}{2}\frac{1}{2\pi\mathrm{j}\tau}\left\{\exp\left[-2\pi\mathrm{j}\tau\left(f_0 - \frac{\Delta f}{2}\right)\right] - \exp\left[-2\pi\mathrm{j}\tau\left(f_0 + \frac{\Delta f}{2}\right)\right]\right\}\right)$$

i.e.

$$\varrho(\tau) = N_0\Delta f\cos(2\pi\tau f_0)\frac{\sin(\pi\tau\Delta f)}{\pi\tau\Delta f}$$

so that we again find

$$\varrho(0) = N_0\Delta f$$

1.6.2 Calculation of the autocorrelation function

Let us consider a white noise ($A^2(f)$ is constant in the useful frequency zone) which has passed through a filter whose transfer function in the Nyquist plane (complex plane) has a circular locus passing through the origin

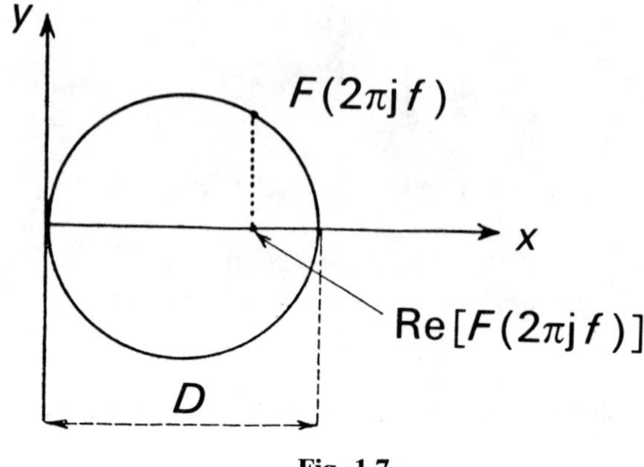

Fig. 1.7

and symmetric with respect to the real axis (Fig. 1.7). This is the case when the filter is an *RC*, *CR*, *LR* or *RL* circuit or an amplifier whose load is a trap circuit.

The spectral density of the filtered noise has the form

$$K_1|F(2\pi jf)|^2$$

where $F(p)$ is the transfer function of the filter, and is the Fourier transform of the autocorrelation function $\varrho(\tau)$ of the filtered noise. The mapping between $K_1|F(2\pi jf)|^2$ and $\varrho(\tau)$ is reciprocal. Furthermore, if $G(p)$ denotes the Laplace transform of $\varrho(t)\,u(t)$, the real part of $G(2\pi jf)$, apart from a factor of 2, is equal to the spectral density $K_1|F(2\pi jf)|^2$ (cf. Section 1.2, point (7)). Now, it is obvious that

$$|F(2\pi jf)|^2 \;=\; D \times \operatorname{Re} F(2\pi jf)$$

Since the relationship between $\varrho(\tau)$ and $\varrho(t)\,u(t)$, like that between $\varrho(t)\,u(t)$ and $G(p)$, is reciprocal, it follows that there is only one function $G(p)$ whose real part is equal to $|F(2\pi jf)|$, apart from a coefficient, which satisfies the problem (i.e. it gives $\varrho(\infty) = 0$ or $p\,G(p) = 0$ in the limit $p = 0$). Therefore, if $F(p)$ is such that $p\,F(p)$ tends to zero as p tends to zero, $F(p)$ is the required function $G(p)$.

1.6.3 Calculation of the spectral density of $v(t)$ (see Example 1.1)

It is clear that the autocorrelation function of $v(t)$ is given by (Fig. 1.8)

$$\varrho_v(\tau) = 0 \qquad \text{for} \quad |\tau| \geqslant 1$$

$$\varrho_v(\tau) = 1 - |\tau| \quad \text{for} \quad |\tau| \leqslant 1$$

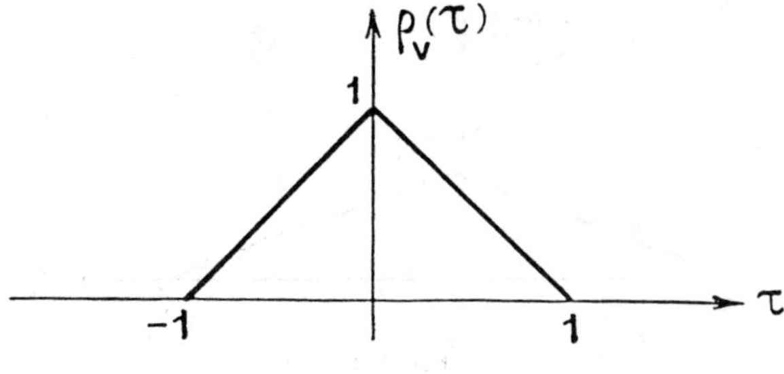

Fig. 1.8

The Laplace transform of $\varrho(t)\,u(t)$ is equal to

$$\frac{1}{p} - \frac{1}{p^2}\,[1 - \exp(-p)]$$

and therefore (cf. Section 1.2, point (7))

$$\begin{aligned} A^2(f) &= 2\,\mathrm{Re}\left\{\frac{1}{2\pi \mathrm{j} f} + \frac{1}{4\pi^2 f^2}\,[1 - \exp(-2\pi \mathrm{j} f)]\right\} \\ &= \frac{1 - \cos(2\pi f)}{2\pi^2 f^2} \\ &= \left[\frac{\sin(\pi f)}{\pi f}\right]^2 \end{aligned}$$

1.6.4 Calculation of the autocorrelation function of $w_1(t)$ (see Examples 1.1 and 1.4)

We have seen in Example 1.4 that the spectral density of $w_1(t)$ can be written as

$$\left[\frac{\sin(2\pi f)}{2\pi f}\right]^2$$

which (see Section 1.6.3) is clearly the Fourier transform of

$$\begin{aligned} \varrho_{w_1}(\tau) &= 0 & \text{for} \quad |\tau| \geqslant 2 \\ \varrho_{w_1}(\tau) &= 0.5 - 0.25|\tau| & \text{for} \quad |\tau| \leqslant 2 \end{aligned}$$

(see Fig. 1.9). We again find that the power of w_1 is equal to 0.5.

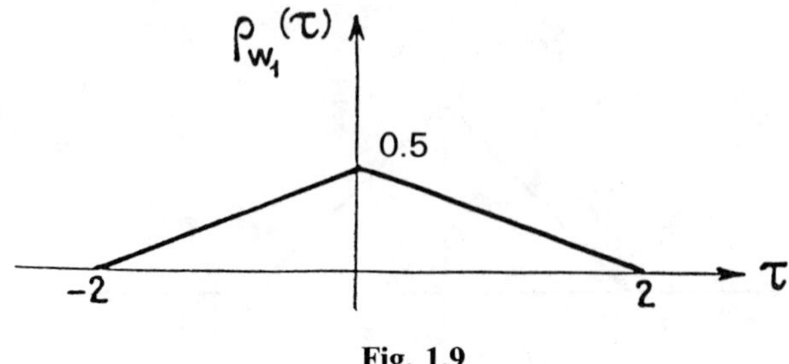

Fig. 1.9

The same result can be obtained in a different manner by writing

$$\varrho_{w_1}(\tau) = \overline{\{\tfrac{1}{2}[v(t) + v(t-1)]\}\{\tfrac{1}{2}[v(t-\tau) + v(t-\tau-1)]\}}$$

$$\begin{aligned}4\varrho_{w_1}(\tau) &= \overline{v(t)\,v(t-\tau)} + \overline{v(t-\tau-1)\,v(t)} + \overline{v(t-1)\,v(t-\tau)} \\ &\quad + \overline{v(t-1)\,v(t-\tau-1)} \\ &= \varrho_v(\tau) + \varrho_v(\tau+1) + \varrho_v(\tau-1) + \varrho_v(\tau)\end{aligned}$$

which gives the same result.

1.6.5 Filtering with an *RC* circuit

A noise voltage $v_1(t)$ is gaussian, its mean value is zero and its r.m.s. value is 1 V. Its spectrum (spectral density) is constant for frequencies below 100 Hz and zero for higher frequencies. It passes through an RC filter with $RC = 1$ s. What is the autocorrelation function of the noise $v_2(t)$ at the output of the filter (Fig. 1.10)?

The transfer function of the filter is given by

$$F(p) = \frac{1}{1+p}$$

(the cut-off frequency is of the order of 0.15 Hz which allows the input noise to be considered as white noise). Thus, according to Section 1.6.2, $1/(1+p)$, apart from a constant factor, is the Laplace transform of $\varrho(t)\,u(t)$. We therefore have

$$\varrho(\tau) = \varrho(0)\exp(-|\tau|)$$

In order to determine $\varrho(0)$, it is useful first to determine the spectral density $A_i^2(f)$ at the input which, for $-100\,\text{Hz} \leqslant f \leqslant 100\,\text{Hz}$, is such that

$$\int_{-100}^{+100} A_i^2(f)\,\mathrm{d}f = (1)^2 = 1$$

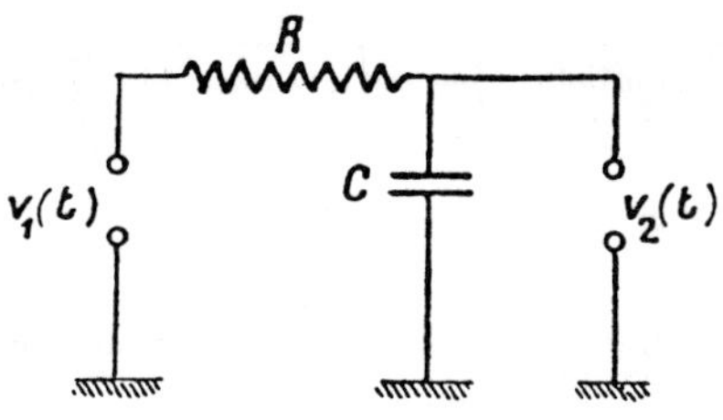

Fig. 1.10

so that

$$A_i^2(f) = \tfrac{1}{200}$$

Thus the spectral density at the output can be written as

$$A_o^2(f) = \frac{1}{200}\frac{1}{|1 + 2\pi j f|^2} = \frac{1}{200(1 + 4\pi^2 f^2)}$$

and the noise power at the output is

$$\varrho(0) = \frac{1}{200}\int_{-100}^{+100} \frac{1}{1 + 4\pi^2 f^2}\,\mathrm{d}f = 2.5 \times 10^{-3}$$

Thus

$$\varrho(\tau) = 2.5 \times 10^{-3} \exp(-|\tau|)$$

1.6.6 Optimization

We wish to measure the vertical velocity of an aerodyne. For this we can either measure its distance X to the ground and differentiate, or measure its vertical acceleration and integrate. We propose using both systems in order to obtain the best result. In practice we do not measure X and $\mathrm{d}^2X/\mathrm{d}t^2$ but X accompanied by white noise with spectral density B and $\mathrm{d}^2X/\mathrm{d}t^2$ accompanied by white noise of spectral density C. The differentiation of X substantially increases the accompanying high frequency noise, and thus a lowpass filter is required to eliminate spectral components above f_0. The integration of $\mathrm{d}^2X/\mathrm{d}t^2$ substantially increases the accompanying low frequency noise and thus a highpass filter which rejects spectral components below f_0 is required (Fig. 1.11). It is now necessary to determine f_0 such that the best result is obtained.

The noise powers at points 1 and 2 in Fig. 1.11 are given by

$$2\int_0^{f_0} |2\pi j f|^2 B\,\mathrm{d}f = B\,\frac{8\pi^2 f_0^3}{3}$$

$$2\int_{f_0}^{\infty} \frac{C}{|2\pi j f|^2}\,\mathrm{d}f = \frac{C}{2\pi^2}\frac{1}{f_0}$$

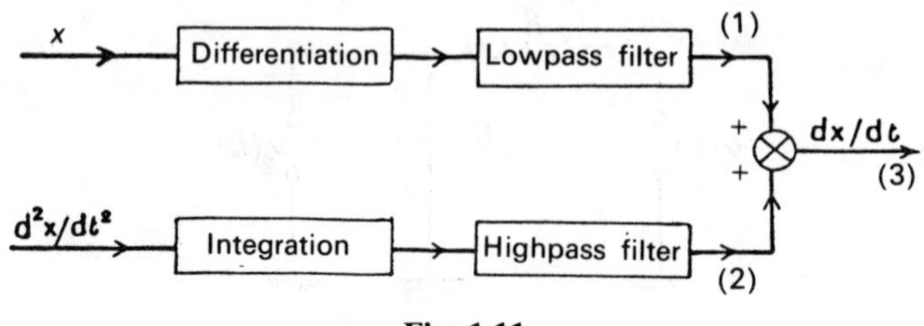

Fig. 1.11

Then the noise power at point 3 is

$$\frac{8\pi^2 B f_0^{\,3}}{3} + \frac{C}{2\pi^2 f_0}$$

The minimum value of this quantity is obtained when

$$(2\pi f_0)^4 = C/B$$

1.7 Sampling theorem

Consider a function $F_1(t)$ which is zero for $t < -T/2$ and $t > T/2$, and which has a Fourier transform $\Phi_1(f)$ given by

$$\Phi_1(f) = \int_{-T/2}^{+T/2} F_1(t) \exp(-2\pi \mathrm{j} f t)\,\mathrm{d}t$$

Thus

$$\mathrm{Re}[\Phi_1(f)] = \int_{-T/2}^{+T/2} F_1(t) \cos(2\pi f t)\,\mathrm{d}t$$

$$\mathrm{Im}[\Phi_1(f)] = -\int_{-T/2}^{+T/2} F_1(t) \sin(2\pi f t)\,\mathrm{d}t$$

Let us now consider a periodic function $f_1(t)$ with period T which, for $-T/2 < t < +T/2$, coincides with $F_1(t)$ (Fig. 1.12). Since it is periodic, we can write

$$f_1(t) = \sum_{-\infty}^{+\infty} A_n \exp\left(\frac{2\pi \mathrm{j} n t}{T}\right)$$

where n is an integer and A_n is the conjugate of A_{-n}. If we write A_n in the form

$$A_n = \alpha_n + \mathrm{j}\beta_n$$

we know that

$$\alpha_n = \frac{1}{T}\int_{-T/2}^{+T/2} F_1(t) \cos\left(2\pi \frac{nt}{T}\right)\mathrm{d}t = \frac{1}{T}\mathrm{Re}\left[\Phi_1\left(\frac{n}{T}\right)\right]$$

$$\beta_n = -\frac{1}{T}\int_{-T/2}^{+T/2} F_1(t) \sin\left(2\pi \frac{nt}{T}\right)\mathrm{d}t = \frac{1}{T}\mathrm{Im}\left[\Phi_1\left(\frac{n}{T}\right)\right]$$

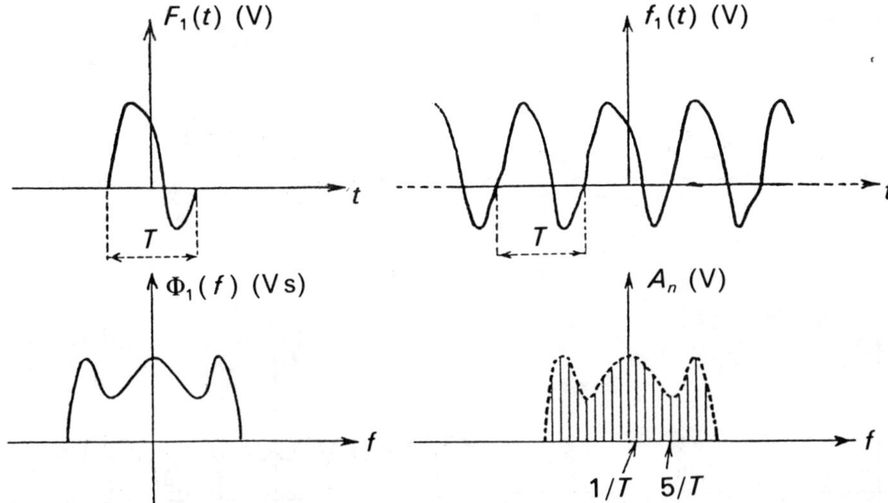

Fig. 1.12 Functions $F_1(t)$ and $f_1(t)$ and their Fourier transforms $\Phi_1(f)$ and A_n.

and hence

$$A_n = \frac{\Phi_1(n/T)}{T}$$

Except for a constant coefficient, the amplitude of the spectrum of the periodic function $f_1(t)$ is equal to the amplitude of the spectrum of $F_1(t)$ at the same frequency as shown in diagrammatic form in Fig. 1.12. (It should be remembered that if A_n is expressed in volts, for example, $\Phi_1(f)$ will be expressed in volt seconds.) Therefore, if we know the spectrum of $F_1(t)$, we can immediately find the spectrum of $f_1(t)$.

Let us now assume that the Fourier transform of $F_1(t)$ is zero for frequencies outside the band $[-\Delta f/2, +\Delta f/2]$, i.e. its spectrum is restricted to the band $\pm\Delta f/2$. It is obvious that the expression

$$f_1(t) = \sum_{-\infty}^{+\infty} A_n \exp\left(\frac{2\pi j n t}{T}\right)$$

contains a finite number of terms given by

$$\frac{\Delta f}{1/T} = T\Delta f$$

Since the conjugate pair (A_n, A_{-n}) provides only two pieces of information (amplitude and phase for example), it can be seen that $f_1(t)$ and therefore

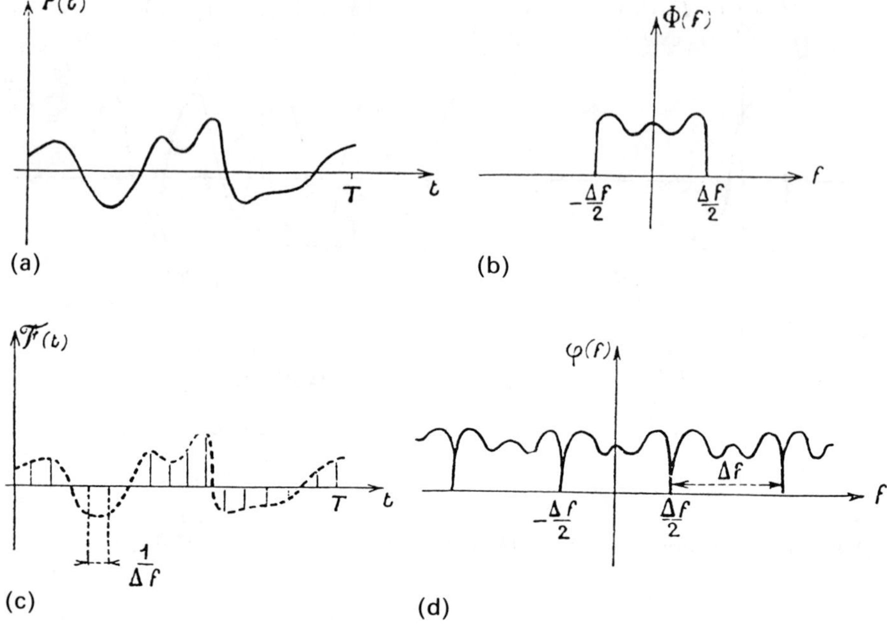

Fig. 1.13 (a) The function $F(t)$; (b) the Fourier transform $\Phi(t)$ of $F(t)$; (c) the function $\mathscr{F}(t)$; (d) the Fourier transform $\varphi(f)$ of $\mathscr{F}(t)$.

$F_1(t)$ depend on only $T\Delta f$ independent parameters, in the same way that a circle in a plane depends on only three parameters and a straight line in a plane depends on only two parameters.

Let us consider an arbitrary function of time $F(t)$ restricted to a duration T and with a spectrum restricted to the band $\pm\Delta f/2$. It is necessary and sufficient to know the values of only $T\Delta f$ parameters in order to describe the function completely. This function and its Fourier transform $\Phi(f)$ are shown in Figs 1.13(a) and 1.13(b) respectively (since $\Phi(f)$ is an imaginary function, Fig. 1.13(b) is only a representative diagram).

Let us now consider the function

$$\mathscr{F}(t) = F(0)\,\delta(t) + F\left(\frac{1}{\Delta f}\right)\delta\left(t - \frac{1}{\Delta f}\right) + F\left(\frac{2}{\Delta f}\right)\delta\left(t - \frac{2}{\Delta f}\right) + \ldots$$

which contains $T\Delta f$ terms. Since the Fourier transform is reciprocal, it is clear that the Fourier transform of $\mathscr{F}(t)$ is a periodic function $\varphi(f)$ with period

$$\frac{1}{1/\Delta f} = \Delta f$$

equal to $\Phi(f)$ for f lying between $-\Delta f/2$ and $+\Delta f/2$ (Figs 1.13(c) and 1.13(d)). If $\mathscr{F}(t)$ is known, $\varphi(f)$ and thus $\Phi(f)$ and hence $F(t)$ are also known. It is therefore sufficient to know the $T\Delta f$ values of $F(t)$ at the sampling times

$$0, \quad \frac{1}{\Delta f}, \quad \frac{2}{\Delta f}, \quad \frac{3}{\Delta f}, \quad \ldots$$

in order to know $F(t)$. Since we also know that $F(t)$ depends on $T\Delta f$ independent parameters, these values are independent. Consequently, the values of the sampling times can be arbitrarily fixed. There exists one and only one function which has these values and a spectrum which is zero outside $\pm\Delta f/2$. It can be seen that T only occurs as an intermediate step in the calculation and that the result is valid regardless of its value. We can therefore state the following extremely important theorem.

Theorem 1.2

Any function $x(t)$, particularly a random function, whose spectrum (Fourier transform or spectral density) is zero outside the interval $[-\Delta f/2, +\Delta f/2]$, is completely defined if the values of $x(t)$ are known at consecutive times separated by intervals of $1/\Delta f$ (sampling times) and these values are independent.

It is important to understand clearly what this theorem means.

Remark 1.3

First, the assumption that a function of time with finite duration has a spectrum of finite width is a mathematical heresy. This is in fact an approximation whose validity increases with increasing $T\Delta f$.

Remark 1.4

If the function $x(t)$ has a zero spectrum outside the frequency interval $[-\Delta f/2, +\Delta f/2]$, it is effectively possible to fix the value of this function arbitrarily at the sampling times, which are separated by $1/\Delta f$, and there is only one method of interpolating between these points in order to obtain a function $x(t)$ whose spectrum is zero for $|f| > \Delta f/2$. This shows that the values are independent at the sampling times.

Remark 1.5

If an extra condition is imposed on the spectrum of $x(t)$, the situation changes. For example, if the shape of the spectrum of $x(t)$ is also given, i.e. if $|\varphi(f)|^2$ is given, apart from a factor, where $\varphi(f)$ is the Fourier transform of $x(t)$, $x(t)$ can still be defined completely if its value is known at the sampling times but it is no longer possible to choose them arbitrarily. In other words, these values are no longer independent.

Remark 1.6

The particular case where the spectrum of a random function $x(t)$ is constant between $-\Delta f/2$ and $+\Delta f/2$ is of interest. This is the case when the noise is white in the band Δf. The autocorrelation function of $x(t)$ (except for a factor) is equal to

$$\frac{\sin(\pi \Delta f t)}{\pi \Delta f t}$$

which implies that it is zero for

$$t = 0 \bmod\left(\frac{1}{\Delta f}\right)$$

The cross-correlation between two sampling times is zero. The sampling values are independent. This implies that, if the values of $x(t)$ chosen at the sampling times are stochastically independent (with zero mean), the random function $x(t)$ with zero spectrum outside the interval $[-\Delta f/2, +\Delta f/2]$, which is produced by interpolation, will have a constant spectral density within this interval. This will be a white noise in the band Δf.

1.8 Tendency to the normal distribution. Gaussian phenomena. Rayleigh distribution

1.8.1 Tendency to the normal distribution

Let us consider a random function with zero mean $x(t)$ where the spectral density is zero outside the band $\pm\Delta f/2$. We can derive from this the random function $Z(t)$ defined by

$$nZ(t) = x(t) + x\left(t + \frac{1}{\Delta f}\right) + x\left(t + \frac{2}{\Delta f}\right) + \cdots + x\left(t + \frac{n}{\Delta f}\right)$$

in which n is very large. $Z(t)$ is the sum of n (very large) independent random variables; its distribution is therefore gaussian (or normal or laplacian).

In practice, $Z(t)$ can be written in another form:

$$Z(t) = \frac{\Delta f}{n}\left[x(t) + x\left(t + \frac{1}{\Delta f}\right) + \cdots + x\left(t + \frac{n}{\Delta f}\right)\right]\left(\frac{1}{\Delta f}\right)$$

$$= \frac{\Delta f}{n}\int_t^{t+n/\Delta f} x(u)\,\mathrm{d}u = \frac{1}{T}\int_t^{t+T} x(u)\,\mathrm{d}u$$

where $T = n/\Delta f$. Thus, regardless of the amplitude distribution of the random function $x(t)$, the random functions

$$\int_t^{t+T} x(u)\,\mathrm{d}u \quad \text{and} \quad \frac{1}{T}\int_t^{t+T} x(u)\,\mathrm{d}u$$

have a gaussian amplitude distribution if $T\Delta f$ is very large (the integration or averaging time is much larger than the reciprocal of the width of the spectrum). This has a very wide-ranging meaning. Let us look at the operation of averaging (or integrating) a function $f(t)$:

$$F(t) = \frac{1}{T}\int_t^{t+T} f(u)\,\mathrm{d}u$$

If

$$f(u) = \int_{-\infty}^{+\infty} \Phi(f)\exp(2\pi \mathrm{j} fu)\,\mathrm{d}f$$

we obtain

$$F(t) = \frac{1}{T}\int_t^{t+T}\mathrm{d}u\int_{-\infty}^{+\infty}\Phi(f)\exp(2\pi \mathrm{j} uf)\,\mathrm{d}f$$

$$= \int_{-\infty}^{+\infty}\Phi(f)\frac{\exp(2\pi \mathrm{j} fT) - 1}{2\pi \mathrm{j} fT}\exp(2\pi \mathrm{j} ft)\,\mathrm{d}f$$

We obtain $F(t)$ by passing $f(t)$ through a filter with transfer function

$$G(p) = \frac{\exp(pT) - 1}{pT}$$

i.e. a lowpass filter, such that only frequencies of less than approximately $1/T$ are retained (3 dB attenuation for $f \approx 1/2T$).

It follows from this that any random function (noise) whose spectrum lies between $-\Delta f/2$ and $+\Delta f/2$, filtered by a lowpass filter with cut-off frequency much less than Δf, gives a random function whose amplitude distribution is gaussian at the output of the filter. In the same manner, any random function whose spectrum lies between $f_0 - \Delta f/2$ and $f_0 + \Delta f/2$ (only positive frequencies are considered) gives a random function with gaussian distribution after filtering by a bandpass filter centered on f_0 where the passband is much less than Δf. This explains why the random functions (noise) encountered in electronics are often gaussian.

1.8.2 Representations of gaussian random functions

Let us consider a gaussian random function $x(t)$ with zero mean value whose spectral density (limited to positive frequencies, i.e. $2A^2(f)$) is equal to N_0 for $f_0 - \Delta f/2 < f < f_0 + \Delta f/2$ and is zero for other values of f ($f_0 \gg \Delta f$). The spectrum of $x(t)$ can be decomposed into a very large number N of contiguous vertical bands of width $\Delta f/N$. In other words, $x(t)$ can be considered as the sum of N independent sinusoidal signals each having a power $N_0 \Delta f/N$:

$$x(t) = \sum_{i=0}^{i=N} \left(\frac{2N_0 \Delta f}{N}\right)^{1/2} \sin(2\pi F_i t + \varphi_i)$$

where F_i is the central frequency of one of the bands and the φ_i are any phases chosen independently for each of the bands. In fact, this representation expresses $x(t)$ as either the sum of a very large number N of independent variables with power

$$\frac{1}{2}\frac{2N_0 \Delta f}{N} = \frac{N_0 \Delta f}{N}$$

i.e. as a gaussian variable with zero mean and power $N_0 \Delta f$, or as having a uniform spectral distribution between $f_0 - \Delta f/2$ and $f_0 + \Delta f/2$.

We now use Example 1.3 to illustrate the application of this representation. Let us consider a random function $y(t)$ which provides a random function $Y(t) = y^2(t)$ at the output of a quadratic detector. $Y(t)$ can also be written as

$$Y(t) = \overline{y^2(t)} + \mathscr{Y}(t) = N_0 \Delta f + \mathscr{Y}(t)$$

and we wish to evaluate the spectral density of $\mathscr{Y}(t)$. In order to do this, we can write

$$y^2(t) = \left[\sum_0^N \left(\frac{2N_0 \Delta f}{N}\right)^{1/2} \cos(2\pi F_i t + \varphi_i)\right]^2$$

It is found that $y^2(t)$ consists of (cf. the binomial theorem) N terms of amplitude $N_0 \Delta f/N$ at zero frequency whose sum is $\overline{y^2(t)}$ and that $\mathscr{Y}(t)$ consists of $N - 1$ terms at frequency $\Delta f/N$, $N - 2$ terms at frequency $2\Delta f/N, \ldots$, one term at frequency $(N - 1)\Delta f/N$ together with terms with frequencies close to $2f_0$ which are of no physical or practical interest because they are generally eliminated in a lowpass filter. The total power of the spectrum at low frequencies (the only ones considered) is equal to half the power of $\mathscr{Y}(t)$ (the other half corresponds to the components close to $2f_0$), i.e. it is given by (see Example 1.3)

$$\frac{2N_0^2 \Delta f^2}{2} = N_0^2 \Delta f^2$$

The shape of the useful spectrum (spectral density) of $\mathscr{Y}^2(t)$ is shown in Fig. 1.14 (the area of the triangle is $2N_0^2 \Delta f(\Delta f/2) = N_0^2 \Delta f^2$).

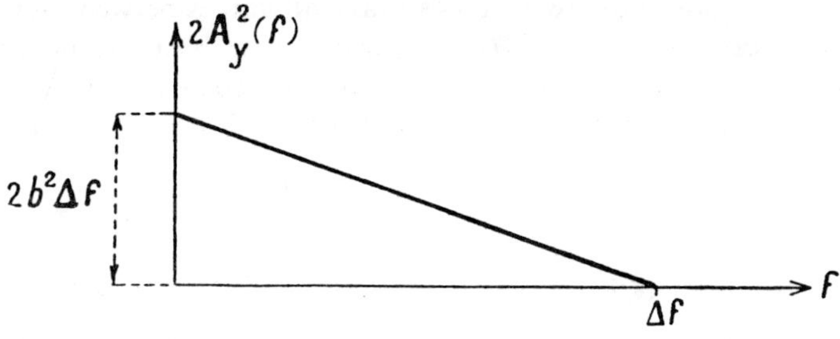

Fig. 1.14

The autocorrelation function of $\mathcal{Y}(t)$ is given by the Fourier transform of $A_{\mathcal{Y}}{}^2(f)$, i.e. by the real part of

$$\frac{2N_0{}^2\Delta f}{2\pi j\tau} + \frac{2N_0{}^2}{4\pi^2\tau^2}[1 - \exp(-2\pi j\tau\,\Delta f)]$$

(see Section 1.2, point (7)), so that

$$\varrho_{\mathcal{Y}}(\tau) = N_0{}^2\Delta f^2\left[\frac{\sin(\pi\tau\,\Delta f)}{\pi\tau\,\Delta f}\right]^2$$

Note The above argument is acceptable to physicists but is not very acceptable to mathematicians (who are quite right, but the above argument shows "how things happen"). The mathematicians arrive at the same result in a more correct manner by proving that, because $y(t)$ is gaussian, the autocorrelation function of $\mathcal{Y}$, i.e. $\varrho_{\mathcal{Y}}(\tau)$, is equal to twice the square of the autocorrelation function of y (see ref. 1, Chapter 9, for the proof of this result):

$$\varrho_{\mathcal{Y}}(\tau) = 2\varrho_y{}^2(\tau)$$

In the present case we have (see Section 1.6.1)

$$\varrho_{\mathcal{Y}}(\tau) = 2N_0{}^2\Delta f^2\left[\frac{\sin(\pi\tau\,\Delta f)}{\pi\tau\,\Delta f}\right]^2\cos^2(2\pi\tau f_0)$$

$$\varrho_{\mathcal{Y}}(\tau) = N_0{}^2\Delta f^2\left[\frac{\sin(\pi\tau\,\Delta f)}{\pi\tau\,\Delta f}\right]^2 + N_0{}^2\Delta f^2\left[\frac{\sin(\pi\tau\,\Delta f)}{\pi\tau\,\Delta f}\right]^2\cos(4\pi\tau f_0)$$

The second term corresponds to frequencies close to $2f_0$ which are generally eliminated. This rigorously proves the result demonstrated above.

Another representation of the same gaussian random function $x(t)$ (with zero mean), which is derived from the one given above, is

$$x(t) = M(t)\sin[2\pi f_0 t + \varphi(t)]$$

where $\varrho(t)$ is a random phase which can take all values between zero and 2π with the same probability and $M(t)$ is a positive variable (variable amplitude of the sine curve randomly modulated in phase) independent of $\varphi(t)$. In order to obtain the corresponding distribution of the amplitude $M(t)$, we can write

$$M^2(t) = x^2(t) + y^2(t)$$

where

$$y(t) = M(t)\cos[2\pi f_0 t + \varphi(t)] = M(t)\sin[2\pi f_0 t + \varphi(t) + \pi/2].$$

It is clear that the amplitude distribution of $y(t)$ is also gaussian and the same as that of $x(t)$, and that $x(t)$ and $y(t)$ are independent (the position of a point in a plane depends on two independent parameters which are either M and θ or x and y).

Denoting the amplitude distributions of x, y and M by $p(x)$, $p(y)$ and $p_M(M)$ (M being the radius vector of a point with cartesian coordinates x and y) and the amplitude distribution of the polar angle φ of this point by $p_\varphi(\varphi)$ (i.e. $p_\varphi(\varphi) = 1/2\pi$), we can write

$$p(x)p(y)\,\mathrm{d}x\,\mathrm{d}y = p_M(M)p_\varphi(\varphi)\,\mathrm{d}M\,\mathrm{d}\varphi$$

$$p(x)p(y)\,M\,\mathrm{d}M\,\mathrm{d}\varphi = p_M(M)\frac{1}{2\pi}\,\mathrm{d}M\,\mathrm{d}\varphi$$

$$p_M(M) = 2\pi Mp(x)p(y)$$

so that, since

$$p(x) = \frac{1}{(2\pi N_0\Delta f)^{1/2}}\exp\left(-\frac{x^2}{2N_0\Delta f}\right)$$

and

$$p_M(M) = 2\pi M\frac{1}{2\pi N_0\Delta f}\exp\left(-\frac{x^2+y^2}{2N_0\Delta f}\right)$$

we have

$$p_M(M) = \frac{M}{N_0\Delta f}\exp\left(-\frac{M^2}{2N_0\Delta f}\right)$$

This type of probability density characterizing a variable with a Rayleigh distribution is often encountered since it characterizes the amplitude distribution of the radius vector of a point whose cartesian abscissa and ordinate have the same gaussian distribution (with zero mean). This result explains the incorrect statement often made that any noise that has been filtered using a narrow passband has a Rayleigh distribution. In fact, its distribution is gaussian, but it can be considered as a sinusoidal signal randomly (and often slowly) modulated in phase, and it is the amplitude of this signal which has a Rayleigh distribution.

It can easily be shown that the mean value of M^2 is twice the power of x either by calculating

$$\int_0^\infty \frac{M^3}{N_0\,\Delta f}\exp\left(-\frac{M^2}{2N_0\,\Delta f}\right)\mathrm{d}M$$

or by writing

$$\overline{x^2(t)} = \overline{M^2(t)\sin^2[2\pi f_0 t + \varphi(t)]}$$

(since $\varphi(t)$ and $M(t)$ are independent) which gives

$$\overline{M^2(t)} = \overline{2x^2(t)} = 2N_0\,\Delta f$$

1.9 Exercises

Exercise 1.1

1. The probability density of a variable x is shown in Fig. 1.15. Find the mean values of (a) x and (b) x^2.

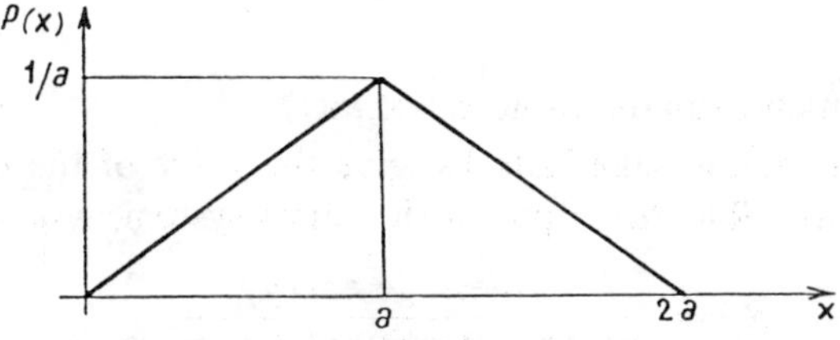

Fig. 1.15

2. A sample of the function $y(t)$ is shown in Fig. 1.16. The value of $y(t)$ varies discontinuously every second, although it remains constant for $n < t < n + 1$ (n is an integer). The values of y on each side of a jump are independent. If y_n is the value of $y(t)$ for $n < t < n + 1$, the absolute value of y_n, i.e. $|y_n|$, has a Poisson distribution and y_n has as many chances of being

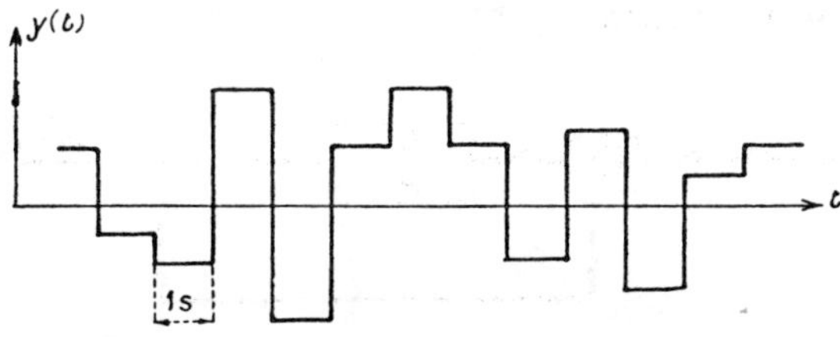

Fig. 1.16

positive as negative; this means that the probability that $|y_n|$ has a noninteger value is zero although the probability that $|y_n|$ has an integer value k is equal to

$$\frac{\lambda^k}{k!}\exp(-\lambda)$$

Give the mean value of $y(t)$, $|y(t)|$, $y^2(t)$ and $y(t)y(t-\tau)$ for $\tau > 1$ and for $\tau < 1$.

Exercise 1.2

The servo system shown in Fig. 1.17 is used for filtering a noise $n(t)$ added to a message $m(t)$. It is assumed that there is no correlation between $m(t)$ and $n(t)$ and that $m(t)$ and $n(t)$ are steady state random functions with spectral density

$$A_m^2(\omega) = \frac{1}{\pi}\frac{a^2A^2}{\pi a^2 + \omega^2}$$

and

$$A_n^2(\omega) = \frac{1}{\pi}B^2 \qquad -\infty < \omega < +\infty$$

respectively.

1. Calculate the mean square value $\overline{m^2}$ of $m(t)$.
2. Assuming that the message lasts 1 s, give the value of the corresponding signal-to-noise ratio R at the input of the servo system, where

$$R = \frac{\text{message energy}}{\text{spectral density of the noise}}$$

3. The error $\varepsilon(t)$ is defined by $\varepsilon(t) = s(t) - m(t)$. Give the expression for the autocorrelation function $\varrho_{\varepsilon\varepsilon}$ of ε as a function of the autocorrelation function $\varrho_{ss}(\tau)$ of $s(t)$, the autocorrelation function $\varrho_{mm}(\tau)$ of $m(t)$ and the cross-correlation functions ϱ_{ms} and ϱ_{sm} of $m(t)$ with $s(t)$ and $s(t)$ with $m(t)$ respectively.
4. Deduce from $\varrho_{\varepsilon\varepsilon}$ the expression giving the spectral density $A_\varepsilon^2(\omega)$ as a function of $A_m^2(\omega)$, $A_n^2(\omega)$ and $H(p)$, where $H(p)$ is the closed-loop transfer function of the servo system.

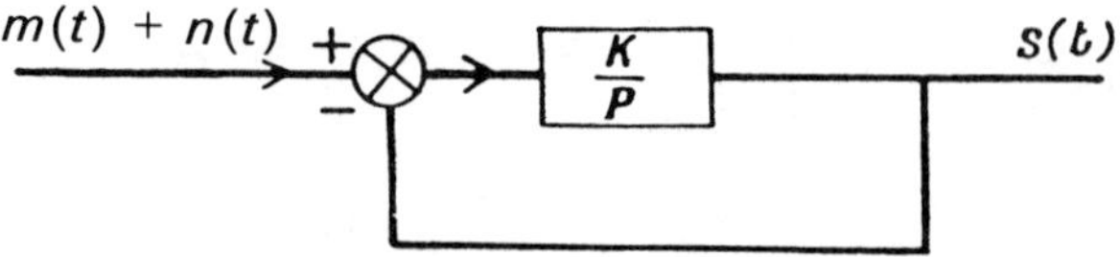

Fig. 1.17

5. Obtain this expression again, simply by using the fact that the servo system is a linear system.

(The method described in 3 and 4 has an educational aim. It goes without saying that an engineer using it instead of the simple method in 5 and without making the effort to give mathematical instruction to the recipients would be extremely incompetent.)

6. Deduce from the expression for the spectral density the mean square value $\overline{\varepsilon^2}$ of the error $\varepsilon(t)$.

7. Calculate the value of K which minimizes this error.

8. For this value of K give the value of the signal-to-noise ratio at the output of the system, i.e. the ratio of the mean square value of the useful output signal to the mean square value of the output noise.

Exercise 1.3

Consider the random function $x(t)$ with zero mean and power equal to unity. Another random function

$$e(t) = x(t)\sin(\omega_0 t)$$

is derived from it.

1. It is assumed that the amplitude distribution of $x(t)$ is gaussian. Give the general expression for the nth-order moment of $e(t)$ for n odd and n even. The mean value of $(\sin t)^{2p}$, where p is an integer, is given by

$$\frac{1}{2} \times \frac{3}{4} \times \frac{5}{6} \times \frac{7}{8} \times \cdots \times \frac{2p-1}{2p}$$

Under the same assumptions, and denoting the autocorrelation function of $x(t)$ by $\varrho_x(\tau)$, give the expression for the autocorrelation function of $y(t) = e^2(t)$.

2. No further assumptions are made about the nature of the amplitude distribution of $x(t)$ but we now assume that $\varrho_x(\tau) = A\exp(-\alpha|\tau|)$. Give the value of A.

3. Calculate the autocorrelation function $\varrho(\tau)$ of $e(t)$ and plot the representative curve of this function.

4. Derive the spectral density of $e(t)$ from this.

5. The signal $e(t)$ is sent to the input of the system shown in Fig. 1.18. Calculate the mean quadratic value of the error i.e.

$$\overline{\varepsilon^2} = \overline{[e(t) - s(t)]^2}$$

It is sufficient to give the result in the form of an integral.

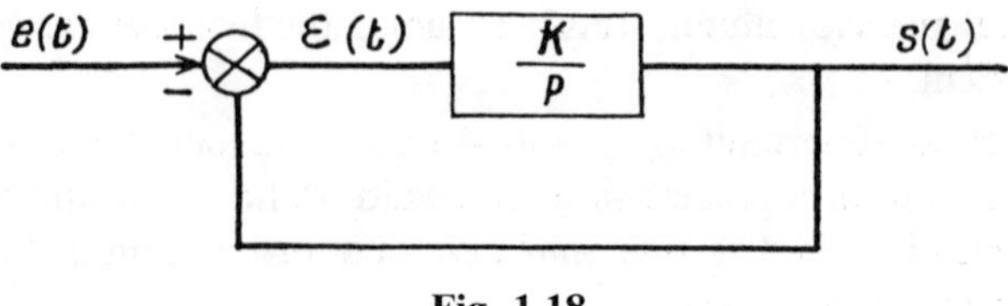

Fig. 1.18

Exercise 1.4

Consider a rectangular search radar reflector (the rectangular shape is chosen in order to simplify the calculations) rotating about a vertical axis situated in its plane of symmetry. The reflector is illuminated by a primary source located at its focus so that we can assume that we are dealing with a uniformly illuminated plane reflector (this assumption of uniform illumination is also made in order to simplify the calculations). All the points of the plane are illuminated in phase except for defects. The antenna rotates at Ωrad s^{-1} opposite a transmitter A located quite a large distance away in the horizontal symmetry plane of the radar antenna and transmitting a pure high frequency sinusoidal wave (Fig. 1.19). The signal received by the antenna is detected directly by a square-law detector and is then amplified in a video amplifier so that the output of this receiver is a voltage V_B proportional to the power received by the antenna at a given time. Finally this video signal is filtered (Fig. 1.20) to give a signal V_S. The aim of this exercise is to study the power of V_S as a function of the nature of the final filter.

1. We consider a trirectangular trihedron rotating with the radar antenna about the vertical axis of the radar Oxyz, where O is the center of the reflector, Ox is a horizontal axis contained in the plane of the reflector and Oy is an axis produced by the normal to the reflector in the direction of radiation (see Fig. 1.21). We denote by $\overline{H(\theta)}$ the value of the electrical field at the level of

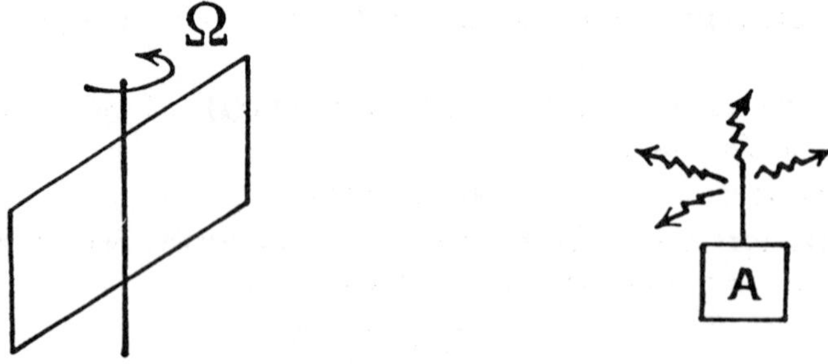

Fig. 1.19

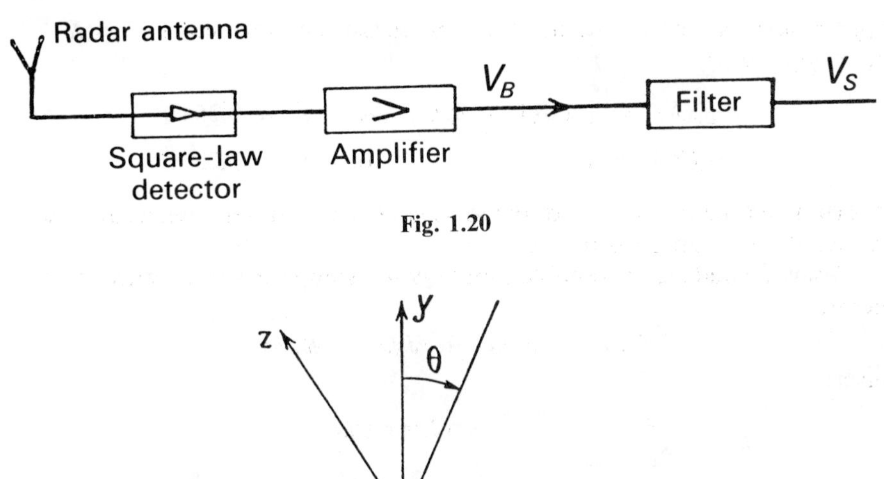

Fig. 1.20

Fig. 1.21

transmitter A as a function of the angle θ between the direction of A (seen from the center O of the reflector) and the direction of the normal to this reflector. It is further assumed (also in order to simplify calculations) that the illumination of a point on the reflector is a function of x only, i.e. $\bar{E}(x)$.

Show that, apart from a factor,

$$\overline{H(\theta)} = \int_{-L}^{+L} \overline{E(x)} \exp\left(\frac{2\pi j x \sin\theta}{\lambda}\right) dx$$

where $2L$ is the horizontal aperture of the radar antenna ($L = 5$ m) and λ is the wavelength under consideration ($\lambda = 0.1$ m). Set $u = (\sin\theta)/\lambda$ and give $H(u)$ in terms of $E(x)$ and vice versa.

2. The assumption that the antenna is uniformly illuminated indicates that $E(x)$ is a complex number with constant modulus A for $-L < x < L$ ($E(x)$ is zero for $x^2 > L^2$). However, the phase $\varepsilon(x)$ is not zero because of defects in the manufacture of the antenna. Rather, it is the sum of a periodic defect

$$\varepsilon_2(x) = \varepsilon_0 \sin\left(\frac{2\pi x}{D}\right)$$

due to the periodic structure of the antenna which consists of vertical panels of width D ($D = 2$ m) and a defect $\varepsilon_3(x)$ which is poorly known and therefore appears as a random function of x. The autocorrelation function $\varrho_3(X)$ of $\varepsilon_3(x)$ is given by

$$\varrho_3(X) = \varrho_3(0)(1 - 2|X|) \quad \text{for} \quad X^2 < 0.25$$

$$\varrho_3(X) = 0 \quad \text{for} \quad X^2 > 0.25$$

It is assumed that $\varepsilon_0 = 0.1$ rad, that ε_3 is gaussian (with zero mean) and that the standard deviation is 0.1 rad.

Show by making reasonable simplifying assumptions that, apart from a factor,

$$H(u) = H_1(u) + H_2(u) + H_3(u)$$

where

$$H_1(u) = \int_{-L}^{+L} \exp(2\pi jxu)\,dx$$

$$H_2(u) = j\int_{-L}^{+L} \varepsilon_2(x)\exp(2\pi jxu)\,dx$$

$$H_3(u) = j\int_{-L}^{+L} \varepsilon_3(x)\exp(2\pi jxu)\,dx$$

Give $H_1(u)$ as a function of u for $-10\,\text{m}^{-1} \leqslant u \leqslant 10\,\text{m}^{-1}$. In a similar manner give $H_2(u)$ as a function of u in the same range of values of u. Finally, give $H_3(u)$ as a function of u in the same range of values of u (assume that $2L$ is large relative to 0.5). Refer to the definition of the spectral density of a random function such as that given in Section 1.3.

3. Use the results obtained above to derive the curve representing the parasitic lobes due to the antenna defects, i.e. the curve representing $H_2(\theta) + H_3(\theta) = H_D(\theta)$ as a function of θ for $+\pi/2 \leqslant \theta \leqslant -\pi/2$.

(*Note.* The lobes corresponding to $u^2 > 10$ are "imaginary" lobes which do not exist physically, at least at a large distance from the antenna.)

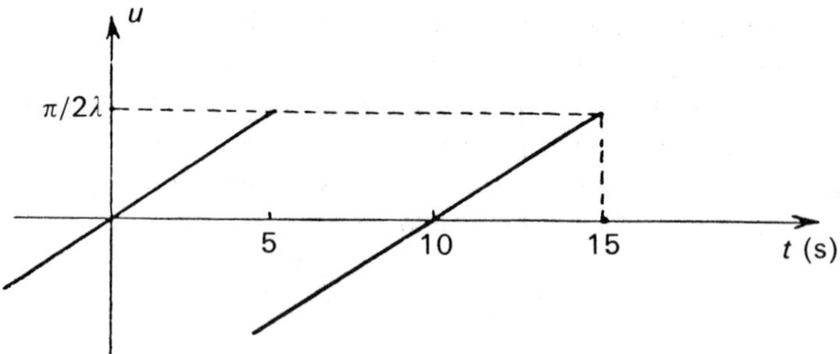

Fig. 1.22

4. It is assumed that the antenna rotates at a constant speed Ω of 6 rev min^{-1}, i.e. $\theta = \Omega t$. Under these conditions $u = (\sin \Omega t)/\lambda$, but in order to simplify the calculations we will assume that the variations in u as a function of time are represented by the curve shown in Fig. 1.22. Show (without performing the calculations) the Fourier transform of V_B.

5. Show the spectral distribution of the power of V_B (spectral density). Compare the powers of the output signals $V_S(t)$ obtained when the final filter allows frequencies to pass in the following ranges: 0.05–0.15 Hz, 0.95–1.05 Hz (tenth harmonic) and 0.95–5.05 Hz.

Chapter 2

Signal and Noise. The Ideal Receiver

2.1 The radar problem

By definition, a radar system is an apparatus for RAdio Detection And Ranging, whose object is to say: "There is something at such and such a distance." The first concepts concerning the electromagnetic detection of targets were rather simple (and they have still not been superseded). A detector consisted of an antenna whose impedance was modified by the proximity of an object which indicated "something abnormal". However, it was soon realized that this information alone was inadequate and attempts were made to determine the position of the "abnormal object". At the same time self-excited oscillator tubes such as magnetrons were being developed which were capable of transmitting high power for a short time. Hence the simple idea was proposed of transmitting brief but very powerful signals and measuring the time between transmission of a signal and reception of the reflection of that signal from the target; this time is proportional to the distance of the target from the transmitter. Since World War II most radar systems have used this principle.

In practice there are a few difficulties. Let us consider a transmitter capable of transmitting a short signal (1 μs) consisting of an extremely powerful (1 MW) high frequency wave. Even if the transmission is concentrated in a reduced solid angle (which will make it possible to obtain a more accurate estimate of the direction of the target) by using a large antenna and if the receiving antenna is also large, the signal reflected from a normal target located at a reasonable distance (a few hundred kilometers) will have a power of the order of 10^{-15} W. It will have the same shape as the transmitted signal but will be very weak (we shall discuss this point further below). The weakness of the received signal is not a problem as it is possible to amplify it. However, this small signal is, unfortunately, always accompanied by background noise originating from atmospheric noise which cannot be reduced to zero or from

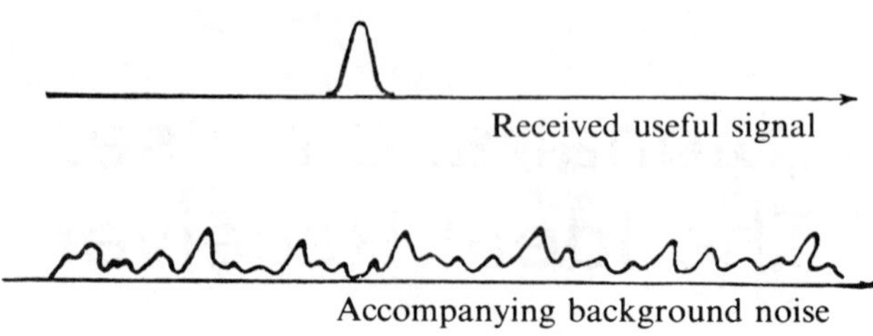

Fig. 2.1

noise coming from the receiver which cannot be reduced to zero or from jamming by an uncooperative enemy (or an incompetent friend). Since this background noise has the effect of concealing the signal, we are interested in filtering it by reducing the passband of the receiver. If this is done, an effective gain is obtained until the passband is approximately equal to the reciprocal of the duration of the signal. If the passband is reduced further, the useful signal is reduced more quickly than the noise. Care is usually taken to avoid doing this, and the best compromise of a passband approximately equal to the reciprocal of the duration τ of the transmitted signal is adopted. Under these conditions the useful signal is no longer the beautiful rectangular signal of the transmission but a rounded signal, and the noise is a sequence of signals with the same shape. This means that it is possible to make a mistake and to identify a noise peak as a signal (false alarm) or assume that a signal is a noise peak (absence of detection) (Fig. 2.1). Therefore, how can we increase the range of a radar system?

It is possible to increase the transmitted power, but this is an approach where we are quickly brought down to earth by considerations such as cost, arcing in the radar etc. The passband of the receiver can be reduced in order to decrease the noise but at the same time it is necessary to increase the duration of the transmitted signal. We then encounter two difficulties with classical radar systems: the longer is the transmitted signal, the less precise is the measurement of distance; two targets, one of which is behind the other, can only be distinguished if their distance increases with the duration of the transmitted signal.

When the problems of obtaining considerable ranges or of obtaining reasonable ranges despite substantial jamming or of obtaining accurate measurements of distance and good distance resolution at reasonable ranges arose it became clear that they could not be solved using classical radar. At this point the mathematical proposals of scientists such as Woodward were considered. Woodward's theories, correctly applied and extended, have led to new types of radar systems which sometimes have qualities (velocity measurement or elimination of parasitic echoes) that classical radar systems did not have. These theories are based on probability calculus and the sampling theorem (see Chapter 1, Section 1.7).

Since the choice of the transmission signal used in classical radar (all-or-nothing amplitude modulation) was not based on any firm foundations, it is no longer necessary, now that a wide range of amplifier tubes is available, to impose such a restriction. It will therefore be assumed that, in general, a signal $S(t)$ is transmitted which may *a priori* take any form and which lasts a time T and has a Fourier transform $\Phi(f)$.

What will be received? In the absence of a target, a noise $n(t) = y(t)$ is received. In the presence of a target, we receive

$$y(t) = kS(t - t_{01}) + n(t)$$

where t_{01} gives the distance to the target.

We must first determine what procedures should be adopted to ensure the minimum error in the reply to the question "Does $y(t)$ contain $kS(t - t_{01})$ or not?" If the answer to this question is in the affirmative, a second question arises: "What is the most probable value of t_{01}?"

Before analyzing this problem, we will examine some simple problems in probability calculus.

2.2 *A priori* and *a posteriori* probabilities

Consider a bridge salon in which two pairs of players X and Y are going to play a game of bridge. A person enters and sits down amongst the spectators: he knows neither players X nor players Y, and the cards have not been distributed. Can he predict the result of the game? He can say that the probability that the pair X will win is $P_1 = 0.5$. This is an *a priori* probability. However, while the cards are being distributed, the other spectators tell him that pair X are world bridge champions and that pair Y are just competent players. He can now say that the probability that X will win is $P_2 = 0.9$. This is an *a posteriori* probability with respect to the previous state of knowledge. Once the cards are distributed, the spectators know more about the game and it appears that pair Y have 37 points in their hands (out of 40); the spectator can deduce from this that the probability that pair X will win the game is no more than $P_3 = 0.2$ (especially if pair Y never bid a slam). Then P_3 is an *a posteriori* probability with respect to P_2, but P_2 is an *a priori* probability with respect to P_3, and so on.

In order to move on to something less simple, let us examine the following game [2]. Mr Cooper has five dice of which four are normal but the

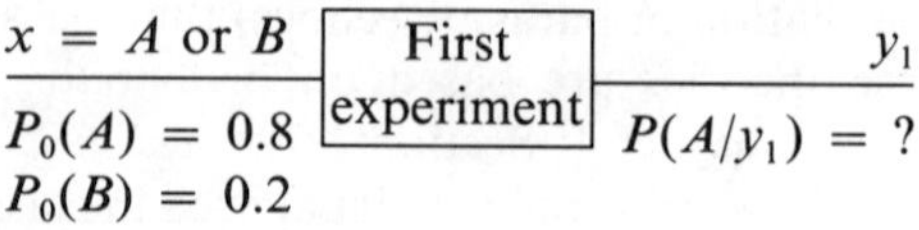

$x = A$ or B	First experiment	y_1
$P_0(A) = 0.8$		$P(A/y_1) = ?$
$P_0(B) = 0.2$		

Fig. 2.2

fifth has three sides with the number six and three sides with the number one. The game between Mr Cooper and Mr Shaw proceeds as follows. Cooper chooses a die at random and asks Shaw if the die is normal or biased, i.e. if $x = A$ (normal die) or $x = B$ (biased die). *A priori*, the probability that $x = A$ is of course 0.8:

$$P_0(A) = 0.8$$

Before replying, Shaw has the right to throw the die twice and see what number comes up each time. He throws once and obtains a number y_1 (Fig. 2.2). If y_1 is 2, 3, 4 or 5, Shaw must logically reply that $x = A$. In other words

$$\text{probability}(x = A) = 1 \text{ if } y_1 = 2$$

$P(A \text{ if } y_1) \equiv P(A/y_1) = 1$ for $y_1 = 2, 3, 4$ or 5 ($P(A/y_1)$ is an *a posteriori* probability with respect to $P_0(A)$). However, if $y_1 = 1$, what is the *a posteriori* probability that $x = A$, i.e. what is $P_1(A)$? The answer is not very difficult. The probability of having $x = A$ and $y_1 = 1$ at the same time ($P(x = A, y_1 = 1)$) is given by

$$(\textit{a priori} \text{ probability of having } x = A)$$
$$\times\ (\text{probability of having } y_1 = 1 \text{ if } x = A)$$

which can also be written as

$$P(x y_1) = P(x) \times P(y_1/x)$$

However, we can also write

$$P(x = A, y_1 = 1) = (\text{probability of having } y_1 = 1)$$
$$\times\ (\text{probability of having } x = A \text{ if } y_1 = 1)$$

or

$$P(x y_1) = P(y_1) \times P(x/y_1)$$

It follows from this that

$$P(x) \times P(y_1/x) = P(y_1) \times P(x/y_1)$$

and hence, if we set $P(y_1) = 1/k_1$,

$$P(x/y_1) = k_1 P(x) \times P(y_1/x)$$

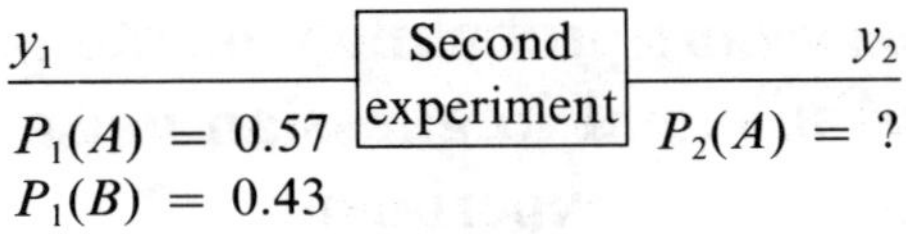

y_1	Second experiment	y_2
$P_1(A) = 0.57$		$P_2(A) = ?$
$P_1(B) = 0.43$		

Fig. 2.3

Thus we can write

$$P_1(A) = k_1 P_0(A) \times P(1/A)$$

$$P_1(B) = k_1 P_0(B) \times P(1/B)$$

where, of course, $P_1(A) + P_1(B) = 1$. Now, it is obvious that

$$P(1/A) = 1/6$$

(if the die is normal, there is one chance out of six of drawing one) and that

$$P(1/B) = 0.5$$

(if the die is biased, there is one chance out of two of drawing one). It follows from this that

$$P_1(A) = k_1 \times 0.8 \times \tfrac{1}{6}$$

$$P_1(B) = k_1 \times 0.2 \times \tfrac{1}{2}$$

and thus

$$1 = k_1 \left(\frac{0.8}{6} + \frac{0.2}{2} \right)$$

$$k_1 = \frac{6}{1.4}$$

Therefore $P_1(A) = 0.57$ and $P_1(B) = 0.43$.

Shaw then throws a second time and another one appears ($y_2 = 1$). For this second experiment, $P_1(A)$ has become the *a priori* probability (Fig. 2.3). This time, we can write

$$P_2(A) = k_2 P_1(A) \times P(1/A)$$

$$P_2(B) = k_2 P_1(B) \times P(1/B)$$

$$P_2(A) = k_2(0.57 \times \tfrac{1}{6}) = 0.095k_2 = 0.31$$

$$P_2(B) = k_2(0.43 \times \tfrac{1}{2}) = 0.215k_2 = 0.69$$

Therefore after two experiments the probability that the die is normal is 0.31.

2.3 *A posteriori* probability of the presence or absence of a signal in gaussian noise after one experiment

We assume that a signal $S(t)$ can take only two values $S(t) = 0$ or $S(t) = V$ with an equal *a priori* probability:

$$P_0(S = 0) = P_0(S = V) = 0.5$$

This signal $S(t)$ arrives accompanied by a gaussian noise $n(t)$. At a given moment, $S(t) + n(t) = y(t)$ is measured and a value y_0 is found. What is the *a posteriori* probability that $S = V$, i.e. $P_1(S = V)$?

Let us assume that $S = 0$. To say that the noise is gaussian is to say that the probability that $y(t)$ lies between y and $y + \mathrm{d}y$ is given by

$$P(y/S = 0)\,\mathrm{d}y = k \exp\left(-\frac{y^2}{2N}\right)\mathrm{d}y$$

where N is the mean value of y^2 if $S = 0$, i.e. the mean value of n^2. However, if it is assumed that $S = V$, it is obvious that

$$P(y/S = V)\,\mathrm{d}y = k \exp\left[-\frac{(y - V)^2}{2N}\right]\mathrm{d}y$$

Then an identical calculation to that in the preceding section gives

$$P_1(V) = k_1 P_0(V) \times P(y_0/S = V)$$

$$P_1(0) = k_1 P_0(0) \times P(y_0/S = 0)$$

or

$$P_1(V) = k_1 \times 0.5 \times \exp\left[-\frac{(y_0 - V)^2}{2N}\right]$$

$$P_1(0) = k_1 \times 0.5 \times \exp\left(-\frac{y_0^{\,2}}{2N}\right)$$

If $y_0 = 0.5V$,

$$P_1(V) = P_1(0) = 0.5$$

and if $y_0 > 0.5$,

$$P_1(V) > P_1(0)$$

We now consider two examples: (i) if $N = 0.1V^2$ and $y_0 = V$,

$$P_1(V) = 0.5k_1 = 1 - 0.66 \times 10^{-2}$$

$$P_1(0) = 0.5k_1 \times 0.66 \times 10^{-2} = 0.66 \times 10^{-2}$$

i.e. there is less than one chance in 100 that $S = 0$; (ii) if $N = 10V^2$ and $y_0 = V$,

$$P_1(V) = 0.5k_1 = 0.515$$

$$P_1(0) = 0.5k_1 \times 0.95 = 0.485$$

i.e. there are almost as many chances that $S = 0$ as that $S = V$.

Note In the first example, it is possible to say that since the "power" of the signal is V^2 and the "power" of the noise is $0.1V^2$, the signal-to-noise ratio is equal to 10 (10 dB). In the second example the signal-to-noise ratio is equal to 0.1 (−10 dB).

2.4 *A posteriori* probability of the presence or absence of a signal in gaussian noise after several successive experiments [2]

We now want to determine whether or not there is a signal under the following conditions: if there is a signal ($x = A$) without noise, we will receive in succession $S_1(A) = 10$, $S_2(A) = -10$, $S_3(A) = 10$; if there is no signal ($x = B$) and no noise, we will receive in succession $S_1(B) = 0$, $S_2(B) = 0$, $S_3(B) = 0$. However, in practice S_1, S_2 and S_3 are each received together with an additional random value, n_1, n_2 and n_3 respectively, obtained by sampling a white gaussian noise whose mean square value is N (i.e. the values of n_1, n_2 and n_3 are independent as in a game of dice where each value sampled has a probability of 1/6 even if it has previously been thrown six times in succession).

Let $P_0(A)$ and $P_0(B)$ be the *a priori* probabilities that $x = A$ and $x = B$ respectively. After $y_1 = S_1 + n_1$ has been received, we can write

$$P_1(A) = k_1 P_0(A) \exp\left\{-\frac{[y_1 - S_1(A)]^2}{2N}\right\}$$

$$P_1(B) = k_1 P_0(B) \exp\left\{-\frac{[y_1 - S_1(B)]^2}{2N}\right\}$$

Substituting for $S_1(A)$ and $S_1(B)$ gives

$$P_1(A) = k_1 P_0(A) \exp\left[-\frac{(y_1 - 10)^2}{2N}\right]$$

$$P_1(B) = k_1 P_0(B) \exp\left(-\frac{{y_1}^2}{2N}\right)$$

After $y_2 = S_2 + n_2$ has been received, we can write

$$P_2(A) = k_2 P_1(A) \exp\left[-\frac{(y_2 + 10)^2}{2N}\right]$$

$$P_2(B) = k_2 P_1(B) \exp\left(-\frac{y_2^2}{2N}\right)$$

Finally, after $y_3 = S_3 + n_3$ has been received, we have

$$P_3(A) = k_3 P_2(A) \exp\left[-\frac{(y_3 - 10)^2}{2N}\right]$$

$$P_3(B) = k_3 P_2(B) \exp\left(-\frac{y_3^2}{2N}\right)$$

In summary

$$P_3(A) = kP_0(A) \exp\left\{-\frac{\Sigma[y_i - S_i(A)]^2}{2N}\right\}$$

$$P_3(B) = kP_0(B) \exp\left(-\frac{\Sigma y_i^2}{2N}\right)$$

However, we can also write

$$P_3(A) = k'P_0(A) \exp\left[-\frac{\Sigma S_i(A)^2}{2N}\right] \exp\left[\frac{\Sigma y_i S_i(A)}{N}\right]$$

$$P_3(B) = k'P_0(B)$$

For example, in the case where $N = 100$, $n_1 = +10$, $n_2 = +7$, $n_3 = -5$ and $x = A$ (which is generally ignored) we receive $y_1 = 20$, $y_2 = -3$, $y_3 = +5$, and we obtain

$$\frac{\Sigma S_i(A)^2}{2N} = +1.5$$

We also find

$$\frac{\Sigma y_i S_i(A)}{N} = \frac{200 + 30 + 50}{100} = 2.8$$

$$P_3(A) = k'P_0(A)\,e^{1.3} = k'P_0(A) \times 3.7$$

$$P_3(B) = k'P_0(B)$$

If, *a priori*, a signal is equally likely to be present or absent

$$P_0(A) = P_0(B) = 0.5$$

and it follows that

$$P_3(A) = 0.79$$

$$P_3(B) = 0.21$$

A posteriori, there are four chances out of five that $x = A$.

There is an obvious difference between the example discussed above and the radar problem. So far it has been assumed that there is only one possible position for the signal. In the case of radar, the location of the object is unknown, i.e. we do not know *a priori* where the signal is, if it exists. Let us look again at the preceding example in order to try and overcome this obstacle. Assume that 10 successive signals are received. If the signals are noise free, we would receive

Position 1

$S_{11} = 10 \quad S_{12} = -10 \quad S_{13} = 10 \quad S_{14} = 0 \; 0 \; 0 \; 0 \quad 0 \quad 0 \quad S_{1\,10} = 0$

Position 2

$S_{21} = 0 \quad S_{22} = 10 \quad S_{23} = -10 \quad S_{24} = 10 \; 0 \; 0 \; 0 \quad 0 \quad 0 \quad S_{2\,10} = 0$

⋮

Position 7

$S_{71} = 0 \quad S_{72} = 0 \quad S_{73} = 0 \quad S_{74} = 0 \; 0 \; 0 \; 10 \; -10 \quad 10 \quad S_{7\,10} = 0$

Position 8

$S_{81} = 0 \quad S_{82} = 0 \quad S_{83} = 0 \quad S_{84} = 0 \; 0 \; 0 \; 0 \quad 10 \; -10 \quad S_{8\,10} = 10$

In other words there are eight possible positions θ from 1 to 8.

Let us assume that the received signal is the sum of a signal at position 2 ($x = A, \theta = 2$) and a gaussian noise ($N = 100$) such that $n_1, n_2, \ldots, n_{10}$ have the following values: 1, 0, 15, 5, -1, 1, 4, 2, -17, 3. The received signal y is given by $y_1 = 1$, $y_2 = 10$, $y_3 = 5$, $y_4 = 15$, $y_5 = -1$, $y_6 = 1$, $y_7 = 4$, $y_8 = 2$, $y_9 = -17$ and $y_{10} = 3$. What response can a receiver give which does not know *a priori* that $x = A$ and $\theta = 2$ but which only knows *a priori* the probabilities $p_0(A\theta)$ that there is a signal at θ and $P_0(B)$ that there is not a signal at θ?

Since y is known, the *a posteriori* probabilities can be written

$$P_1(A\theta) = k_1 P_0(A\theta) \exp\left(-\frac{\Sigma_i S_{\theta i}^2}{2N}\right) \exp\left(+\frac{\Sigma_i y_i S_{\theta i}}{N}\right)$$

$$P_1(B) = k_1 P_0(B)$$

(In fact, it is sufficient to perform the calculation over the duration of the signal since $S_\theta = 0$ outside the signal.) Then the *a posteriori* probability for which there is a signal at any position can be written as

$$P_1(A) = k_1 \sum_\theta P_0(A\theta) \exp\left(-\frac{\Sigma_i S_{\theta i}^2}{2N}\right) \exp\left(+\frac{\Sigma_i y_i S_{\theta i}}{N}\right)$$

The following results are obtained by calculation:

$$-\frac{\Sigma_i S_{\theta i}^2}{2N} = -1.5 \quad \text{independent of } \theta$$

$$\frac{\Sigma_i y_i S_{1i}}{N} = -0.4 \quad \frac{\Sigma y S_2}{N} = 2 \quad \frac{\Sigma y S_3}{N} = -1.1$$

$$\frac{\Sigma y S_4}{N} = 1.7 \quad \frac{\Sigma y S_5}{N} = 0.2 \quad \frac{\Sigma y S_6}{N} = -0.1$$

$$\frac{\Sigma y S_7}{N} = -1.5 \quad \frac{\Sigma y S_8}{N} = 2.2$$

The corresponding exponentials are

0.22		
0.67	7.4	0.33
5.5	1.2	0.92
0.3	9.1	

If, in addition, it is assumed that

$$P(A\theta) = 0.5/8 \quad \text{independent of } \theta$$

and

$$P(B) = 0.5$$

we obtain

$$P_1(A) = k_1 \times (0.5/8) \times 0.22(0.67 + 7.4 + \cdots + 9.1)$$

$$= k_1 \times (0.5/8) \times 0.22 \times 25.4 = 0.35k_1 = 0.41$$

$$P_2(B) = 0.5k_1 = 0.59$$

Thus when we know the values of y, the nature of the signal (if it exists), the value of N and the *a priori* probabilities defined above, we can deduce the following probabilities for the location of the signal:

no signal	0.57
signal at position 1	$(0.41 \times 0.67)/25.4 = 0.01$
signal at position 2	0.12
signal at position 3	0.005
signal at position 4	0.09
signal at position 5	0.02
signal at position 6	0.01
signal at position 7	0.005
signal at position 8	0.15

Note If we had been sure *a priori* that there was a signal and that the eight positions were *a priori* equally probable, the experiment would have led to the following conclusion:

there is a probability of 0.29 that there is a signal at position 2
there is a probability of 0.37 that there is a signal at position 8

We now assume that the experiment was carried out in the presence of less noise ($N = 25$ instead of $N = 100$) and the noise samples were divided by 2, so that the received signal (signal at position 2 plus noise) becomes

$$0.5 \quad 10 \quad -2.5 \quad 12.5 \quad -0.5 \quad 0.5 \quad 2 \quad 1 \quad -8.5 \quad 1.5$$

We then find

$$\frac{\Sigma_i \, S_{\theta i}{}^2}{2N} = -6$$

$$\frac{\Sigma y S_1}{N} = -4.8 \quad \frac{\Sigma y S_2}{N} = 10 \quad \frac{\Sigma y S_3}{N} = -6.2$$

$$\frac{\Sigma y S_4}{N} = 5.4 \quad \frac{\Sigma y S_5}{N} = 0.4 \quad \frac{\Sigma y S_6}{N} = -0.2$$

$$\frac{\Sigma y S_7}{N} = -3 \quad \frac{\Sigma y S_8}{N} = 4.4$$

and the corresponding exponentials are

2.5×10^{-3}		
0.007	22×10^3	0.002
0.22×10^3	1.5	0.8
0.05	0.08×10^3	

Then

$$P_1(A) = k_1 \times (0.5/8) \times 2.5 \times 10^{-3}(22\,300) = 3.5k_1 = 0.875$$

$$P_1(B) = k_1 \times 0.5 = 0.125$$

and we can deduce the following probabilities for the location of the signal:

no signal	0.125
signal at position 1	5×10^{-7}
signal at position 2	0.86
signal at position 3	8×10^{-8}
signal at position 4	0.01
signal at position 5	6×10^{-5}
signal at position 6	3×10^{-5}
signal at position 7	2×10^{-6}
signal at position 8	0.003

Finally we reduce the noise even further ($N = 6.25$) such that the noise samples are again divided by 2 and the received signal (signal at position 2 plus noise) becomes

$$0.25 \quad 10 \quad -6.25 \quad 11.25 \quad -0.25 \quad 0.25 \quad 1 \quad 0.5 \quad -4.25 \quad 0.75$$

This time we obtain

$$-\frac{\Sigma_i S_{\theta i}{}^2}{2N} = -24$$

$$\frac{\Sigma y S_1}{N} = -26 \quad \frac{\Sigma y S_2}{N} = 44 \quad \frac{\Sigma y S_3}{N} = -28$$

$$\frac{\Sigma y S_4}{N} = 19 \quad \frac{\Sigma y S_5}{N} = 1 \quad \frac{\Sigma y S_6}{N} = 0$$

$$\frac{\Sigma y S_7}{N} = -6 \quad \frac{\Sigma y S_8}{N} = 9$$

with the corresponding exponentials

$$3.7 \times 10^{-11}$$

0	1.3×10^{19}	0
2×10^{8}	3	0
0	10^{4}	

Then

$$P_1(A) = k_1 \times (0.5/8) \times 3.7 \times 10^{-11} \times 1.3 \times 10^{19} = 1 - 2 \times 10^{-8}$$

and

$$P_1(B) = k_1 \times 0.5 = 2 \times 10^{-8}$$

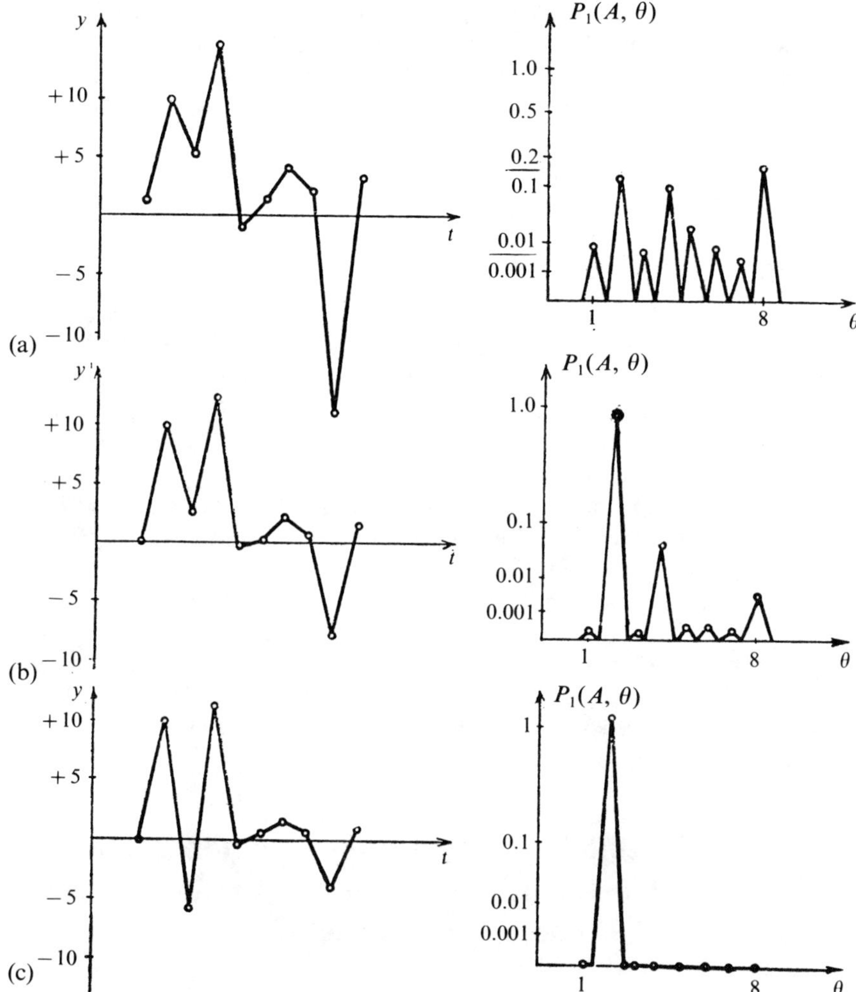

Fig. 2.4 Received signal as a function of time (left-hand side) and the probability that the signal exists as a function of the position θ (right-hand side) for the same signal with three different signal-to-noise ratios $\Sigma S_\theta^2/2N$: (a) 1.5; (b) 6; (c) 24.

Thus the probability that there is no signal or that there is a signal at a position other than 2 is less than 10^{-7}.

These results are illustrated in Fig. 2.4, from which the following conclusions can be drawn: for a low signal-to-noise ratio, all positions have probabilities of the same order and the most probable position is not the true position; when the signal-to-noise ratio is multiplied by 4 (increased by 6 dB) three positions are significantly more probable than the others, and the most probable position is the true position; finally, when the signal-to-noise ratio is increased by a further 6 dB, there is only one probable position which is the true position.

Before starting our analysis of the radar problem, the following mathematical points arising from the preceding discussion should be noted. Once the expressions

$$\frac{\Sigma y S_1}{N}, \ldots, \frac{\Sigma y S_\theta}{N}$$

have been calculated, no further reference to y is necessary. In order to calculate $P_1(A\theta)$, which is the aim of the operation, we only need to know these expressions and the *a priori* probabilities of the energy of the useful signal related to the energy of the interfering signal. In the final analysis the information obtained from a knowledge of y (which makes it possible to obtain *a posteriori* probabilities from *a priori* probabilities) is derived from a knowledge of only the curve $\Sigma y S_\theta / N$ as a function of θ, and $\Sigma S_\theta{}^2/2N$ which characterizes the signal-to-noise ratio.

It should be noted that searching for the maximum of $\Sigma y S_i / N$, i.e. searching for the correlation between y and each of the S_i, is only the mathematical consequence of the fact that we are seeking the minimum of $\Sigma(y - S_i)^2/2N$, i.e. we are looking for the most probable value of S_i by applying the law of least squares and looking for the value of S_i which gives the smallest residue.

2.5 Probability of the presence or absence of a continuous signal in continuous gaussian noise. The ideal receiver [2]

In the previous section we expressed in mathematical form the *a posteriori* probabilities of the existence of a signal at a given position under the following conditions: the signal is received at successive moments in time and the noise associated with each measurement is independent of the noise of the preceding measurement; the signal can only occupy a finite number of positions. The radar problem is less simple: we wish to determine whether a signal of the form $S(t - t_{01})$ exists in the accompanying noise $n(t)$ under the following conditions†: $S(t)$ is known but occurs in the form of a continuous

† In practice the signal takes the form $kS(t - t_{01})$ (see Section 2.1). However, for simplicity we let $k = 1$. It can easily be shown that this simplification does not affect the conclusions.

signal; $n(t)$ is known only in terms of its statistical properties and is also continuous; t_{01} can take all values t_0 in a given interval. In the classical method, we try to replace the continuous functions $S(t)$ and $n(t)$ by successive samples so that we can examine successive and independent experiments.

Suppose that the signal $S(t)$ lasts for a time T and occupies a frequency band Δf ($\Phi(f)$ is zero for $|f| > \Delta f/2$) (i.e. $S(t)$ is assumed to be a video signal). Suppose further that the noise $n(t)$ accompanying the possible signal $S(t - t_{01})$ is gaussian with a mean power N and occupies the same frequency band Δf in a uniform manner. We define the *a priori* probability that there is a signal $S(t - t_0)$ (signal at t_0) by $p(t_0)$. In this case $p(t_0)$ is a probability density. In fact we are measuring $y(t)$ which is the sum of $n(t)$ and a possible signal $S(t - t_0)$. What is the *a posteriori* probability distribution $p(t_0/y)$ if $y(t)$ is known?

It has been shown (see Section 1.7) that if $y(t)$ is sampled at consecutive time intervals separated by $1/\Delta f$, the number of independent measurements made is a maximum. Therefore let $t_1, t_2, \ldots, t_n$ be the $T\Delta f$ sampling times and $y(t_1), y(t_2), \ldots, y(t_n)$ be the corresponding values of $y(t)$. We can then write, as in Section 2.4,

$$p(t_0/y) = k_1 p(t_0) \exp\left\{-\frac{\Sigma[y(t_i) - S(t_i - t_0)]^2}{2N}\right\}$$

$$p(t_0/y) = k_2 p(t_0) \exp\left[-\frac{\Sigma S(t_i - t_0)^2}{2N}\right] \exp\left\{\frac{\Sigma[y(t_i)\, S(t_i - t_0)]}{N}\right\}$$

$$p(t_0/y) = k_2 p(t_0) \exp\left[-\frac{\Sigma S(t_i - t_0)^2 (1/\Delta F)}{2N/\Delta F}\right] \exp\left[\frac{\Sigma y(t_i)\, S(t_i - t_0)(1/\Delta F)}{N/\Delta F}\right]$$

Since $1/\Delta f$ is the interval between two consecutive t_is, this expression can be replaced by

$$p(t_0/y) = k_2 p(t_0) \exp\left[-\frac{1}{2N_0}\int_T S(t - t_0)^2\, \mathrm{d}t\right] \exp\left[\frac{1}{N_0}\int_T y(t)\, S(t - t_0)\, \mathrm{d}t\right] \tag{2.1}$$

where $N_0 = N/\Delta f$ represents the noise power density per unit bandwidth, i.e. per hertz.

Remark 2.1

It should be remembered that in the derivation of this expression it is assumed that the noise $n(t)$ is gaussian. This is usually a valid assumption for the following reasons: the internal noise of a radar receiver is gaussian; noise has maximum entropy (i.e. it jams most efficiently) if it is gaussian; if the (jamming) noise is not gaussian, good antijamming equipment usually has precisely the effect of making it gaussian. However, if the noise $n(t)$ is definitely not gaussian, the theory given above is no longer valid and care must be taken if the conclusions are used.

Remark 2.2

Two successive approximations have been made: first it is assumed that the spectrum of the signal $S(t)$ is limited, and secondly we pass from the sum to the integral. The validity of these approximations increases with increasing $T\Delta f$. Attention should be drawn to the fact that, even if $T\Delta f$ is small, the two approximations compensate one another exactly: the final result given in eqn (2.1) is rigorous even if $T\Delta f$ is not large [3] (see Section 2.6).

Remark 2.3

Finally it has been assumed that the noise $n(t)$ is white in the band Δf. If this is not true (i.e. if the noise is colored), it is easy to allow for this [3] (see Section 2.6 and Chapter 3, Remark 3.1).

In conclusion, it is clear that $y(t)$ is involved only in the expression

$$\int_T y(t)\, S(t - t_0)\, \mathrm{d}t$$

Thus any equipment (receiver) which calculates the average product of the received signal and the assumed signal over the duration of the received signal contains all the information on $y(t)$.

We are now able to define the ideal receiver.

Definition 2.1

If we know the probability that the target exists (see the discussion of the probability of false alarms in Chapter 3, Section 3.3) and it is satisfactory to compare the various possible positions t_0 in order to find the most probable position, it is easy to see that, since the integral $\int_T S(t - t_0)^2\, \mathrm{d}t$ is independent of t_0, eqn (2.1) can be replaced by

$$p(t_0/y) = k_3 p(t_0) \exp\left[\frac{1}{N_0}\int_T y(t)\, S(t - t_0)\, \mathrm{d}t\right] \tag{2.2}$$

If we assume further that $p(t_0)$ does not vary much as a function of t_0 compared with the variations in the other factor, which is a valid assumption in most practical applications, we can find the maxima of $p(t_0/y)$, i.e. the most probable positions, by replacing eqn (2.2) by

$$p(t_0/y) = K \exp\left(\frac{1}{N_0}\int_T y(t)\, S(t - t_0)\, \mathrm{d}t\right) \tag{2.3}$$

This means that the most probable positions are those for which

$$C(t_0) = \frac{1}{N_0} \int_T y(t)\, S(t - t_0)\, dt \tag{2.4}$$

is a maximum. Thus it can be concluded that it is necessary to correlate the received signal with the transmitted signal such that an expression is obtained whose magnitude represents the possibility that a target is present and whose maxima correspond to the most probable positions of that target. A receiver is *ideal* when it performs this correlation between the received signal and the transmitted signal.

Remark 2.4

It is generally valid to assume that $p(t_0)$ varies very little, particularly in surveillance radars. If $p(t_0)$ is known approximately as a function of t_0, it can easily be allowed for. This is generally done in tracking radars where $p(t_0)$ is assumed to be constant over a given range of t_0 and zero elsewhere (this is ensured in practice by using a "range gate" in the receiver).

2.6 More rigorous definition of the ideal receiver [3]

In this section we give a more rigorous justification of the concept of the ideal receiver for a useful signal $S(t)$ with finite duration. In this case we do not assume that the spectrum of $S(t)$ has finite width. (Although the logical location of this section is here at the end of Chapter 2, it is recommended that Chapter 3 should be read first.)

2.6.1 Preamble

It has already been established that, if a measurement y is the sum of a centered random gaussian variable n with variance σ^2 and a possible signal U whose *a priori* probability of existence (event A) is $P_0(A)$, the probability $P(A/y)$ that U is present in the measured value of y can be written

$$P(A/y) = kP_0(A) \text{ (distribution of } y \text{ if } A)$$

Similarly, the probability $P(B/y)$ that U is absent (event B) from the measured value of y can be written

$$P(B/y) = kP_0(B) \text{ (distribution of } y \text{ if } B)$$

where $P_0(B)$ is the *a priori* probability of the absence of U and $P_0(A) + P_0(B) = 1$. Therefore we have

$$P(A/y) = kP_0(A)\frac{1}{\sigma(2\pi)^{1/2}}\exp\left[\frac{(y-U)^2}{2\sigma^2}\right]$$

$$P(B/y) = kP_0(B)\frac{1}{\sigma(2\pi)^{1/2}}\exp\left(\frac{y^2}{2\sigma^2}\right)$$

or

$$P(A/y) = k_1\exp[V(y)]$$

$$P(B/y) = k_2$$

where

$$\frac{k_2}{k_1} = \frac{P_0(B)}{P_0(A)}$$

and

$$V(y) = \frac{Uy}{\sigma^2} - \frac{U^2}{2\sigma^2}$$

We shall assume that, in practice, all of $P_0(A)$ is ignored (or, equivalently, that there is little information on $P_0(A)$).

In this case the probability that the signal U is present in y increases with increasing $V(y)$. What generally interests the user is that it is possible to fix a probability of false alarm (probability of stating incorrectly that there is a signal U in y). Now, if the signal U is absent, it is clear that y is a zero-mean gaussian variable with variance σ^2 so that $V(y)$ is a gaussian variable with a well-known distribution. It follows that, if the condition is made that $V(y)$ must be higher than a threshold K_1 in order to declare U present, the probability of false alarm is fixed (Neyman–Pearson criterion). This is what has usually been done in deciding to declare "signal present" for $V(y) > K_1$; the lower is the fixed probability of false alarm, the larger is K_1.

It is clear that if several independent measurements can be carried out in order to see if the signal is present, and each time the presence of the signal corresponds to the existence of a U_i accompanied by a zero-mean gaussian n_i (with $\overline{n_i^2} = \sigma_i^2$), we have

$$P(A/y) = k_1'\Pi_i\{\exp[V_i(y_i)]\} = k_1'\exp[\Sigma V_i(y_i)]$$

where

$$\sum V_i(y_i) = \sum\frac{U_i y_i}{\sigma_i^2} - \sum\frac{U_i^2}{2\sigma_i^2}$$

and the target will be declared if

$$\sum\frac{U_i y_i}{\sigma_i^2} - \sum\frac{U_i^2}{2\sigma_i^2} > K_1$$

which means that the target will be declared if

$$\Sigma \frac{U_i y_i}{\sigma_i^2} > K$$

where

$$K = K_1 + \frac{\Sigma U_i^2}{2\sigma_i^2}$$

In the absence of a signal, $\Sigma U_i y_i/\sigma_i^2$ is a zero-mean gaussian variable with variance $\Sigma(U_i^2/\sigma_i^2)$. In the presence of a signal, it is a gaussian variable with the same variance and mean value $\Sigma(U_i^2/\sigma_i^2)$, so that the probability P_f of false alarm is given by (see Chapter 3, Section 3.3)

$$P_f = \frac{1}{[2\pi\Sigma(U_i^2/\sigma_i^2)]^{1/2}} \int_K^{+\infty} \exp\left[\frac{-v^2}{2\Sigma(U_i^2/\sigma_i^2)}\right] dv$$

and the probability P_d of detection is given by

$$P_d = \frac{1}{[2\pi\Sigma(U_i^2/\sigma_i^2)]^{1/2}} \int_{K-\Sigma(U_i^2/\sigma_i^2)}^{+\infty} \exp\left[\frac{-v^2}{2\Sigma(U_i^2/\sigma_i^2)}\right] dv$$

2.6.2 Assumptions

The signal $S(t)$ transmitted by radar is assumed to have duration T, i.e. $S(t) = 0$ for $t < 0$ and $t > T$, and finite energy, i.e. $\int_0^T S^2(t)\,dt$ is finite. Therefore its Fourier transform never vanishes definitely in a rigorous manner.

The noise $n(t)$ accompanying the possible useful signal $S(t - t_{01})$ is assumed to be gaussian and steady state but not necessarily white. The steady state condition implies that it is possible to define for $n(t)$ an autocorrelation function of the form $\varrho(\tau)$ which represents the mathematical expectation (mean value) of $n(t)\,n(t - \tau)$:

$$\varrho(\tau) = E[n(t)\,n(t - \tau)]$$

The fact that the noise is not white means that the Fourier transform of $\varrho(\tau)$ (i.e. the spectral density of $n(t)$) is a nonconstant function (real and even) of the frequency f. Denoting the spectral density of $n(t)$ by $N^2(f)$, we have

$$\varrho(\tau) = \int N^2(f) \exp(2\pi j f \tau)\,df$$

and the mean power of the noise is equal to $\varrho(0)$, i.e. $\int N^2(f)\,df$.

2.6.3 Properties of $S(t)$, $n(t)$ and $S(t - t_0)$

The set E of possible functions $S(t)$ is a vector space over R (because we can define $S_1(t) + S_2(t)$, we can define $-S_1(t)$ . . . in the same way that we can define $\alpha S_1(t)$ $\forall \alpha \in R$ and $\alpha S_1(t) \in E$). For this vector space, it is possible to define the scalar product of any two vectors $S_1(t)$ and $S_2(t)$ by the expression $\int_0^T S_1(t)\, S_2(t)\, \mathrm{d}t$. If this scalar product exists, the space E is a Hilbert space $L^2[0, T]$. The noise slice $n(t)$ between zero and T belongs to the same Hilbert space.

The problem of knowing whether a signal $S(t - t_0)$ is present in the received signal $y(t)$ (accompanied by a noise $n(t)$) reduces to examining the time slice between t_0 and $t_0 + T$, i.e. the Hilbert space $L[t_0, t_0 + T]$ which is closely related to the Hilbert space $L^2[0, T]$. We shall call this new Hilbert space $E(t_0)$ and associate with it the scalar product

$$\int_{t_0}^{t_0+T} S_1(t)\, S_2(t)\, \mathrm{d}t$$

In order to pass to the case of independent measurement examined in Section 2.6.1, we shall attempt to express any function $x(t)$ of $E(t_0)$, in particular $S(t - t_0)$, $n(t)$ between t_0 and $t_0 + T$, and $y(t)$ between t_0 and $t_0 + T$, in the form $x(t) = x_i X_i(t)$. The functions $X_i(t)$ are independent because they form a basis of orthonormal vectors for $E(t_0)$. It is clear that this basis must be chosen so that it is possible to introduce the shape of the autocorrelation function $\varrho(\tau)$ of the noise.

2.6.4 Search for a basis of orthonormal vectors over $E(t_0)$

Consider the mapping which transfers a vector $x(t)$ of $E(t_0)$ to another vector $y(t)$ of the same set, defined by

$$y(t) = A[x(t)]$$

where

$$y(t) = K(t)_{t_0,t_0+T} \int_{t_0}^{t_0+T} \varrho(t - u)\, x(u)\, \mathrm{d}u$$

$K(t)_{t_0,t_0+T}$ denotes the function which is zero for $t < t_0$ and $t > t_0 + T$ and unity for $t_0 \leqslant t \leqslant t_0 + T$. The mapping A is linear because

$$\forall x(t),\ y(t) \in E(t_0) \ \text{ and } \ \forall \alpha, \beta \in R$$

we have

$$A[\alpha x(t) + \beta y(t)] = \alpha A[x(t)] + \beta A[y(t)]$$

The mapping is symmetric, which is easily verified because $\varrho(\tau)$ is an even function.

This implies that it is possible to find a basis of eigenvectors $V_i(t)$ for $E(t_0)$, associated with eigenvalues λ_i. These eigenvectors are orthogonal and such that

$$A[V_i(t)] \equiv \lambda_i V_i(t) \text{ with } \lambda_i \in R$$

The set of functions $V_i(t)$ defined in this way constitutes the required orthonormal vector basis, i.e. any function $x(t)$ of $E(t_0)$ can be put in the form

$$x(t) = \Sigma x_i V_i(t)$$

where the x_i are defined by

$$x_i = \int_{t_0}^{t_0+T} x(t)\, V_i(t)\, \mathrm{d}t$$

and the functions $V_i(t)$ are independent. The orthogonality of $V_i(t)$ and $V_k(t)$ means that

$$\int_{t_0}^{t_0+T} V_i(t)\, V_k(t)\, \mathrm{d}t = 0 \text{ if } i \neq k$$

2.6.5 Definition of an ideal receiver

$S(t - t_0)$ can be written as $\Sigma U_i V_i(t)$, $y(t)$ between t_0 and $t_0 + T$ can be written as $\Sigma y_i V_i(t)$, and $n(t)$ between t_0 and $t_0 + T$ can be written as $\Sigma n_i V_i(t)$. The calculation of $\Sigma U_i y_i/\sigma_i^2$ is the ideal calculation to carry out and, as noted in Section 2.6.1, it involves making the expression

$$\Gamma(t_0) = \sum \frac{U_i y_i}{\sigma_i^2}$$

explicit. In order to do this, it is first necessary to evaluate $\sigma_i^2 = E(n_i^2)$ (the mathematical expectation of n_i^2). It can be proved that

$$\sigma_i^2 = E(n_i^2) = \lambda_i$$

In fact, it is known that the mathematical expectation of $n(t_0)\, n(t_0 + \tau)$ is equal to $\varrho(\tau)$. Let

$$\varrho(\tau) = E[n(t_0)\, n(t_0 + \tau)]$$

Now

$$n(t_0) = \Sigma n_i V_i(t_0)$$

$$n(\tau_0 + \tau) = \Sigma n_i V_i(t_0 + \tau) \quad \text{for } \tau \leqslant T$$

and

$$E[n(t_0)\, n(t_0 + \tau)] = \Sigma E(n_i^2)\, V_i(t_0)\, V_i(t_0 + \tau) \quad \text{for } 0 \leqslant \tau \leqslant T$$

However, in addition,

$$\int_{t_0}^{t_0+T} \varrho(u - t_0)\, V_i(u)\, \mathrm{d}u \;=\; \lambda_i V_i(t_0)$$

which means that

$$\varrho(t - t_0) \;=\; \Sigma \lambda_i V_i(t_0)\, V_i(t) \quad \text{for } t_0 \leqslant t \leqslant t_0 + T$$

and therefore that

$$\varrho(\tau) \;=\; \Sigma \lambda_i V_i(t_0)\, V_i(t_0 + \tau) \quad \text{for } 0 \leqslant \tau \leqslant T$$

It therefore follows that

$$\sigma_i^2 \;=\; E(n_i^2) \;=\; \lambda_i$$

Thus the ideal receiver must calculate

$$\Gamma(t_0) \;=\; \sum \frac{U_i y_i}{\lambda_i}$$

and it is necessary to declare detection if $\Gamma(t_0) > K$. The threshold K depends on the accepted probability of false alarm.

Let us now define a function $R(t)$ such that

$$\begin{aligned} R(t - t_0) \;&=\; \sum \frac{U_i}{\lambda_i} V_i(t) \quad \text{for } t_0 \leqslant t \leqslant t_0 + T \\ R(t) \;&=\; 0 \quad \text{for } t < t_0 \text{ and } t > t_0 + T \end{aligned} \tag{2.5}$$

Of course the function $R(t - t_0)$ belongs to the set $E(t_0)$. The scalar product of $R(t - t_0)$ and the section of $y(t)$ lying between t_0 and $t_0 + T$ can clearly be written

$$\int_{t_0}^{t_0+T} R(t - t_0)\, y(t)\, \mathrm{d}t \equiv \int_{-\infty}^{+\infty} R(t - t_0)\, y(t)\, \mathrm{d}t$$

or

$$\sum \frac{U_i}{\lambda_i} y_i \equiv \Gamma(t_0)$$

Thus $\Gamma(t_0)$ is the cross-correlation function of the received signal with a reference $R(t)$ defined by eqns (2.5). It can be defined alternatively as follows. Since

$$\int_{t_0}^{t_0+T} \varrho(t - u)\, R(u - t_0)\, \mathrm{d}u \equiv \int_{t_0}^{t_0+T} \varrho(t - u) \sum \frac{U_i}{\lambda_i} V_i(u)\, \mathrm{d}u$$

$$= \sum \left[\frac{U_i}{\lambda_i} \int_{t_0}^{t_0+T} \varrho(t - u)\, V_i(u)\, \mathrm{d}u \right] \;=\; \Sigma U_i V_i(t) \;=\; S(t - t_0)$$

we have

$$\int_{t_0}^{t_0+T} \varrho(t - u)\, R(u - t_0)\, \mathrm{d}u = S(t - t_0)$$

or, setting $\tau = u - t_0$ and $\theta = t - t_0$,

$$\int_0^T \varrho(\theta - \tau)\, R(\tau)\, \mathrm{d}\tau = S(\theta)$$

or, using more classical notation

$$\int \varrho(t - \tau)\, R(\tau)\, \mathrm{d}\tau = S(t) \tag{2.6}$$

The ideal receiver therefore calculates the cross-correlation function of the received signal with a reference $R(t)$ related to the signal $S(t)$ and to the autocorrelation function of the noise by eqn (2.6).

2.6.6 White noise

In the case of white noise, $\varrho(\tau)$ is equal to the Dirac delta function $\delta(\tau)$ (apart from a factor). In this case

$$\int \varrho(t - \tau)\, R(\tau)\, \mathrm{d}\tau = R(t)$$

apart from a factor, and therefore

$$R(t) \equiv S(t)$$

apart from a factor, and we have

$$\Gamma(t_0) = C(t_0) = \int y(t)\, S(t - t_0)\, \mathrm{d}t$$

2.6.7 Colored noise (approximate proof)

We denote the Fourier transforms of $S(t)$, $R(t)$ and $\varrho(t)$ by $\Phi(f)$, $\mathscr{R}(f)$ and $N^2(f)$ respectively ($N^2(f)$ is the spectral density of the noise $n(t)$). According to the convolution theorem

$$S(t) = \int R(\tau)\varrho(t - \tau)\, \mathrm{d}\tau \equiv \int \mathscr{R}(f)\, N^2(f) \exp(2\pi \mathrm{j} f t)\, \mathrm{d}f$$

which shows that

$$\Phi(f) = \mathscr{R}(f)\, N^2(f)$$

and

$$\mathscr{R}(f) = \frac{\Phi(f)}{N^2(f)}$$

In the case of colored noise, the matched filter will have a transfer function $\mathscr{R}^*(f)$, i.e. $\Phi^*(f)/N^2(f)$ (see Chapter 3, Remark 3.1).

Chapter 3

Performance of Radar Systems Equipped with Ideal Receivers

3.1 Two procedures for making an ideal receiver

We explained in Chapter 2 why the ideal receiver has to calculate the expression

$$C(t_0) = \frac{1}{N_0}\int_T y(t)\, S(t - t_0)\, \mathrm{d}t$$

The first procedure which can be used to obtain $C(t_0)$ effectively consists of applying this formula exactly by forming the product of $y(t)$ (received signal) with as many values of $S(t - t_0)$ as desired (these values are obtained from a suitably delayed sample (reference) of the transmission signal) and then integrating this product over the time T. This type of receiver is known as a correlation receiver. Radar systems using noise transmissions or the pulse Doppler radar systems described in Chapter 4, Sections 4.1 and 4.2, use correlation receivers.

$C(t_0)$ can be obtained in a different manner by using the convolution theorem (see Chapter 1, Section 1.5). If we denote the Fourier transform of the received signal $y(t)$ by $\mathcal{Y}(f)$, we can write the identity

$$C(t_0) = \frac{1}{N_0}\int_T y(t)\, S(t - t_0)\, \mathrm{d}t \equiv \frac{1}{N_0}\int_{\Delta f} \mathcal{Y}(f)\, \Phi(-f)\, \exp(2\pi \mathrm{j} f t_0)\, \mathrm{d}f \tag{3.1}$$

Thus, if we let $y(t)$ pass through a filter matched to the transmitted signal $S(t)$ with transfer function $\Phi(-p/2\pi\mathrm{j})$, we obtain at the output of the filter the expression $C(t)$ multiplied by a constant factor, i.e. $C(t_0)$ in real time.

Can such a filter exist?

It is easy to show that, if such a filter exists, its response to a Dirac delta function is $S(-t)$, i.e. if $S(t)$ begins at $t = 0$, the response of a filter matched to a Dirac delta function also situated at $t = 0$ is terminated at the moment of excitation. This is obviously physically absurd. Thus a matched filter cannot exist. However, there is nothing to prevent the design of a filter which gives a signal of the form $S(T_1 - t)$ as a response to the Dirac delta function $\delta(t)$ under the same conditions, since T_1 is longer than the duration T of the signal $S(t)$, i.e. the filter is equivalent to a delay line giving a delay T_1 followed by a matched filter. The passage of $y(t)$ through such a filter, which will still be called "matched" by extension, no longer gives $C(t)$ but $C(t - T_1)$ at the output. This is not a problem because T_1 is known.

Another way of characterizing a matched filter (apart from the delay T_1) is based on the facts that the amplitude of its transfer function (its "response") is equal to $|\Phi(f)|$ and the phase of its transfer function is opposite to the argument of $\Phi(f)$. It can thus be seen that if a transmission signal $S(t)$ is applied to the input of a matched filter an output signal is obtained whose Fourier transform has a modulus of $|\Phi(f)|^2$ with zero phase. In other words, the matched filter removes the phase (or frequency) modulation of a signal $S(t)$ passing through it. A receiver consisting of a matched filter is therefore also an ideal receiver.

Approximately matched filter receivers are used in well-designed classical radar (see Chapter 4, Section 4.4) and pulse compression radar (see Chapter 4, Section 4.5).

Remark 3.1

If the noise $n(t)$ in the band Δf is not white, i.e. if the spectral density $N^2(f)$ of the noise varies as a function of f, certain changes must be made which are described simply below (for a more rigorous description see Chapter 2, Section 2.6). Let us assume that the useful signal and the noise pass through a whitening filter whose transfer function is $\mathcal{N}(f)$ where

$$|\mathcal{N}(f)|^{-2} = N^2(f)$$

The output of this whitening filter will produce a white noise (with a spectral density of unity) and a modified useful signal with a Fourier transform $\Phi(f)\mathcal{N}(f)$. The matched filter with this modified useful signal must have a transfer function equal to the conjugate of $\mathcal{N}(f)\Phi(f)$, i.e. $\Phi^*(f)\mathcal{N}^*(f)$. Therefore the total reception filter must have a transfer function

$$\mathcal{N}(f)[\Phi^*(f)\mathcal{N}^*(f)] = \frac{1}{N^2(f)}\Phi^*(f)$$

which clearly shows that the receiver must consist of a filter with a transfer function $1/N^2(f)$ and a normal matched filter (or ideal receiver) in series. Finally, the fact that the noise is colored makes it necessary to incorporate what is sometimes incorrectly called a whitening filter in the receiver, i.e. a filter with a transfer function $1/N^2(f)$.

3.2 The output signal of the ideal receiver

Let us assume that there is a target at t_{01}. The received signal $y(t)$ can be written as (see Chapter 2, Section 2.1)

$$y(t) = kS(t - t_{01}) + n(t) \tag{3.2}$$

and the output of the ideal receiver can be written as

$$C(t_0) = C_u(t_0) + C_p(t_0) \tag{3.3}$$

where

$$C_u(t_0) = \frac{1}{N_0} \int_T kS(t - t_{01}) S(t - t_0) \mathrm{d}t \tag{3.4}$$

$$C_p(t_0) = \frac{1}{N_0} \int_T n(t) S(t - t_0) \mathrm{d}t \tag{3.5}$$

Thus the output signal of the ideal receiver is the sum of a useful signal $C_u(t_0)$, which would be obtained in the absence of noise, and a parasitic signal $C_p(t_0)$, which would be obtained in the absence of a target.

3.2.1 Properties of the useful signal

$C_u(t_0)$ is the autocorrelation function $\varrho(\tau)$ of the useful signal $S(t)$, where $\tau = t_0 - t_{01}$, multiplied by a factor kT/N_0:

$$C_u(\tau) = \frac{kT}{N_0} \frac{1}{T} \int_T S(u) S(u - \tau) \mathrm{d}u$$

where $u = t - t_{01}$, and

$$C_u(t_0) = \frac{kT}{N_0} \varrho(t_0 - t_{01}) \tag{3.6}$$

Therefore $C_u(t_0)$ is maximum for $t_0 = t_{01}$, which is reassuring because in this way it is found that in the absence of noise ($n(t)$ and therefore $C_p(t_0)$ are zero) $C(t_0)$ (and therefore $p(t_0/y)$) is maximum for $t_0 = t_{01}$. In the absence of noise, the ideal receiver indicates that the most probable position of the target is its actual position.

The shape of $C_u(t_0)$ depends only on $|\Phi(f)|^2$ (apart from a factor), since the Fourier transform of $\varrho(\tau)$ is equal to $|\Phi(f)|^2$ (apart from a factor). It follows that the shape of the useful output signal of a radar system with an ideal receiver depends only on the modulus of the Fourier transform of the transmitted signal and not on its phase. For example, two radar systems transmitting signals $S_1(t)$ and $S_2(t)$ with very different lengths could give the same useful signal at the output of the receiver if their spectra differ only by the phases of their components.

The value of $C_u(t_0)$ for $t_0 = t_{01}$, i.e. the amplitude of the useful signal at the target position, can easily be obtained:

$$C_u(t_{01}) = \frac{1}{N_0 k} \int_T [kS(t - t_{01})]^2 \, dt$$

$[kS(t - t_{01})]^2$ is the power of the received signal and its integral over the duration T of the signal is the energy E of the received signal, so that the amplitude of the useful signal is given by

$$C_u(t_{01}) = \frac{1}{k} \frac{E}{N_0} = \frac{R}{k} \tag{3.7}$$

In this expression R is the signal-to-noise energy ratio of the received signal:

$$R = \frac{\text{energy of the received signal}}{\text{spectral density of the accompanying noise}} \tag{3.8}$$

R is a dimensionless ratio which is usually expressed in decibels.

Note It was assumed in Chapter 2, Section 2.5, that $S(t)$ has a video spectrum so that in practice the maximum of $C(t_0)$ is unique (see Section 3.4).

3.2.2 Properties of the parasitic signal

The parasitic signal $C_p(t_0)$ can also be written

$$C_p(t_0) = \frac{1}{N_0} \sum n(t_i) S(t_i - t_0) \frac{1}{\Delta f}$$

where the t_i are sampling times separated by $1/\Delta f$. Apart from the factor $1/N_0 \Delta f$, this is the sum of $T\Delta f$ independent terms which are the products of a known value and a gaussian random variable $n(t_i)$, with zero mean, whose power (variance) $N = \overline{n^2(t_i)}$ is the power of the noise in the band Δf. $C_p(t_0)$ is therefore also a gaussian random variable with zero mean whose variance $\overline{C_p^2(t_0)}$ is the sum of the variances of each of the terms:

$$\overline{C_p^2(t_0)} = \frac{1}{N_0^2 \Delta^2 f} \sum N S^2(t_i - t_0) = \frac{1}{N_0} \sum S^2(t_i - t_0) \frac{1}{\Delta f}$$

$$\overline{C_p^2(t_0)} = \frac{1}{N_0} \int_T S^2(t - t_0) \, dt = \frac{1}{N_0 k^2} \int [kS(t - t_0)]^2 \, dt$$

$$\overline{C_p^2(t_0)} = R/k^2 \tag{3.9}$$

3.2.3 Conclusions

The output signal of the ideal receiver is, apart from the factor $1/k$, the sum of a function with a maximum amplitude R at $t_0 = t_{01}$ (the useful signal) and a gaussian random function with mean value zero and standard deviation $R^{1/2}$ (parasitic signal). It is also, apart from the factor $R^{1/2}/k$, the sum of a function with a maximum amplitude $R^{1/2}$ at $t_0 = t_{01}$ (useful signal) and a gaussian random parasitic signal with mean value zero and a standard deviation of unity. If $R \ll 1$, $C(t_0)$ will have many maxima (t_{01} will not necessarily correspond to a maximum) and $p(t_0/y)$ will also have many maxima (Fig. 3.1). However, if $R \gg 1$, $C(t_0)$ will almost certainly have a very pronounced maximum close to t_{01} (Fig. 3.2).

Remark 3.2
At the output of the receiver the power of the useful signal is

$$C_u^2(t_{01}) = \frac{R^2}{k^2}$$

and the power of the parasitic signal is

$$\overline{C_p^2(t_0)} = \frac{R}{k^2}$$

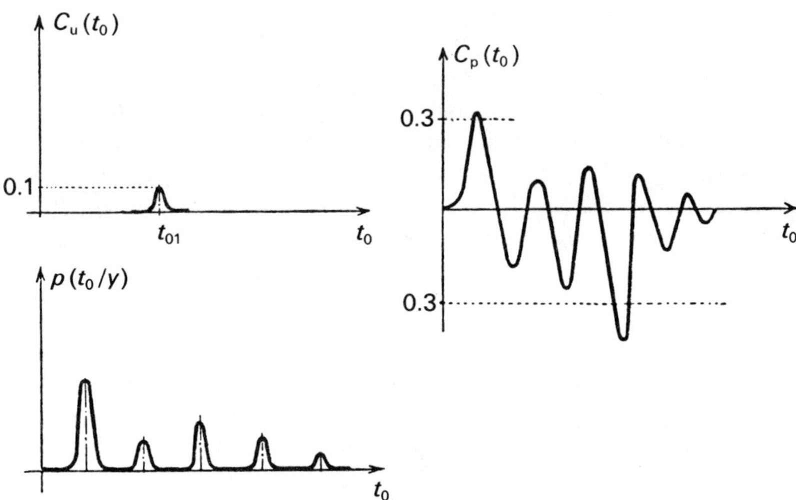

Fig. 3.1 $C_u(t_0)$, $C_p(t_0)$ and $p(t_0/y)$ for $R = 0.1$.

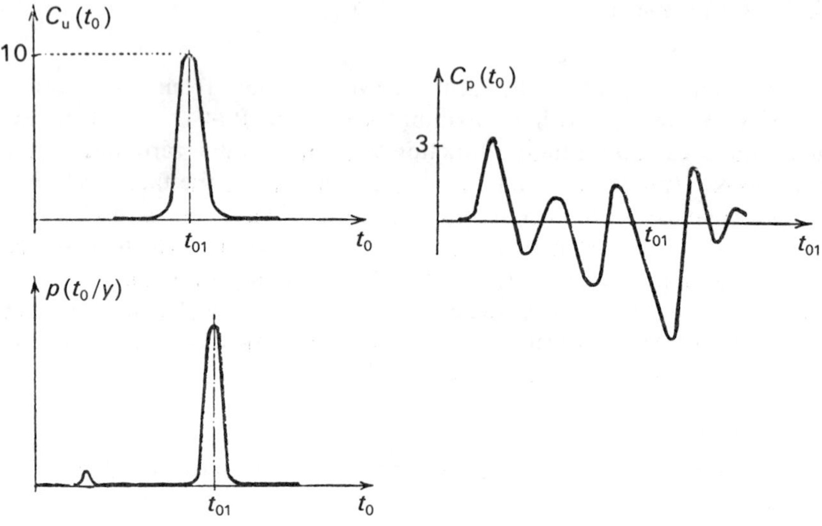

Fig. 3.2 $C_u(t_0)$, $C_p(t_0)$ and $p(t_0/y)$ for $R = 10$.

Therefore the signal-to-noise power ratio is equal to R. Thus we can say that the signal-to-noise power ratio at the output of an ideal receiver is equal to the signal-to-noise energy ratio at the input.

Remark 3.3

Classical theories made use of the signal-to-noise power ratio SN at the input of the receiver. Modern theories of radar replace this by the signal-to-noise energy ratio, which is the only signal-to-noise ratio which has any physical meaning and the only one which will be used in what follows. When the transmission signal $S(t)$ (and therefore the received useful signal) has a constant power P throughout its duration T, we can write

$$\mathrm{SN} = \frac{P}{N} = \frac{E}{T}\frac{1}{N_0 \Delta f} = \frac{R}{T\Delta f}$$

If $T\Delta f \approx 1$ (classical radar), SN $\approx R$. If $T\Delta f$ is very large, $R \gg$ SN, so that it is sometimes said that the ideal receiver improves the signal-to-noise ratio in the ratio $T\Delta f$, which can sometimes reach values of 10^5–10^6 as will be seen later. It is also possible to consider a spectral density N_0' equal to $N/(\Delta f/2)$ where N is the noise power and $\Delta f/2$ is the bandwidth of the positive frequencies occupied by this noise (which is assumed to be video). Under these conditions, R can be written as

$$R = \frac{E}{N/\Delta f} = \frac{2E}{N_0{}'}$$

3.3 Probability of false alarm and detection. Range ambiguity

3.3.1 Probability of false alarm

The calculations carried out in Chapter 2, in Section 2.5, and in Section 3.2 do not differ from those carried out in Chapter 2, Section 2.4. Let us therefore assume that a signal $S(t - t_0)$ exists at a certain location t_{01} and that its spectrum is a video spectrum centered on zero frequency. If the noise is significant, an *a posteriori* probability distribution $p(t_0/y)$ with a large number of peaks of the same size is generally found, i.e. no information is obtained. If the noise power decreases, the number of peaks also decreases, but it is still possible to identify the target position wrongly. Finally, when the noise power is very small, the probability distribution $p(t_0/y)$ has only a single peak close to t_{01} (see Figs 3.1, 3.2 and 3.3 and compare with Fig. 2.4). In other words, if $R \ll 1$, it is very likely that the target will be located incorrectly and that a noise peak will be identified as the target, whereas if $R \gg 1$ there is a good chance of locating the target correctly and not making a gross error in the position of t_{01}.

In most applications a "detection threshold" is defined as follows. The threshold is fixed so that in the absence of the useful signal (e.g. for t_0 very different from t_{01}) $C_p(t_0)$, which is then equal to $C(t_0)$, has a very low probability of exceeding it. Only values of t_0 for which $C(t_0)$ exceeds this threshold are considered. Thus when $C(t_0)$ exceeds the threshold we say that there is a target, but when $C(t_0)$ does not exceed it we say that there is no target.

However, $C_p(t_0)$ may exceed the threshold and a target will be declared even though there is nothing there, so that a false alarm is obtained. The probability that, at any value t_0, $C_p(t_0)$ exceeds the threshold is called "the false alarm probability". The higher is the false alarm probability, the greater is the chance of detecting phantom targets. In the case of a rocket with proximity radar, this means that there will be a greater chance of making it explode unnecessarily. In the case of a search radar, a larger number of phantom targets will appear on the screen. The tolerable false alarm

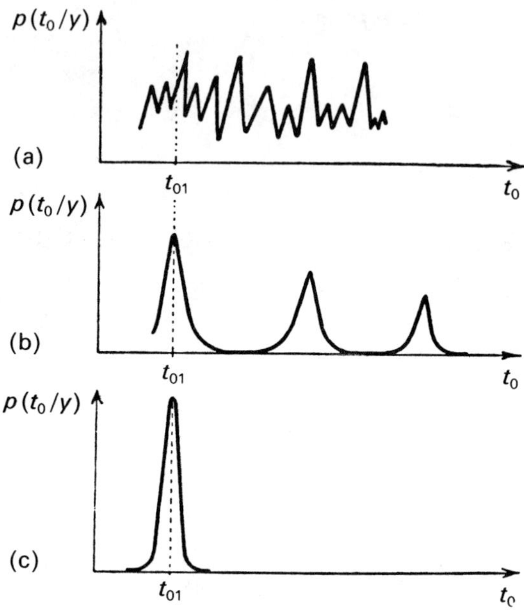

Fig. 3.3 Probability distribution $p(t_0/y)$ for (a) large noise power, (b) medium noise power and (c) small noise power.

probability (or the average number of "false blips" per second) is determined by operational or technical considerations (computer saturation etc.) and thus fixes the detection threshold. If this threshold is compared with the standard deviation of the parasitic signal $C_p(t_0)$, the false alarm probability P_f is given by the probability that a gaussian random function with zero mean and a standard deviation of unity exceeds the base threshold K:

$$P_f = \frac{1}{(2\pi)^{1/2}} \int_K^{+\infty} \exp\left(-\frac{v^2}{2}\right) dv \qquad (3.10)$$

For $P_f = 10^{-3}$, we obtain $K = 3.1$.

3.3.2 Detection probability

Under these conditions, it is interesting to calculate, as a function of R for the false alarm probability P_f, the probability that $C(t_0)$ exceeds the threshold when there is a target, i.e. the probability that the target is detected

Table 3.1 Detection probabilities for $P_f = 10^{-3}$

R		P_d
0.1	(−10 dB)	10^{-3} [a]
1	(0 dB)	0.02 (low)
4	(6 dB)	0.15
10	(10 dB)	0.5
16	(12 dB)	0.85
25	(14 dB)	0.98
36	(15.5 dB)	0.999

[a] The "signal" $C_u(t_0) = 0.1$ is so weak relative to the noise $C_p(t_0)$ that it has no impact: whether there is a signal or not, there is almost the same probability that $C(t_0)$ will exceed the threshold level.

at t_{01} when it is there. This is known as the detection probability, and is the probability that the sum of $R^{1/2}$ and a random phenomenon with zero mean and a standard deviation of unity is greater than K:

$$C_u(t_{01}) + C_p(t_{01}) > K$$

It can easily be found using a table of error functions Θ (see Appendix B) that the detection probabilities for a false alarm probability P_f of 10^{-3} are as given in Table 3.1 (see also Fig. 3.4 and the first fold-out graph at the end of the book). More generally, we have

$$P_d = \frac{1}{(2\pi)^{1/2}} \int_{K-R^{1/2}}^{+\infty} \exp\left(-\frac{v^2}{2}\right) dv \tag{3.11}$$

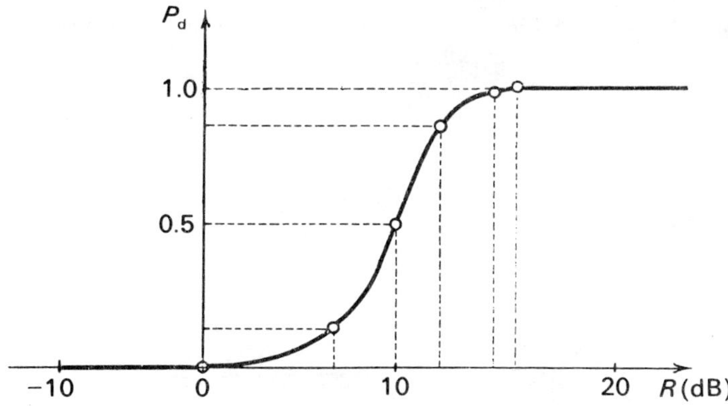

Fig. 3.4 Detection probability P_d for a false alarm probability P_f of 10^{-3} as a function of R.

The possible existence of ambiguity peaks makes it necessary to take precautions so that targets can only be detected if the value of R is sufficiently high (in this case of the order of magnitude of 13 dB if we want a detection probability of 0.9). A detection probability of 0.5 is obtained when the probability that $C_p(t_0)$, which is a gaussian random phenomenon with zero mean, is larger than $K - R^{1/2}$ is 0.5, which is clearly obtained for

$$K - R^{1/2} = 0 \qquad R = K^2$$

It can be seen that for a given P_f (a given K), P_d is an increasing function of R.

3.3.3 Range ambiguity

3.3.3.1 First property of ambiguity

The function $C_u(t_0)$ always has its highest peak at $t_0 = t_{01}$, but it can decrease more or less rapidly when t_0 moves away from t_{01} and may present other less significant maxima for values of t_0 other than t_{01} (Fig. 3.5). It has been demonstrated above that if R is very large $C(t_0)$ is maximum for t_0 very close to t_{01}. However, as R becomes smaller and $C_u(t_0)$ becomes larger for a given value of t_0, the chance that a value of t_0 other than t_{01} will be assigned as the position of the target (position of the highest peak of $C(t_0)$) increases. In other words, the larger is $C_u(t_0)$ (for $t_0 \neq t_{01}$), the greater is the chance at a given value of R of finding a maximum for $C(t_0)$ at this point and of taking this (incorrect) value as the correct value.

Consequently, the amplitude of the function $C_u(t_{01} + \theta)$ for a value θ gives an indication of the possibility of making an error θ in the determination of the position of the target. It is customary to call the function $C_u(t_{01} + \theta)$, normalized so that its maximum is equal to unity, the "ambiguity". The

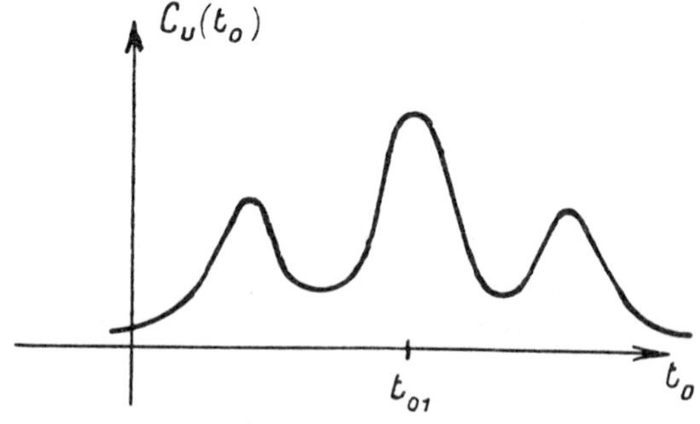

Fig. 3.5

range ambiguity $\mathscr{A}(\theta)$ is then defined by

$$\mathscr{A}(\theta) = \frac{\int_T S(t)\, S(t - \theta)\, \mathrm{d}t}{\int_T S^2(t)\, \mathrm{d}t} \tag{3.12}$$

Note Electronics engineers usually employ imaginary notation (sometimes wrongly and sometimes justified) and write $\exp(\mathrm{j}\omega t)$ instead of $\cos(\omega t)$. If this is the case, the above expression should be replaced by

$$\mathscr{A}(\theta) = \frac{\int_T S(t)\, S^*(t - \theta)\, \mathrm{d}t}{\int |S(t)|^2\, \mathrm{d}t} \tag{3.13}$$

Care should also be taken in the interpretation of calculations in similar cases.

3.3.3.2 Second property of ambiguity

The first property of ambiguity covers the case (ideal but unfortunately rare) where we are dealing with a single target. Let us suppose that we are dealing with a primary target located at t_{01} and a parasitic target located at $t_{01} + \theta$ which gives a signal p^2 times more powerful than that of the primary target, i.e. the primary target will give the useful signal $C_u(t_0)$ and the parasitic target will give the signal $pC_u(t_0 - \theta)$ whose value at the position t_{01} of the useful target is given by

$$pC_u(t_{01} - \theta) = pC_u(t_{01} + \theta)$$

or

$$p\mathscr{A}(\theta)\, C_u(t_{01})$$

In other words, at the position of the primary target the parasitic target gives a signal whose amplitude is $p|\mathscr{A}(\theta)|$ times that of the useful signal, or the parasitic signal has a power at the position of the primary target whose ratio with respect to that of the useful signal is $p^2|\mathscr{A}^2(\theta)|$, i.e. the ratio of the powers of the echoes of the targets multiplied (weighted) by the square of the ambiguity at the relative distance θ of the parasitic target (with respect to the primary target). The lower is the ambiguity at the position of the parasitic target, the less the latter disturbs the primary target. The ambiguity $\mathscr{A}(\theta)$ gives the positions (peaks of $\mathscr{A}(\theta)$) where parasitic targets cause most problems. They cause least interference if they are located in the troughs of $\mathscr{A}(\theta)$.

3.4 Revision of the results when the useful signal is a microwave signal

For convenience in radar technology (implementation of antennas, propagation) it is necessary to use a signal $S(t)$ occupying a frequency spectrum which is very narrow relative to one of the frequencies of this spectrum, e.g. a spectrum lying between 10 000 and 10 001 MHz. In this case (see Chapter 1, Section 1.6.1), the autocorrelation function of the useful signal has a very large number of maxima (in the given example approximately 10 000 separated from each other by 10^{-10} s. In other words $C_u(t_0)$ can be written as

$$\gamma_u(t_0)\cos(2\pi f t_0 + \varphi_1)$$

where f is fixed (of the order of 10 000 MHz in our example) and $\gamma_u(t_0)$ and φ_1 vary slowly as a function of t_0 (relative to the rapid variation of $\cos(2\pi f t_0 + \varphi_1)$). This also means that, even if R is quite large, the *a posteriori* distribution $p(t_0/y)$ will have a very large number of peaks (separated by approximately 10^{-10} s in the given example) because $C(t_0)$ is also of the form

$$C(t_0) = \gamma(t_0)\cos(2\pi f t_0 + \varphi)$$

where $\gamma(t_0)$ is the (positive) envelope of $C(t_0)$ which contains relatively crude information on the position of the target and φ contains information on this position accurate to the fourth or fifth decimal place. As in practice such an accuracy can never be achieved, it makes no sense to keep the information given by φ and to carry a distribution with a large number of neighboring peaks (it is also an illusion to think of measuring the position of an airplane or a missile with an accuracy to 0.1 mm).

A radar system, even if it is ideal, may therefore destroy this information on φ. This is done by assuming that all values of φ are equiprobable and by integrating with respect to φ, which gives

$$p(t_0/y) = k'p(t_0)\int_0^{2\pi} \exp[\gamma(t_0)\cos(2\pi f_0 t + \varphi)]\,d\varphi$$

$$p(t_0/y) = k'p(t_0)\,I_0[\gamma(t_0)]$$

where I_0 is the modified Bessel function

$$I_0 \approx \exp(x)(2\pi x)^{-1/2}\left(1 + \frac{1}{8x} + \ldots\right)$$

A receiver detecting the signal $C(t_0)$ with a detection characteristic of the form $I_0(x)$ can therefore still be considered as an ideal receiver. In addition, when R is very large (which has been shown to be essential) $I_0[\gamma(t_0)]$ behaves as $\exp[\gamma(t_0)]$. Under these conditions, the calculations and formulae presented in this chapter can be applied without reservation (remember the definition of R in Remark 3.3 as the ratio of twice the energy E of the received signal to the spectral density N_0 of the noise with respect to the band Δf of the positive frequencies occupied by the signal).

Remark 3.4

A simple and useful physical interpretation of this result consists in noting that there are two ways of defining the signal-to-noise power ratio at the output of an ideal receiver supplying $C(t_0)$ when the signal $S(t)$ is a microwave signal (or the signal on the carrier, e.g. the intermediate frequency). The first method (used by information theoreticians) consists in defining this ratio, which is denoted SN_A, in the spirit of what has just been said, as the ratio of the square of the maximum amplitude of $C_u(t_0)$, i.e. $\gamma_u{}^2(t_{01})$, to the noise power. The second method (used by laboratory technicians) consists in defining this ratio, which is denoted SN_B, in the normal manner, i.e. as the ratio of the average power $\frac{1}{2}\gamma_u{}^2(t_{01})$ of $C_u(t_0)$ around t_{01} to the noise power. Therefore $SN_A = 2SN_B$.

Since $SN_A = 2E/N_0$ and therefore $SN_B = E/N_0$ we can make the connection between the video signal $S(t)$ considered up to now and the signal on the carrier $S(t)$.

3.5 Accuracy of range measurement

Let us recall the assumptions which have already been made: the useful signal $S(t - t_{01})$ lasts for a time T during which the measurement is carried out and occupies a spectrum of width Δf, and the noise $n(t)$ accompanying it is gaussian and occupies the same frequency bandwidth ΔF. The problem is now to determine with what accuracy it is possible to specify the value of t_{01} when there is a target such that $S(t - t_{01})$ exists.

If $R \gg 1$ at the point where it is possible to neglect $C_p(t_0)$ completely around $C_u(t_0)$ in $C(t_0)$ and if in addition the *a priori* distribution $p(t_0)$ varies only slightly with t_0 in the area surrounding t_{01}, which is always the case, the maximum of $p(t_0/y)$ corresponds to the maximum of $C(t_0)$ and thus to the maximum of $C_u(t_0)$ and therefore is situated at t_{01} since $C_u(t_0)$ is, apart from a constant factor, the autocorrelation function of $S(t - t_{01})$.

When $R \gg 1$ but it is no longer permissible to neglect $C_p(t_0)$ completely with respect to $C_u(t_0)$, it is logically accepted that the value of t_{01} is the value of t_0 which makes $C(t_0)$ a maximum and this time there is an error due to the noise $C_p(t_0)$. Therefore it is necessary to examine the behavior of $C(t_0)$ around its maximum and, before that, the behavior of $C_u(t_0)$ around $t_0 = t_{01}$. Setting $\tau = t_0 - t_{01}$, we have (see Section 3.2.1)

$$C_u(\tau) = \frac{k}{N_0}\int_T S(u)\,S(u - \tau)\,\mathrm{d}u$$

which can be written

$$C_u(\tau) = C_u(0) + C_u''(0)\,\frac{\tau^2}{2}$$

for small τ. Since $C_u'(0) = 0$ (see Chapter 1, Section 1.5.4)

$$C_u''(\tau) = -\frac{k}{N_0}\int_T S'(u)\,S'(u-\tau)\,du$$

and

$$C_u''(0) = -\frac{k}{N_0}\int_T S'^2(t)\,dt = -\frac{k}{N_0}\int_{-\Delta f/2}^{+\Delta f/2} |\Psi(f)|^2\,df$$

where $\Psi(f)$ is the Fourier transform of $S'(t)$, the first derivative of $S(t)$, whose Fourier transform $\Phi(f)$ is zero for $|f| > f/2$ (cf. Chapter 1, Section 1.51). Since $\Psi(f) = 2\pi j f\Phi(f)$, we have

$$C_u''(0) = -\frac{4\pi^2 k}{N_0}\int_{\Delta f} f^2|\Phi(f)|^2\,df = -\frac{4\pi^2}{N_0 k}\int_{\Delta f} f^2k^2|\Phi(f)|^2\,df$$

We recall (eqn (3.7)) that $C_u(0) = R(k)$ and that

$$\int_T k^2S^2(t)\,dt = E = \int_{\Delta f} k^2|\Phi(f)|^2\,df$$

We set, by definition,

$$\int_{\Delta f} f^2|\Phi(f)|^2\,df = B^2\int_{\Delta f} |\Phi(f)|^2\,df \tag{3.14}$$

In this way we define a second-order moment B of the spectrum $|\Phi(f)|^2$ (using terminology borrowed from probability theory). B can be described as the radius of gyration of an area limited by the frequency axis and the curve $|\Phi(f)|^2$.

Finally, we have the expression

$$C_u(\tau) = \frac{R}{k}(1 - 2\pi^2B^2\tau^2)$$

and

$$C(\tau) = \frac{R}{k}(1 - 2\pi^2B^2\tau^2) + C_p(\tau)$$

where

$$C_p(\tau) = \frac{1}{N_0}\int_T n(u + t_{01})\,S(u-\tau)\,du$$

The maximum of $C(\tau)$ is given by $C'(\tau) = 0$:

$$-\frac{4\pi^2B^2\tau R}{k} + C_p'(\tau) = 0$$

It is obtained for τ satisfying the expression

$$\tau = \frac{k}{4\pi^2B^2R}\,C_p'(\tau)$$

If the noise $n(t)$ were negligible, it would be obtained for $\tau = 0$. The value of τ obtained in this way therefore represents the error made in measuring the range t_{01}.

In order to obtain more information, it is necessary to know a little more about $C_p'(\tau)$. This can be written (using standard notation)

$$C_p'(\tau) = -\frac{1}{N_0}\int_T n(u + t_{01})\, S'(u - \tau)\, du$$

$$C_p'(\tau) = -\frac{1}{N_0 \Delta f}\sum n(t_i + t_{01})\, S'(t_i - \tau)$$

According to the arguments of Section 3.2.2, $C_p'(\tau)$ acts as a gaussian random variable with zero mean and variance

$$\overline{C_p'(\tau)} = \frac{1}{N_0{}^2 \Delta f^2}\sum N S'^2(t_i - \tau) = \frac{1}{N_0}\sum S'^2(t_i - \tau)\frac{1}{\Delta f}$$

$$\overline{C_p'(\tau)} = \frac{1}{N_0}\int_T S'^2(t - \tau)\, dt = \frac{1}{N_0}\int_T S'^2(t)\, dt$$

$$\overline{C_p'(\tau)} = \frac{1}{N_0}\int_{\Delta f} |\Psi(f)|^2\, df = \frac{4\pi^2}{N_0 k^2} B^2 E = \frac{4\pi^2 B^2 R}{k^2}$$

Thus the error τ made in measuring t_{01} acts as a gaussian error with zero mean and standard deviation

$$(\overline{\tau^2})^{1/2} = \frac{k}{4\pi^2 B^2 R}\frac{2\pi B}{k} R^{1/2}$$

$$= \frac{1}{2\pi B R^{1/2}} \tag{3.15}$$

The above formula is known as Woodward's formula.

Note In order to obtain the standard deviation of the measurement of the range of the target to the radar, it is obviously sufficient to multiply the result by $c/2$, where c is the velocity of light. For example, 1 μs gives 150 m.

If R is sufficiently large, we will make a gaussian error in the determination of t_{01} with a standard deviation given by eqn (3.15) where $R = E/N_0$ is the ratio of the energy of the signal in the measurement time to the spectral density of the noise and B is the second-order moment of the spectrum of the transmitted useful signal.

Remark 3.5

The transmitted signal is a microwave signal whose spectrum is centered on a frequency f_0. The preceding formula can be applied provided that B is defined by

$$\int_{\Delta f} (f - f_0)^2 |\Phi^2(f)| \, df = B^2 \int_{\Delta f} |\Phi^2(f)| \, df \tag{3.16}$$

and R is defined by $R = 2E/N_0$.

Remark 3.6

The value of B depends only on the modulus of $\Phi(f)$ and not on its argument (phase of the spectrum) (see Section 3.2.1).

Remark 3.7

If the spectrum is bell shaped, its width at 3 dB is given by $2B$. If, however, the spectrum of $S(t)$ is rectangular in the interval Δf ($|\Phi(f)| =$ constant), we have the relationship

$$2B = \frac{\Delta f}{\sqrt{3}}$$

Remark 3.8

If the transmitted signal $S(t)$ is periodic with period T_R (repetition period), i.e. if it is formed by a regular sequence of elementary signals, it is obvious that the received signal $S(t - t_{01})$ is identical with the signal $S(t - t_{01} - T_R)$. In other words, t_{01} is measured accurately, but theoretically it is possible to be wrong about T_R by a whole number of T_Rs. Moreover, in this case the spectrum of $S(t)$ is a spectrum of lines separated by $1/T_R$ (see Chapter 1, Section 1.7). The transmission of a periodic signal $S(t)$ therefore creates an ambiguity in the measurement of t_{01}. This is often a completely theoretical ambiguity: since the order of magnitude of t_{01} is generally known, the values of $t_{01} + kT_R$ (k is an integer) are eliminated *ipso facto*. This ambiguity is also encountered in the diagram of the ambiguity function $\mathscr{A}(\theta)$ associated with $S(t)$. Since $\Phi(F)$ consists of lines with a regular spacing of $1/T_R$, the same is true of $|\Phi(f)|^2$, and therefore $\mathscr{A}(\theta)$, with Fourier transform $|\Phi(f)|^2$ apart from a factor, is a periodic function of θ whose period is

$$T_R = \frac{1}{1/T_R}$$

In practice, for $S(t)$ to be truly periodic its duration T must be infinite, which is impossible. Therefore $S(t)$ will never be completely periodic and neither will $\mathscr{A}(\theta)$, which in practice will consist of a peak at $\theta = 0$ surrounded by peaks for $\theta = kT_R$ which will have a height close to unity. Thus there is still a very real ambiguity.

3.6 Range resolution

3.6.1 General considerations

Although what has just been presented is more or less reliable (there were a few inconsistencies in the arguments used in arriving at eqn (3.15)), this section should be considered as an introductory treatment which deserves to be put on a sounder basis. The problem to be examined is as follows. This time, we are dealing with two targets situated in the same direction but at neighboring ranges. To what extent can we see two targets and measure the position of each of them accurately?

It is important to realize that a single target can be located very accurately if R is very large, and with infinite accuracy provided that R is infinitely large. However, if two targets are very close together, it is not possible to distinguish them even if R is extremely large. It is quite easy to understand why this should be the case. Let us suppose that R is very large and that there are two identical neighboring targets, one at t_{01} and the other very close to it at t_{02}. Then the target at t_{01} will give

$$C_{1u}(t_0) \quad = \quad R[1 \; - \; 2\pi^2 B^2(t_0 \; - \; t_{01})^2]$$

and the target at t_{02} will give

$$C_{2u}(t_0) \quad = \quad R[1 \; - \; 2\pi^2 B^2(t_0 \; - \; t_{02})^2]$$

The two targets together will therefore have a total C_u given by

$$C_{1u}(t_0) + C_{2u}(t_0) = 2R\left[1 - 2\pi^2 B^2\left(\frac{t_{01} - t_{02}}{2}\right)^2 - 2\pi^2 B^2\left(t_0 - \frac{t_{01} + t_{02}}{2}\right)^2\right]$$

This expression has only a single maximum at

$$t_0 \quad = \quad \frac{t_{01} + t_{02}}{2}$$

In other words, if the two targets are sufficiently close for the limited expansions used to be valid, only one target is seen and the two targets cannot be resolved even if R is extremely large.

In order for the two targets to be resolved (when they are in the same direction), it is necessary that, for $t_0 = t_{02}$, $C_{1u}(t_0)$ (of the first target) is almost zero. In other words, two targets separated by θ are resolved only if $C_u(\tau)$ for $\tau = t_{02} - t_{01} = \theta$, i.e. $C_u(\theta)$, is very small so that $C_u(\tau) + C_u(\tau - \theta)$ has two distinct maxima. In this way an interval θ_{min} can be defined, above which two identical targets are resolved and below which they are no longer resolved.

Remark 3.9

When a periodic signal with repetition period T_R is transmitted, a target at range t_{01} and another at $t_{01} + kT_R$ are equally confused by the radar (this difficulty is often purely theoretical).

3.6.2 First general rule

Regardless of how the resolving power is defined in terms of range, it is a function of and only of the shape of the useful signal at the output of the ideal receiver. Thus the resolving power in terms of range depends only on $|\Phi(f)|^2$, i.e. the modulus of the transmitted spectrum and not its phase.

3.6.3 Analysis of the problem

3.6.3.1 Classical radar

Let us first consider the example of an ideal radar transmitting a non-frequency-modulated square-wave signal of duration T in the presence of two identical targets with the same radial velocity whose radial range differs by θ. Figure 3.6 shows the envelope of the output signal of the radar for $\theta < T$ and $\theta > T$ depending on whether the signals received from the two targets are in phase or have opposite phase (the autocorrelation function of the transmitted signal is a triangle with base $2T$).

This example shows that, as soon as θ is larger than T, two maxima will be obtained on the envelope of the output signal corresponding to the true positions of the targets irrespective of the phase of the received signals. However, it can also be seen that, if θ is less than T, one maximum will be

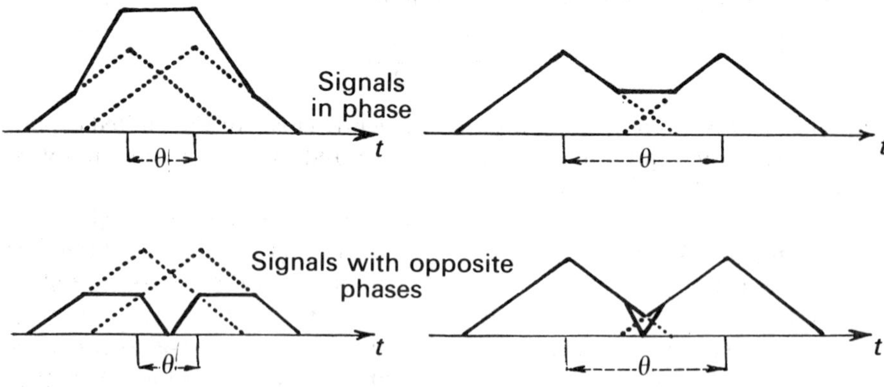

Fig. 3.6

obtained if the signals are in phase and two maxima will be obtained if they have opposite phase. In this case the maxima are flat and do not allow the positions of each of the targets to be obtained. Thus we can say that if $\theta < T$ the signals cannot be separated, and if $\theta > T$ the signals can be separated.

It is important to understand that, although this illustration is quite realistic in practice, it is pessimistic in theory. If, because the noise is negligible, the signals on the left-hand side of Fig. 3.6 are effectively received and it is known that the individual signals are triangular with duration $2T$, in theory it is possible to deduce from an examination of the received signals that two targets are present at particular positions. (It is sufficient to establish that the signal lasts for more than $2T$ in order to deduce that there is a target which is present from the start time plus T and another at the end time minus T.) This remark, which refers to angular measurements, also makes it possible to see that the classical concepts of resolving power in optics are obsolete and pessimistic (see Chapter 6, Section 6.8), although they are based on the same fundamental idea as that which we have attempted to use here for range measurement.

Having made this reservation, it is usually necessary to accept that, when the output signal of the radar only contains one maximum in the presence of a single target, two targets will be resolvable if the sum of the envelopes of the signals (radar output) corresponding to each of the targets has two maxima. However, if the sum of the envelopes of the two signals has only one maximum, the two targets cannot be resolved.

This idea can be exploited by defining the resolution ambiguity $A^2(\theta)$. We consider two targets of which the larger, which is taken as a reference, is at a range assumed to be the origin of the θs and has a signal energy of unity and the smaller is at a range θ with respect to the first. Then the resolution ambiguity is equal to A^2 at the range θ when, for a signal energy of the second target equal to A^2, the latter can just be separated from the first ($A^2 < 1$). In other words, if a target at range θ has a radar cross-section greater than A^2 times the radar cross-section of the reference target (see Chapter 5), it is resolvable; otherwise it is not resolvable.

3.6.3.2 Non-frequency-modulated gaussian signal

Let us consider a transmitted signal of the form

$$S(t) = \exp\left(-\frac{\pi t^2}{T^2}\right)$$

($S(t)$ is in fact equal to $\exp(-\pi t^2/T^2)\exp(j\omega_0 t)$ but $\exp(j\omega_0 t)$ is ignored because we are interested in the envelope of the signal.) A reference target will give a signal

$$C_u(t) = \exp\left(-\frac{\pi t^2}{2T^2}\right)$$

Another target at range θ and with an equivalent surface A^2 times that of the reference will give

$$A \exp\left[-\frac{\pi(t-\theta)^2}{2T^2}\right]$$

For a given A, it is a question of determining when

$$\left|\exp\left(-\frac{\pi t^2}{2T^2}\right)\right| + A\left|\exp\left[-\frac{\pi(t-\theta)^2}{2T^2}\right]\right|$$

has two maxima. It is found (after a relatively complex calculation which is strictly classical and involves several differentiations and changes of variable) that this result is obtained for

$$A = 1 \quad \text{when} \quad |\theta| > 1.1T$$

$$A = 0.5 \quad \text{when} \quad |\theta| > 1.4T$$

$$A = 0.1 \quad \text{when} \quad |\theta| > 1.9T$$

$$A = 10^{-2} \quad \text{when} \quad |\theta| > 2.3T$$

$$A = 10^{-3} \quad \text{when} \quad |\theta| > 2.6T$$

Table 3.2 shows $A^2(\theta)$ as a function of $|\theta|$. The data in this table were obtained using the expression

$$A^2(\theta) = \left[1 - \frac{1}{u^2} + \left(1 - \frac{2}{u^2}\right)^{1/2}\right]^2 u^4 \exp\left[-2u^2\left(1 - \frac{2}{u^2}\right)^{1/2}\right]$$

where

$$u^2 = \frac{\pi\theta^2}{2T^2} \quad \text{for} \quad u^2 > 2$$

and

$$A^2(\theta) = 1 \quad \text{for} \quad u^2 < 2$$

The data are also shown in Fig. 3.7 (the broken curve represents $\mathscr{A}(\theta)$). Figure 3.8 shows the same curves but with $A^2(\theta)$ expressed in decibels.

Table 3.2

$\lvert\theta\rvert$	$\leqslant 1.1T$	$1.4T$	$1.9T$	$2.3T$	$2.6T$
$A^2(\theta)$	1	0.25	10^{-2}	10^{-4}	10^{-6}

3.6.3.3 Non-frequency-modulated square-wave pulse with duration T

It can easily be found that, for any A, two targets (at the same radial speed) cannot be resolved (with the accepted definition) if $|\theta| < T$ and can

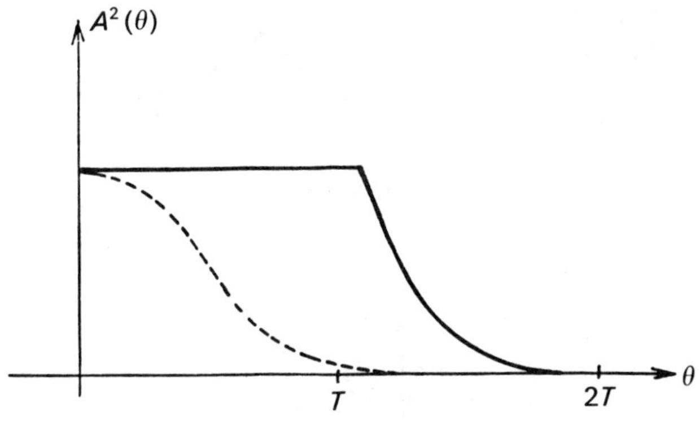

Fig. 3.7

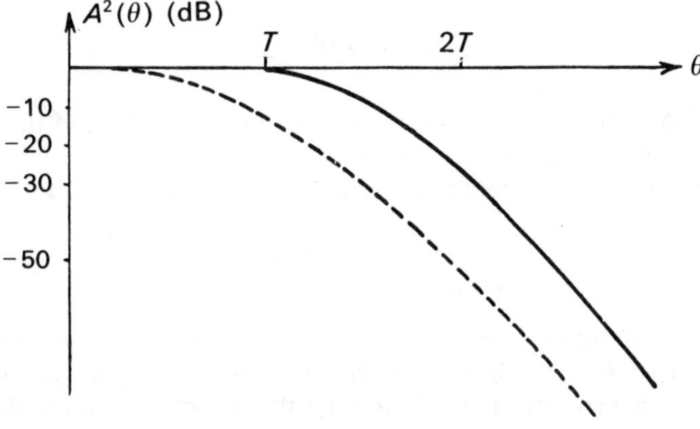

Fig. 3.8

always be resolved if $|\theta| > T$ (Fig. 3.9). Thus

$$A^2(\theta) = 1 \text{ if } |\theta| < T$$

$$A^2(\theta) = 0 \text{ if } |\theta| > T$$

(see Fig. 3.10 where $\mathscr{A}^2(\theta)$ is shown as a broken curve).

3.6.4 Approximate formula

The value of θ above which $A^2(\theta)$ decreases can be considered as the range resolving power of a radar. Let θ_{min} be this value: then two targets

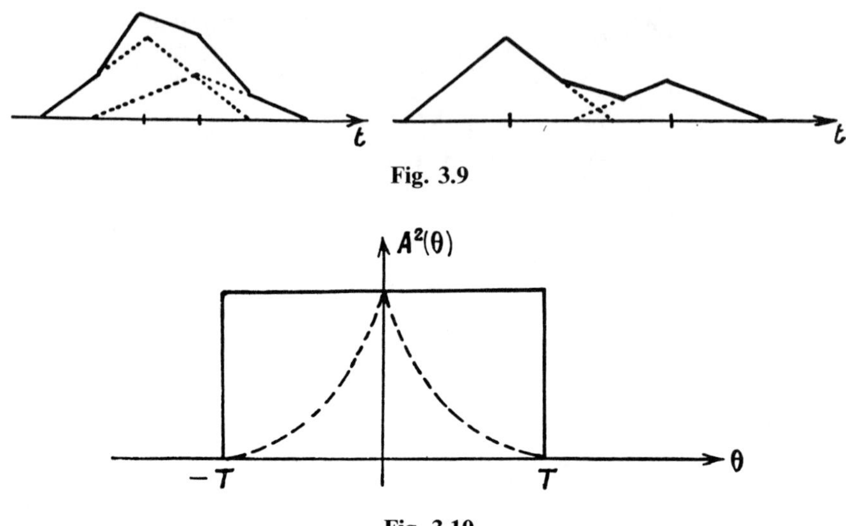

Fig. 3.9

Fig. 3.10

which are closer than θ_{min} cannot be distinguished regardless of their amplitudes. It is convenient, when there is no time to carry out the complete calculation, to use an approximate formula which gives θ_{min} with reasonable accuracy in many cases:

$$\theta_{min} = \int |\mathscr{A}^2(\theta)| \, d\theta \tag{3.17}$$

This definition of the resolving power of a radar, although it is not unreasonable, may seem rather arbitrary; none the less it is very useful. Classical transformations can also be used to write this expression in an alternative form†:

$$\theta_{min} = \frac{\int |\Phi(f)|^4 \, df}{\left[\int |\Phi(f)|^2 \, df\right]^2} \tag{3.18}$$

where $\Phi(f)$ is the Fourier transform of the transmitted useful signal. $1/\theta_{min}$ is also called the frequency span of the transmitted useful signal. For example, if the transmitted useful signal has a square-wave spectrum of width Δf,

$$\theta_{min} = \frac{1}{\Delta f}$$

† The results summarized in Chapter 1 are sufficient to prove this result because the function $\mathscr{A}(t)$ has a Fourier transform

$$\frac{|\Phi(f)|^2}{\int |\Phi(f)|^2 \, df}$$

3.7 Accuracy of the measurement of radial speed

Up to now we have assumed that the received signal $S(t - t_{01})$ is identical (apart from a factor) with the transmitted signal delayed by t_0. This is only exact if the target is fixed or has zero radial speed. If, however, the target has a nonzero radial velocity V_R and, as is frequently the case†, the width Δf of the transmitted spectrum is very small compared with the central frequency f_0 (corresponding to a wavelength λ), the signal received from the target is obtained by shifting the spectrum of $S(t - t_{01})$ by a frequency f_D, called the Doppler frequency, which is given by

$$f_D = \frac{2V_R}{\lambda} \tag{3.19}$$

In practice, the case which has just been discussed is the ideal case where either $f_D = 0$ or f_D is known, which makes it possible to shift the spectrum of the received signal artificially by $-f_D$, thus returning to the previous problem. This is the case when we are measuring the position of a target whose velocity is known or whose velocity is sufficiently low. (In practice, this means that we know the frequency f_D with an accuracy to better than approximately $1/T$.)

The problem which now has to be examined is to determine with what accuracy f_D can be measured if the position of the target is known with an accuracy to better than $1/\Delta f$. We proceed exactly as in the previous problem. If a spectrum $\Phi(f)$ is sent, it returns shifted by f_D and immersed in the noise. We now propose to evaluate the theoretical error with which f_D can be measured. All the arguments which have been made remain valid provided that time and frequency are exchanged. Starting from the envelope $\sigma(t)$ of the transmitted useful signal, we arrive at the definition of a time T_f in the following manner. The origin of the time is defined such that

$$\int_T t\sigma^2(t)\,dt = 0$$

(at the energy center of gravity of the signal) and T_f is defined by

$$\int_T t^2\sigma^2(t)\,dt = T_f^2 \int_T \sigma^2(t)\,dt \tag{3.20}$$

With a square-wave signal of duration T, we thus have

$$2T_f = \frac{T}{\sqrt{3}} \qquad T_f = \frac{T}{2\sqrt{3}}$$

† It is possible to make radar systems transmitting a signal occupy a spectrum width Δf of the order of the central frequency f_0, but the study of such systems lies outside the scope of this section.

The standard deviation of the best measurement which can be made of f_D is given by

$$\frac{1}{2\pi T_f R^{1/2}} \tag{3.21}$$

which corresponds to a standard deviation of the measurement of V_R equal to

$$\frac{\lambda}{4\pi T_f R^{1/2}} \tag{3.22}$$

In particular, if we have a square-wave signal of duration T, the standard deviation of the measurement of the radial velocity is given by

$$\frac{\lambda\sqrt{3}}{2\pi T R^{1/2}} \tag{3.23}$$

For example, if $\lambda = 0.1\,\mathrm{m}$, $T = 5 \times 10^{-3}\,\mathrm{s}$ and $R = 20$, it is theoretically possible to measure the radial velocity with an accuracy to $1.2\,\mathrm{m\,s^{-1}}$.

Remark 3.10

The Doppler velocity of targets can be measured using an alternative procedure which involves deriving successive positions of the target. The accuracy can be obtained using similar but different formulae.

3.8 (Radial) velocity ambiguity

In Section 3.3 we defined a function of the range θ called the ambiguity function $\mathscr{A}(\theta)$, and we define here in a similar manner an ambiguity function of the Doppler frequency F by using the expression

$$a(F) = \frac{\int_{\Delta f} \Phi(f)\,\Phi^*(f - F)\,\mathrm{d}f}{\int |\Phi(f)|^2\,\mathrm{d}f} \tag{3.24}$$

which is a corollary to expression (3.13) and by recalling that

$$\int |\Phi(f)|^2\,\mathrm{d}f \equiv \int |S(t)|^2\,\mathrm{d}t$$

This function is maximum and equal to unity for $F = 0$. It also has the following two properties.

(1) The amplitude of $a(F)$ gives an indication of the possibility of making an error F in the determination of the Doppler frequency (radial velocity) of a target whose distance is known. The larger is $a(F)$ for a given value F, the greater is the chance of making an error of F in the measurement of the Doppler frequency.

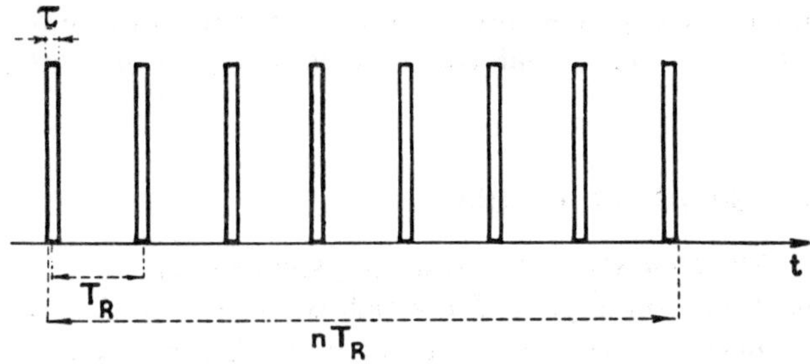

Fig. 3.11

(2) When two targets are at the same range, and the primary target has a Doppler frequency f_D and the parasitic target has a Doppler frequency $F_D + F$ and is p^2 times more energetic than the primary target, the response of the parasitic target at the frequency F_D has a power of $p^2|a^2(F)|$ times that of the response of the primary signal. In other words, the smaller is $|a(F)|$, the less the parasitic target interferes with the primary target.

Example 3.1

Let us consider a transmitted signal formed, as in the case of classical radar systems, from a certain number n of elementary regularly spaced square-wave signals, each lasting a time τ and separated by a time T_R, where τ is very small compared with T_R (Fig. 3.11). Such a signal is the equivalent in the time domain of a line spectrum in the frequency domain. Now, it has been seen in Section 3.5 that when the signal spectrum consists of lines separated by $1/T_R$ an error of kT_R can be made in the measurement of t_{01} and two targets separated by kT_R are confused. In a similar manner, when the transmitted signal is as shown in Fig. 3.11 we can conclude that an error of k/T_R can be made in f_D and hence two targets whose Doppler frequencies differ by k/T_R are confused. This is included in the function $a(F)$ by the existence of ambiguity peaks for $F = k/T_R$.

3.9 Resolving power in terms of (radial) velocity

We now return to the treatment in Section 3.6, *mutatis mutandis*.

3.9.1 General rule

The resolving power in terms of (radial) velocity, i.e. the ability to resolve two targets at the same distance because they have sufficiently different

Doppler frequencies, depends only on $\sigma(t)^2$, i.e. the envelope of the transmitted signal (the shape of the signal which is assumed to be centered on zero frequency).

3.9.2 Analysis of the problem

3.9.2.1 Non-frequency-modulated gaussian signal

A resolution ambiguity $A_1{}^2(F)$, which is a function of F, will be defined as follows. Consider two targets at the same distance. One of them (the larger) is taken as a reference at a Doppler frequency assumed to be the origin for F, and the other (the smaller) is at a Doppler frequency F with respect to the first. Then the resolution ambiguity is equal to $A_1{}^2$ at the frequency F, when the two targets can just be resolved ($A_1{}^2 < 1$), if the signal energy of the second target is $A_1{}^2$ times that of the first.

If we now consider the expression

$$\Gamma(f_0) = \int \Phi(f)\,\Phi^*(f-f_0)\,\mathrm{d}f = \left[\int \Phi^*(f)\,\Phi(f-f_0)\,\mathrm{d}f\right]^*$$

which is the corollary of $C_u(\tau)$, $A_1{}^2(F)$ is defined as being such that

$$|\Gamma(f)| + A_1|\Gamma(f-F)|$$

has just two maxima.

We now return to the example in Section 3.6.3.2, where

$$\sigma(t) = \exp\left(-\frac{\pi t^2}{T^2}\right)$$

Then $\Phi(f)$ is given by

$$\Phi(f) = k_1 \exp[-\pi(Tf)^2]$$

and therefore $\Gamma(f)$ is given by

$$\Gamma(f) = k_2 \exp\left[-\frac{\pi}{2}(Tf)^2\right]$$

Hence $A_1{}^2(F)$ is given by

$$A_1{}^2(F) = \left[1 - \frac{1}{v^2} + \left(1 - \frac{2}{v^2}\right)^{1/2}\right]^2 v^4 \exp\left[-2v^2\left(1 - \frac{2}{v^2}\right)^{1/2}\right]$$

where

$$v^2 = \frac{\pi}{2}(TF)^2 \quad \text{for } v^2 > 2$$

and

$$A_1{}^2(F) = 1 \quad \text{for } v^2 < 2$$

i.e.

$$A_1{}^2(F) = A^2(FT^2)$$

3.9.2.2 Non-frequency-modulated square-wave pulse with duration T

If we calculate the expression $\Gamma(f)$ for this case, we find that it is equal to the Fourier transform of a square-wave signal of duration T:

$$\Gamma(f) = \frac{\sin(\pi f T)}{\pi f T}$$

Because of this it has its highest peak equal to unity for $f = 0$ and an infinite number of secondary peaks with (approximately)

$$|f| = \frac{1}{2T} + \frac{2k}{T} \qquad k > 1$$

and approximate amplitude $1/\pi T f$, so that $|\Gamma(f)| + A_1|\Gamma(f - F)|$ always has a large number of peaks. The definition given above then becomes invalid. In fact, we can state that two targets whose Doppler frequencies differ by F are resolvable if $\Gamma(f) + A_1\Gamma(f - F)$ has a maximum around $f = 0$ and another maximum around $f = F$ which are higher than the neighboring maxima (Fig. 3.12). It is then found that, for $A_1 = 1$, resolution is possible if F is larger than approximately $1.1/T$. It is also found that for $F \gg 1/T$ resolution is possible for

$$A_1^2 > \frac{1}{(\pi F T)^2}$$

which gives the following approximate values for $A_1^2(F)$:

$$A_1^2(F) = 1 \quad \text{for } F < \frac{1.1}{T}$$

$$A_1^2(F) = \frac{1}{(\pi F T)^2} \quad \text{for } F > \frac{1.1}{T}$$

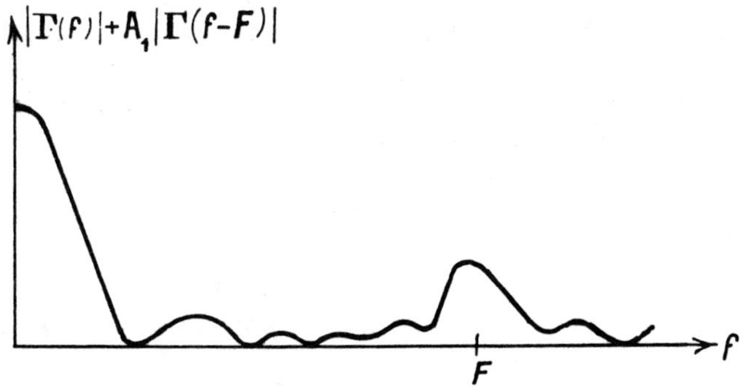

Fig. 3.12

3.9.3 Approximate formula

The value of F above which $A_1(F)$ decreases can be considered as the Doppler frequency resolving power of the radar. Denote this value by $F_{\min}$. Then two targets which are closer than $F_{\min}$ in terms of Doppler frequency (and at the same radial distance) cannot be resolved no matter what their amplitudes are.

In many cases $F_{\min}$ can be obtained with reasonable accuracy using the following approximate formula (do not use this formula when the signal $S(t)$ is quasi-periodic):

$$F_{\min} = \int |a^2(F)| \, \mathrm{d}F \tag{3.25}$$

This formula can also be written as

$$F_{\min} = \frac{\int_T |\sigma(t)|^4 \, \mathrm{d}t}{\left[\int_T |\sigma(t)|^2 \, \mathrm{d}t\right]^2} \tag{3.26}$$

which is a corollary to eqn (3.18).

For example, if $S(t)$ is a square-wave signal with duration T, we find $F_{\min} = 1/T$. Thus two targets at the same range can be distinguished if their radial velocities differ by $V_{\min} = \lambda/2T$. Numerically, if $\lambda = 0.1\,\mathrm{m}$ and $T = 5 \times 10^{-3}\,\mathrm{s}$, $V_{\min} = 10\,\mathrm{m\,s^{-1}}$.

3.10 Range–velocity ambiguity

3.10.1 General considerations

In the previous sections we have examined the following: the likelihood of making an error in the measurement of the range of a target whose Doppler frequency is assumed to be known or of making an error in the measurement of the Doppler frequency of a target whose range is known; the interference with a primary target produced by a parasitic target either at the same Doppler frequency and different range or at the same range with a different Doppler frequency. This is all expressed mathematically in the ambiguity functions $\mathscr{A}(\theta)$ and $a(f)$.

In practice the problem is more general: both the range and Doppler frequency have to be measured; the interference with a primary target caused by a parasitic target which has neither the same range nor the same Doppler frequency has to be evaluated. It is clear that similar arguments have to be used.

What is $\mathscr{A}(\theta)$? When it is suitably normalized, it is the response at $t_0 = \theta$ and $f_D = 0$ of a target at $t_0 = 0$ with a Doppler frequency $f_D = 0$. This time, we consider the suitably normalized response $\mathscr{A}(\theta, F)$ for $t_0 = \theta$ and $f_D = F$ of a target at $\theta = 0$ and $f = 0$, i.e. the correlation of the received signal $S(t)$ of a target situated at $t_0 = 0$ with $f_D = 0$, with the reference $S(t)$ shifted by time θ and frequency F:

$$\mathscr{A}(\theta, F) = \frac{\int_T S(t)\, S^*(t - \theta) \exp(2\pi j F t)\, dt}{\int |S(t)|^2\, dt} \tag{3.27}$$

What is $a(F)$? When it is suitably normalized, it is the output of the filter matched to a signal situated at $t_0 = 0$ and at $f_D = F$, to which a signal corresponding to a target situated at $t_0 = 0$ and $f_D = 0$ is applied. This time we consider the suitably normalized output of the filter matched to a signal situated at $t_0 = \theta$ and $f_D = F$ excited by a signal corresponding to a target situated at $t_0 = 0$ and $F_D = 0$:

$$a(\theta, F) = \frac{\int \Phi(f)\, \Phi^*(f - F) \exp(2\pi j F \theta)\, df}{\int |\Phi(f)|^2\, df} \tag{3.28}$$

These two lines of argument lead to the same result because $|\mathscr{A}(\theta, F)|$ is equal to $|a(\theta, F)|$. In practice, $|\mathscr{A}(\theta, F)|$ is generally used for obvious reasons, or $|\mathscr{A}(\theta, F)|^2$ is used for reasons discussed later. $\mathscr{A}(\theta, F)$ (or its derivatives $|\mathscr{A}(\theta, F)|$ and $|\mathscr{A}^2(\theta, F)|$) is the two-dimensional ambiguity function. This ambiguity has two properties.

(1) The amplitude of $\mathscr{A}(\theta, F)$ indicates the likelihood of simultaneously making an error of θ in range and F in Doppler frequency in the determination of the two coordinates of a target. The closer $|\mathscr{A}(\theta, F)|$ is to unity, the more likely is a simultaneous error in θ and F.

(2) If two targets lie in the same direction, with the primary target at a range t_0 with a Doppler frequency f_D and the parasitic target at a range $t_0 + \theta$ with a Doppler frequency $f_D + F$, and the parasitic target is p^2 times more energetic than the primary target, then the parasitic target gives a response at range t_0 and Doppler frequency f_D with a power $p^2|\mathscr{A}^2(\theta, F)|$ times larger than that of the primary signal. In other words, the smaller is $|\mathscr{A}(\theta, F)|$, the less the parasitic target interferes with the primary target.

$|\mathscr{A}^2(\theta, F)|$ represents the interference with the primary target caused by other targets as a function of their relative position (θ, F).

The mathematical properties of the ambiguity function can be summarized as follows.

(1) It has a maximum for $\theta = 0$ and $F = 0$ and the maximum is equal to unity.

(2) It is symmetric with respect to this point:

$$|\mathscr{A}(\theta, F)| = |\mathscr{A}(-\theta, -F)|$$

(3) The volume contained by the ambiguity $|\mathscr{A}^2(\theta, F)|$ is constant:

$$\iint |\mathscr{A}^2(\theta, F)|\,d\theta\,dF = 1 \tag{3.29}$$

In other words, it is impossible to reduce the ambiguity at a point without seeing it increase elsewhere.

We recall that we have defined (in an approximate manner) the resolving power in terms of the pure range θ_{min} (possibility of distinguishing two targets at the same velocity by their different ranges) and the Doppler frequency F_{min} (possibility of distinguishing two targets at the same range by their different Doppler frequencies) by

$$\theta_{min} = \int |\mathscr{A}^2(\theta, 0)|\,d\theta \tag{3.17'}$$

and

$$F_{min} = \int |\mathscr{A}^2(0, F)|\,dF \tag{3.25'}$$

It should be noted that eqn (3.29) does not imply that the product of θ_{min} and F_{min} defined by eqns (3.17′) and (3.25′) is constant. In particular, signals can exist such that the contours of $|\mathscr{A}^2|$ are ellipses with the θ and F axes being the major axes (Fig. 3.13). Obviously, in this case the volume of the ambiguity is equal to unity and the product $\theta_{min} f_{min}$ is very small when, for example, the ambiguity contains a sharp central peak at zero ($\theta = 0, F = 0$) and a plateau around this peak where the ambiguity decreases slowly on moving away from zero. For example, if the ambiguity is a revolution around zero of the form $\mathscr{A}^2(\varrho)$ (with $\varrho^2 = \tau^2 + F^2$), the integral

$$F_{min} = \theta_{min} = 2\int_0^\infty \mathscr{A}^2(\varrho)\,d\varrho$$

may be very small, whereas the integral

$$\int_0^\infty 2\pi\varrho\mathscr{A}^2(\varrho)\,d\varrho$$

is equal to unity. In this case, the area where the ambiguity $|\mathscr{A}^2|$ is greater than 0.8, for example, may be very small, whereas the area where the ambiguity $|\mathscr{A}^2|$ is greater than 10^{-6} would be extremely large. This is why the use of the term ambiguity area instead of ambiguity volume may give rise to false interpretation.

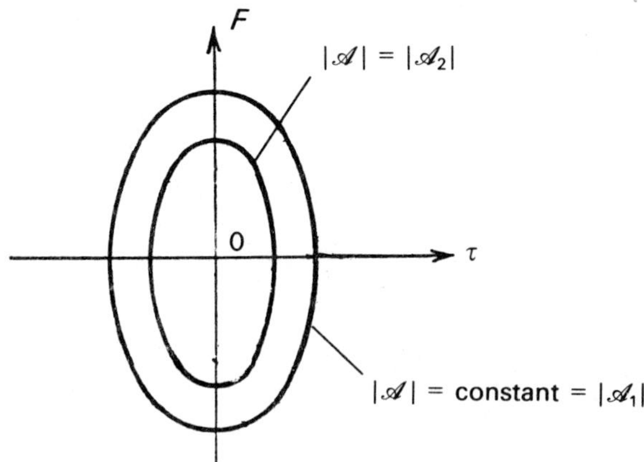

Fig. 3.13

3.10.2 Non-frequency-modulated square-wave pulse with duration T

Consider the case where the transmitted signal has a constant amplitude over its duration T and is not frequency modulated. These conditions hold for a classical radar system transmitting signals which are completely matched at the receiver (by correlation for example). In this case, neglecting the carrier frequency, we can write (see Section 3.4)

$$S(t) = \frac{u(t) - u(t-T)}{T^{1/2}}$$

where $u(t)$ is the unit level of the Laplace transform $1/p$, and

$$\mathscr{A}(\theta, F) = \int [u(t) - u(t-T)][u(t-\theta) - u(t-\theta-T)] \exp(2\pi \mathrm{j} Ft) \frac{\mathrm{d}t}{T}$$

where $|\mathscr{A}^2(\theta, F)|$ is an even function of θ. Obviously it is sufficient to calculate $\mathscr{A}(\theta, F)$ for $\theta > 0$.

For $\theta > T$, the product $S(t)\, S^*(t - \theta)$ is identically zero and therefore $\mathscr{A}^2$ is also identically zero. For $\theta < T$, we have

$$\mathscr{A}(\theta, F) = \frac{1}{T}\int [u(t-\theta) - u(t-T)] \exp(2\pi \mathrm{j} Ft)\, \mathrm{d}t$$

$$= \frac{1}{T} \frac{\exp[\pi \mathrm{j} F(\theta + T)] \sin[\pi F(T-\theta)]}{\pi F}$$

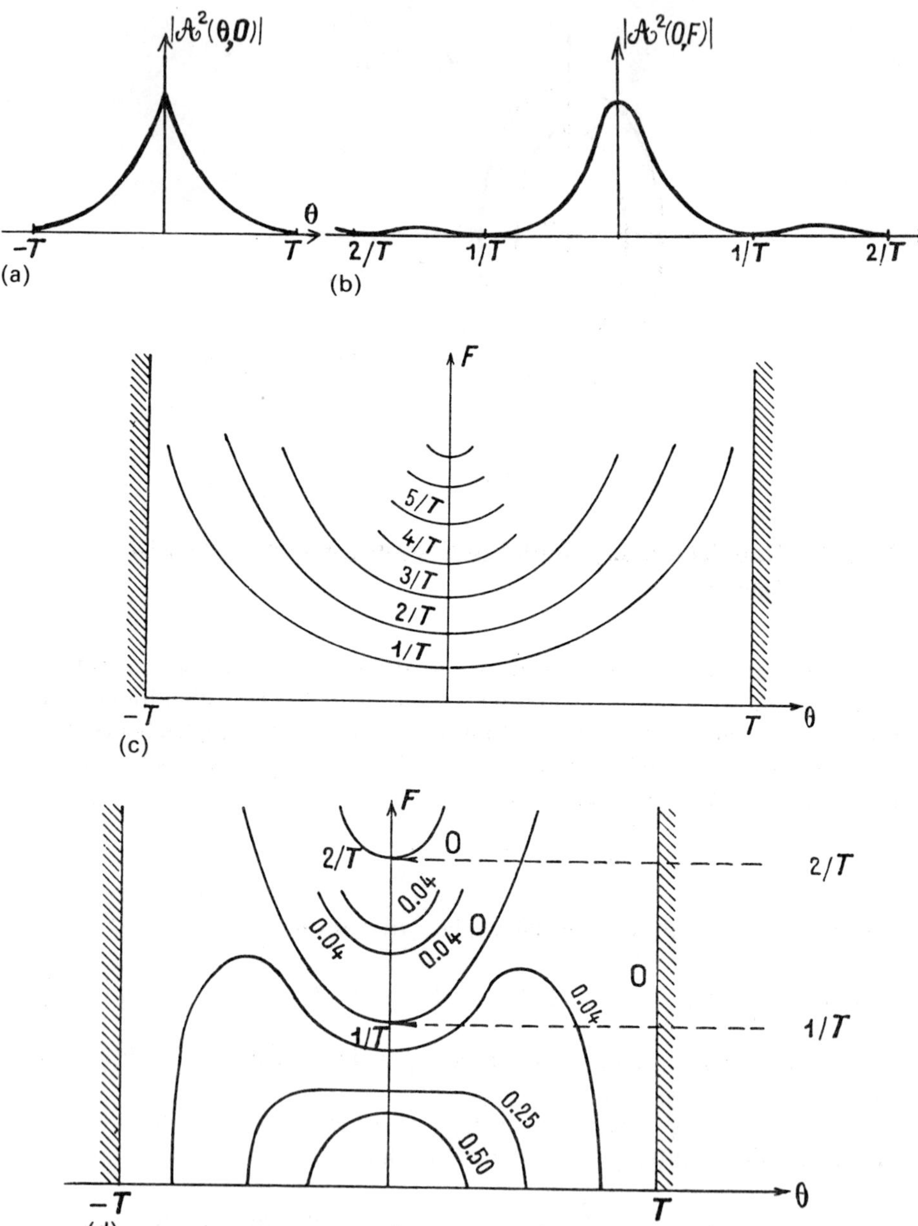

Fig. 3.14 (a) $|\mathscr{A}^2(\theta, 0)|$ as a function of θ; (b) $|\mathscr{A}^2(0, F)|$ as a function of F; (c) locus of points where $|\mathscr{A}^2| = 0$ in the plane (θ, F); (d) locus of points where $|\mathscr{A}^2|$ takes values of 0.5, 0.25 and 0.04 in the plane (θ, F).

and hence

$$|\mathscr{A}^2(\theta, F)| = \frac{[\sin \pi F(T - \theta)]^2}{\pi^2 F^2 T^2}$$

Figure 3.14 shows the behavior of $|\mathscr{A}^2(\theta, F)|$ under various conditions. We find that

$$\theta_{\min} = 2\int_0^{\infty} \frac{(T - \theta)^2}{T^2}\,\mathrm{d}\theta = \frac{2T}{3}$$

$$F_{\min} = 2\int_0^{\infty} \left[\frac{\sin(\pi F T)}{\pi F T}\right]^2 \mathrm{d}F = \frac{1}{T}$$

and hence

$$\theta_{\min} F_{\min} = 2/3$$

3.10.3 Non-frequency-modulated gaussian signal

The spectrum used in classical radar systems is limited in practice. The useful transmission signal of a classical radar system can be represented in an equally valid manner by writing $S(t)$ in the form

$$S(t) = K' \exp\left[-\pi\left(\frac{t}{T}\right)^2\right]$$

The Fourier transform of $S(t)$ is

$$\Phi(f) = k \exp[-\pi(Tf)^2]$$

(The width of the signal at 3 dB is of the order of $2T/3$ and the width at 3 dB of the spectrum $\Phi(F)$ is of the order of $2/3T$.) Then

$$\mathscr{A}(\theta, F) = K'^2 \int \exp\left[-\pi\left(\frac{t}{T}\right)^2\right] \exp\left[-\pi\left(\frac{t-\theta}{T}\right)^2\right] \exp(2\pi \mathrm{j} F t)\,\mathrm{d}t$$

$$= K'^2 \exp(\pi \mathrm{j} F\theta) \exp\left[-\frac{\pi}{2}\left(\frac{\theta}{T}\right)^2\right] \int \exp\left(-\frac{2\pi u^2}{T^2}\right) \exp(2\pi \mathrm{j} F u)\,\mathrm{d}u$$

where $u = t - \theta/2$. Then

$$\mathscr{A}(\theta, F) = \exp\left[-\frac{\pi}{2}\left(\frac{\theta}{T}\right)^2\right] \exp\left[-\frac{\pi}{2}(TF)^2\right] \exp(\pi \mathrm{j} F\theta)$$

and hence

$$|\mathscr{A}^2(\theta, F)| = \exp\left[-\pi\left(\frac{\theta}{T}\right)^2\right]\exp[-\pi(TF)^2]$$

The loci of the points where $|\mathscr{A}^2|$ is constant are formed by ellipses whose major axes are the θ and F axes. We find that

$$\theta_{\min} = 2\int_0^\infty \exp\left[-\pi\left(\frac{\theta}{T}\right)^2\right]\mathrm{d}\theta = T$$

$$F_{\min} = 2\int_0^\infty \exp[-\pi(TF)^2]\,\mathrm{d}F = \frac{1}{T}$$

and hence

$$\theta_{\min}F_{\min} = 1$$

3.10.4 Gaussian signal with linear frequency modulation

A gaussian signal with linear frequency modulation is a simple mathematical example of the transmission signal of a pulse compression radar (in fact, pulse compression radar systems tend to use square-wave frequency-modulated signals (see Chapter 4)). The transmission signal can be written in the form

$$S(t) = \exp\left[-\pi\left(\frac{t}{T}\right)^2\right]\exp\left(-\mathrm{j}\frac{\pi}{K}t^2\right)$$

i.e. the frequency of the signal varies linearly as a function of time with a gradient $\mathrm{d}f/\mathrm{d}t = 1/K$. In this case

$$\mathscr{A}(\theta, F) = k\int \exp\left[-\pi\left(\frac{t}{T}\right)^2\right]\exp\left[-\mathrm{j}\left(\frac{\pi}{K}\right)t^2\right]\exp\left[-\pi\left(\frac{t-\theta}{T}\right)^2\right]$$

$$\times \exp\left[\mathrm{j}\left(\frac{\pi}{k}\right)(t-\theta)^2\right]\exp(2\pi\mathrm{j}Ft)\,\mathrm{d}t$$

$$= k\exp\left(-\frac{\pi\theta^2}{2T^2}\right)\exp(\pi\mathrm{j}F\theta)\int\exp\left[-2\pi\left(\frac{u}{T}\right)^2\right]$$

$$\times \exp\left[-2\pi\mathrm{j}u\left(\frac{\theta}{K-F}\right)\right]\mathrm{d}u$$

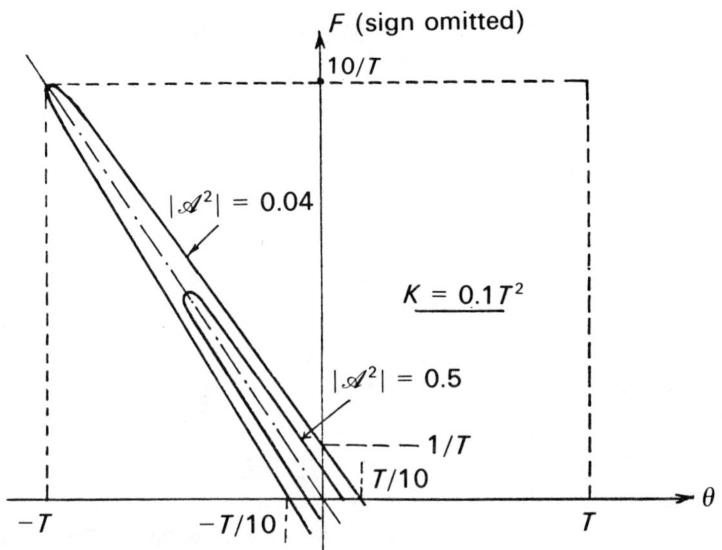

Fig. 3.15

where $u = t - \theta/2$. Then

$$\mathscr{A}(\theta, F) = \exp\left(-\frac{\pi\theta^2}{2T^2}\right)\exp(\pi jF\theta)\exp\left[-\left(\frac{\pi}{2}\right)\left(F - \frac{\theta}{K}\right)^2 T^2\right]$$

and hence

$$|\mathscr{A}^2(\theta, F)| = \exp\left(-\frac{\pi\theta^2}{T^2}\right)\exp\left[-\pi T^2\left(F - \frac{\theta}{K}\right)^2\right]$$

The loci of the points where $|\mathscr{A}^2|$ is constant are formed by ellipses described by the equation

$$\frac{\theta^2}{T^2} + \left(F - \frac{\theta}{K}\right)^2 T^2 = \text{constant}$$

i.e. the diameter in the direction of the F axis is the straight line with equation $F = \theta/K$ (Fig. 3.15). In other words, the ambiguity is concentrated on the straight line $F = \theta/K$. This is in good agreement with the results found in Chapter 4.

We also find

$$\theta_{\min} = \frac{T}{(1 + T^4/K^2)^{1/2}}$$

T^2/k is approximately equal to the compression ratio ϱ, which is defined in Chapter 4, and is significantly greater than unity. Then

$$\theta_{\min} \approx T/\varrho$$

$$F_{\min} = 1/T$$

The product $\theta_{\min} F_{\min}$ does not always make physical sense since the ambiguity is concentrated in a direction which does not always coincide with the θ or F axes.

3.10.5 Square-wave pulse trains

As an example Figs 3.16–3.20 show the function $|\mathscr{A}^2(\theta, F)|$ for a train of three square-wave signals of duration τ with a repetition period of T_R. The shaded areas in Fig. 3.20 correspond to areas where the ambiguity is greater than 0.25 and the contours correspond to the loci $|\mathscr{A}^2| = 0.1$.

3.10.6 Square-wave pulse with linear frequency modulation

Figure 3.21 shows the ambiguity function $|\mathscr{A}(\theta, F)|$ of a square-wave pulse with linear frequency modulation corresponding to a $T\Delta F$ product (compression ratio) of 20. The origin $\theta = 0$, $F = 0$ is in the middle of the "background" and the θ axis is parallel to the edge in front of the box cover.

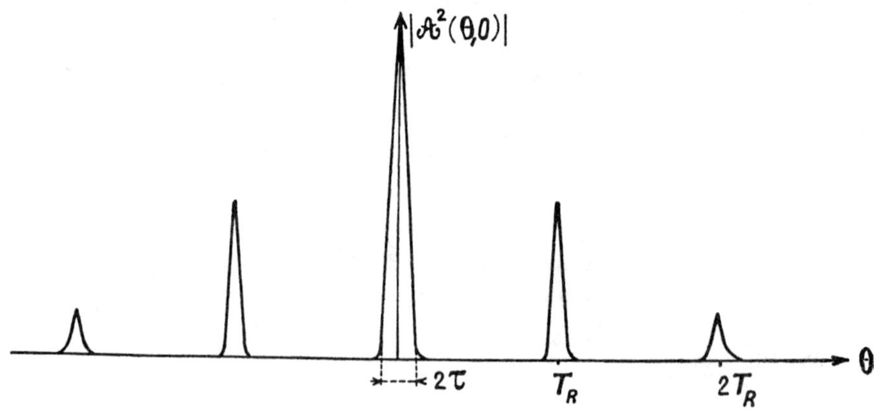

Fig. 3.16 $|\mathscr{A}^2(\theta, 0)|$ as a function of θ.

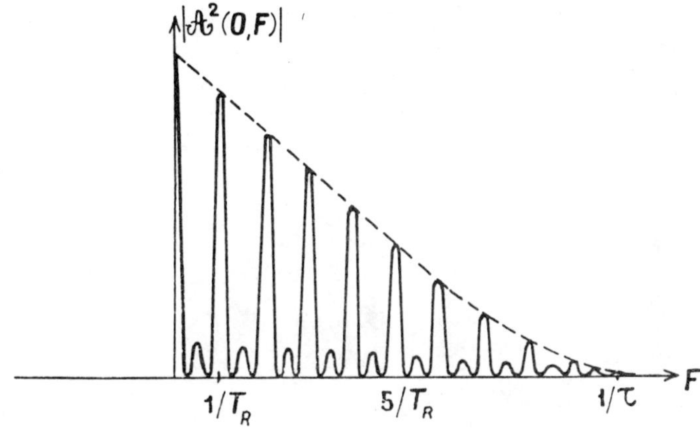

Fig. 3.17 $|\mathscr{A}^2(0, F)|$ as a function of F.

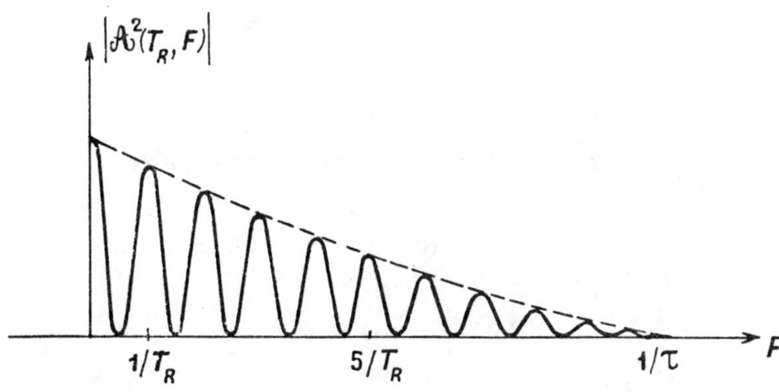

Fig. 3.18 $|\mathscr{A}^2(T_R, F)|$ as a function of F.

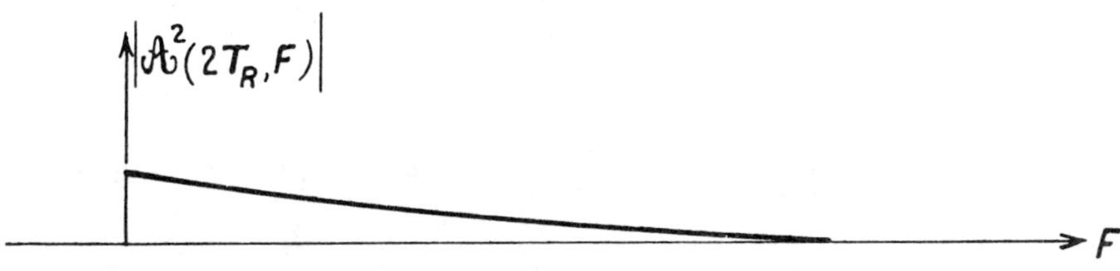

Fig. 3.19 $|\mathscr{A}^2(2T_R, F)|$ as a function of F.

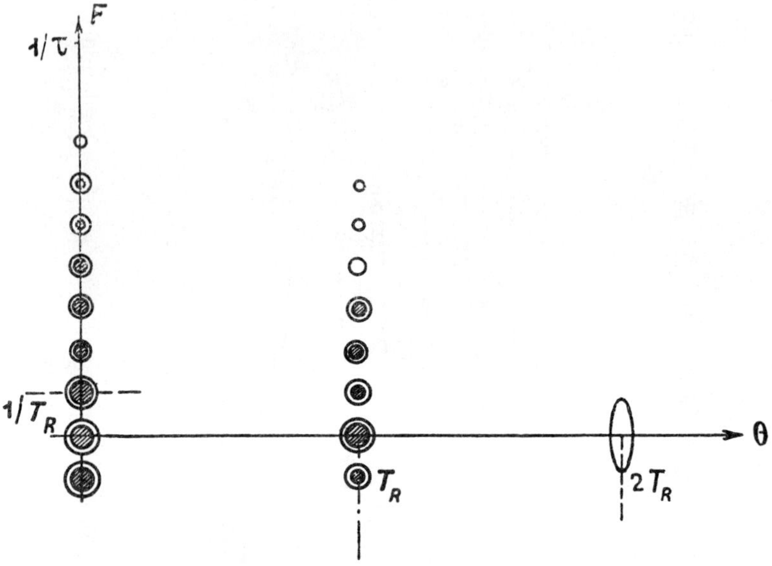

Fig. 3.20 Contours of $\mathscr{A}^2$.

3.11 Range–velocity resolution

The resolution ambiguity $A^2(\theta, F)$ can be defined by generalizing the statements in Sections 3.6.3 and 3.9.2, and its value is such that

$$|\mathscr{A}(t, f)| + A|\mathscr{A}(t - \theta, f - F)|$$

has just two maxima.

3.11.1 Non-frequency-modulated gaussian signal (see Section 3.10.3)

A value of A such that

$$\exp\left[-\frac{\pi}{2}(Tf)^2\right]\exp\left[-\frac{\pi}{2}\left(\frac{t}{T}\right)^2\right] + A\exp\left\{-\frac{\pi}{2}[T(f - F)]^2\right\}$$
$$\times \exp\left[-\frac{\pi}{2}\left(\frac{t - \theta}{T}\right)^2\right]$$

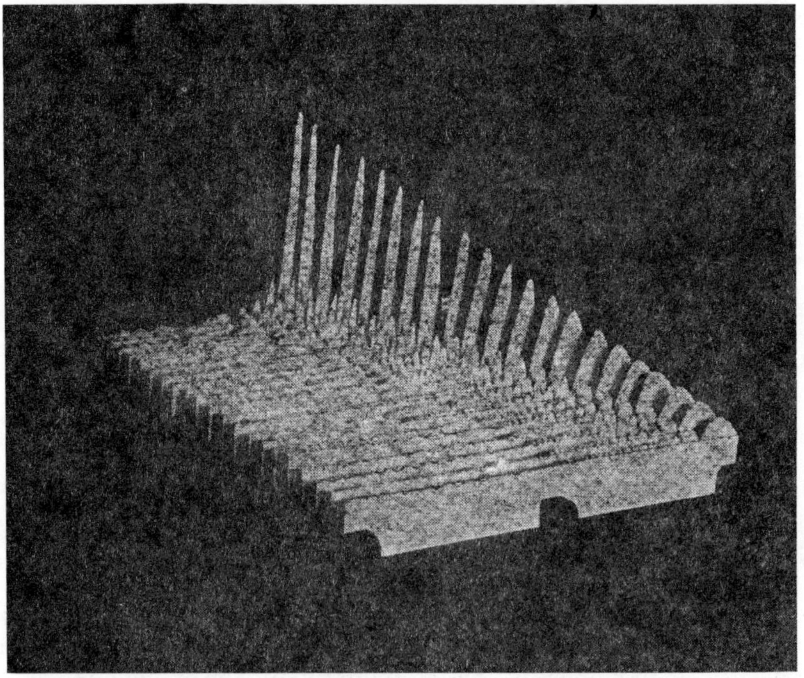

Fig. 3.21

has just two maxima is required. This result can be found (in a rigorous manner here and in general to an excellent approximation) by determining the condition for which the intersection of the surface

$$|\mathscr{A}(t, f)| + A|\mathscr{A}(t - \theta, f - F)|$$

with the plane $f = (F/\theta)\, t$ has just two maxima, i.e. in this case when

$$\exp\left[-\frac{\pi}{2}\left(\frac{T^2F^2}{\theta^2} + \frac{1}{T^2}\right)t^2\right] + A \exp\left[-\frac{\pi}{2}\left(\frac{T^2F^2}{\theta^2} + \frac{1}{T^2}\right)(t - \theta)^2\right]$$

has two maxima. It can be seen that

$$A^2(\theta, F) = A^2\left(\left[\theta^2 T^2\left(\frac{T^2F^2}{\theta^2} + \frac{1}{T^2}\right)\right]^{1/2}, 0\right)$$

$$A^2(\theta, F) = A^2([\theta^2 + T^4F^2]^{1/2}, 0)$$

and the results shown in Fig. 3.22 are obtained. The hatched area corresponds to $A^2 = 1$.

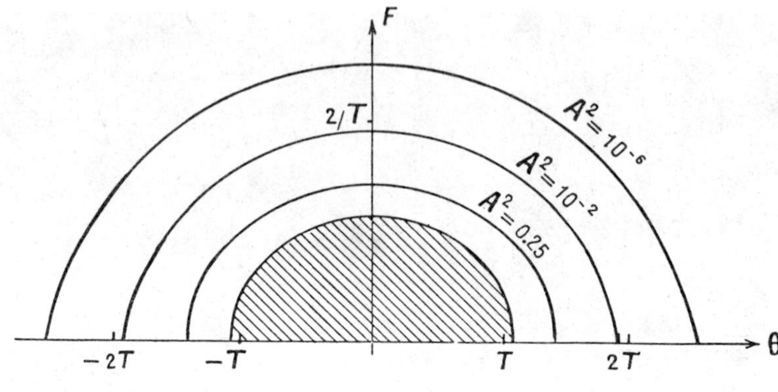

Fig. 3.22

A point in the plane θ, F where $A^2(\theta, F) = A_0^{\,2}$ is such that two targets whose ranges differ by θ and whose Doppler frequencies differ by F can be separated if the ratio of the radar cross section of the smallest target to the radar cross section of the largest target is greater than $A_0^{\,2}$. The contour enclosing the hatched area is a special contour; it corresponds to the impossibility of resolution regardless of the level of the two targets. The hatched area is called the total ambiguity area or, more simply, the ambiguity area.

3.11.2 Gaussian signal with linear frequency modulation

In an identical manner to that used above, it is found that

$$A_2^{\,2}(\theta, F) = A_1^{\,2}\left(\left[\theta^2 + T^4\left(F - \frac{\theta}{K}\right)^2\right]^{1/2}\right)$$

The ambiguity found in 3.10.4 is given and the curves shown in Fig. 3.23 are obtained.

3.11.3 Square-wave pulse which is not frequency modulated

When $F(T - |\theta|)$ is sufficiently large relative to unity, it can be seen, using a similar approach to that in Section 3.9.2.2, that there is resolution if

$$A^2 > \frac{1}{(\pi FT)^2}$$

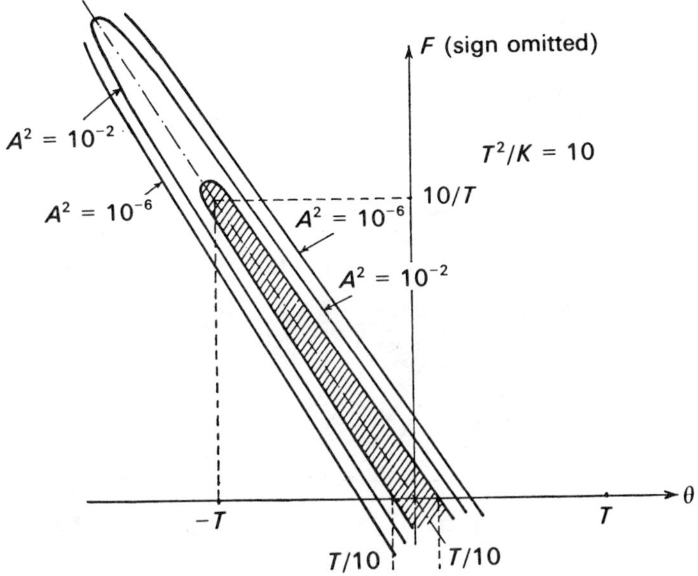

Fig. 3.23

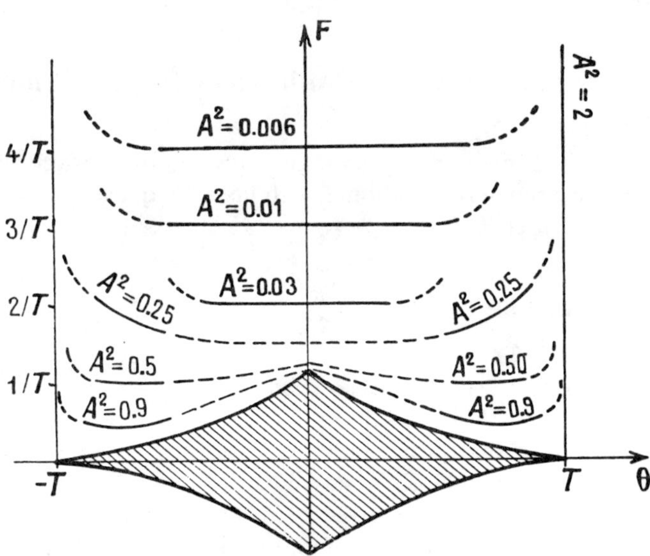

Fig. 3.24

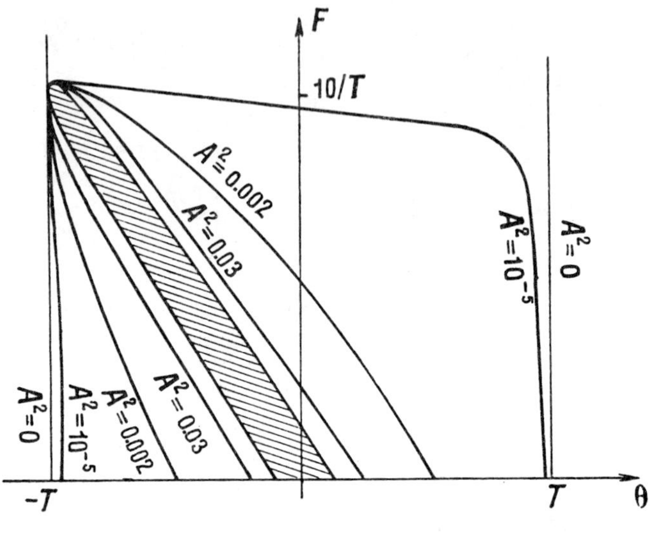

Fig. 3.25

The contour $A^2(\theta, F) = 1$ is defined as in Section 3.11.1 by the intersection of the surface $|\mathscr{A}(t, f)| + |\mathscr{A}(t - \theta, f - F)|$ with the plane $f = (F/\theta)\,t$. $A^2(\theta, F)$ is shown approximately in Fig. 3.24.

3.11.4 Square-wave pulse with linear frequency modulation

Figure 3.25 gives an approximate representation of the ambiguity A^2 of a square-wave signal of duration T with linear frequency modulation between $f_0 - 5/T$ and $f_0 + 5/T$.

3.11.5 Random signal

Let us consider a transmission signal formed from a noise with constant power (random phase modulation) occupying a rectangular spectrum of width Δf when a square-wave pulse of duration T is extracted from this noise (Fig. 3.26). The product $T\Delta f$ is assumed to be very large (e.g. 10^6) but not infinite, i.e. T is very large relative to $1/\Delta f$ but is finite. Under these conditions, the expression for $C_u(\theta)$ is very similar to the autocorrelation function of the spectral density of the source noise (rectangular spectral density) but it is not rigorously identical with it. The discrepancy has two different sources.

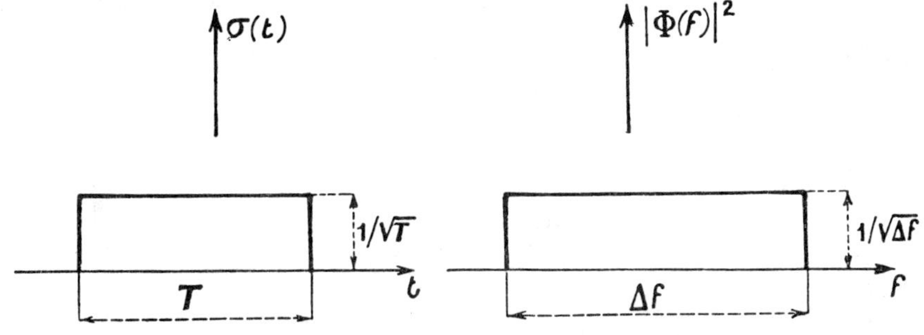

Fig. 3.26

(1) The autocorrelation function should be multiplied by $|(T - \theta)/T|$ in order to allow for the fact that the signals $S(t)$ and $S^*(t - \theta)$ only coexist for the time $T - \theta$ which enables $C_u(\theta) = 0$ to be found for $\theta \geqslant 0$.

(2) $|T/(T - \theta)|C_u(\theta)$ is only equal to the autocorrelation function of $S(t)$ if $T - \theta$ is infinitely large relative to $1/\Delta f$. Since this is certainly not the case, we must take $|T/(T - \theta)|C_u(\theta)$ to be equal to the autocorrelation function of $S(t)$ plus a corrective term due to the finite nature of $T - \theta$ which becomes dominant for large θ.

More generally, because $T - \theta$ is finite, the expression $\mathscr{A}(\theta, F)$ is equal to what it would be if T were infinite multiplied by $|(T - \theta)/T|$ with the addition of a corrective term. This corrective term can be calculated by assuming that the point (θ, F) is sufficiently far from the center of coordinates for the assumption that $S(t)$ and $S^*(t - \theta)\exp(-2\pi jFt)$ are independent to be valid. It is then given by

$$\sigma = \int S(t)\,S^*(t - \theta)\exp(-2\pi jFt)\,dt$$

$$\sigma = \sum S(t_i)\,S^*(t_i - \theta)\exp(-2\pi ijFt_i)\frac{1}{\Delta f}$$

where the t_i are the $(T - \theta)\Delta f$ sampling times (separated by $1/\Delta f$).

The real part of σ is a random variable. It is gaussian because $(T - \theta)\Delta f$ is large (except when θ is extremely close to T) and has a zero mean because $S(t_i)$ and $S^*(t_i - \theta)\exp(-2\pi jFt_i)$ are assumed to be independent. It is characterized by its variance

$$(T - \theta)\Delta f\left(\frac{1}{\Delta f^2}\frac{1}{T^2}\frac{1}{2}\right) = \frac{T - \theta}{2T^2\Delta f}$$

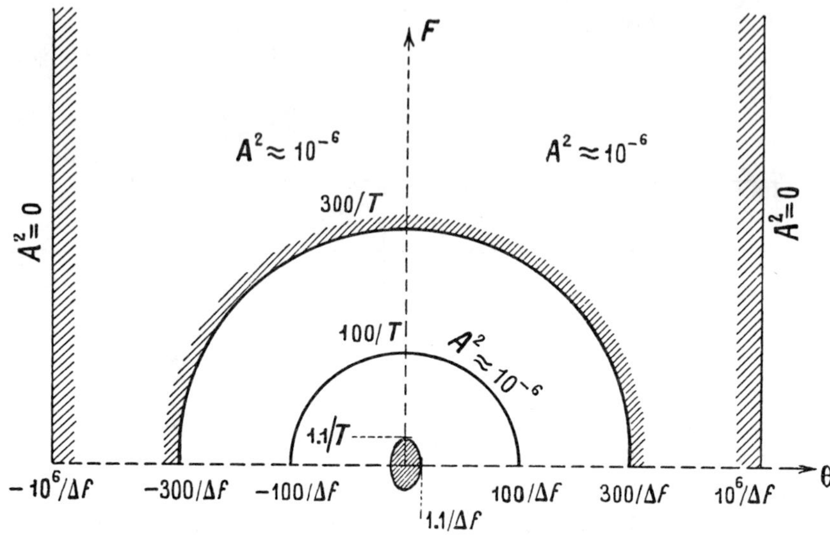

Fig. 3.27

The same arguments apply to the imaginary part of σ, so that $|\sigma|$ has a Rayleigh distribution and the mean value of $|\sigma^2|$ is given by

$$\overline{|\sigma^2|} = \left|\frac{T-\theta}{T}\right| \frac{1}{T\Delta f}$$

It can therefore be seen that a target located at a position (θ, F), which would easily allow it to be distinguished from the reference target if T were infinite, runs the risk of not being distinguished because of the term σ. It is logical to accept that, on average and in terms of order of magnitude, it can only be resolved if

$$A^2 > \left|\frac{T-\theta}{T}\right| \frac{1}{T\Delta f}$$

$A^2(\theta, F)$ is represented quite well by Fig. 3.27 (where $T\Delta f = 10^6$). With the approximations made, Fig. 3.27 also represents the ambiguity $|\mathscr{A}^2(\theta, F)|$. The ambiguity surface consists of a central peak of small volume surrounded by a plateau whose height is of the order of $1/T\Delta f$.

Remark 3.11

It appears from the foregoing that if a random signal burst is transmitted, the ambiguity can be reduced to as low a level as desired by increasing $T\Delta f$ to infinity. In fact, there is an upper limit on $T\Delta f$. In the presence of a target with radial velocity V_R the Doppler deviation is not constant over the entire range of the spectrum but differs by $2V_R\,\Delta f/c$, where c is the velocity of light, between the end frequencies of the spectrum used. It is therefore not possible to make a consistent measurement for too long a time T because the following condition must hold:

$$\frac{1}{T} > \frac{2V_R\,\Delta f}{c}$$

and hence

$$T\,\Delta f < \frac{c}{2V_R}$$

This result can also be obtained by stating that the time T must be less than the time taken by the target to travel the distance corresponding to approximately $1/\Delta f$, i.e. the distance $c/2\Delta f$. This condition gives

$$V_R T < \frac{c}{2\,\Delta f}$$

and hence

$$T\,\Delta f < \frac{c}{2V_R}$$

Therefore the higher the speed of the target, the more difficult it is to locate its position. This statement is sometimes called Heisenberg's principle applied to radar [4]. The following numerical examples show the limitations on $T\,\Delta f$: $V_R = 50\,\mathrm{m\,s^{-1}}$, $T\,\Delta f < 3 \times 10^6$; $V_R = 6000\,\mathrm{m\,s^{-1}}$, $T\,\Delta f < 2.5 \times 10^4$.

Note Part of the problem is due to the fact that in the foregoing it has been assumed that the Doppler effect gives a translation of the frequency spectrum whereas it has the effect of producing a similarity transformation of the frequency scale. When this property is applied in radar systems (the theory of such systems has yet to be developed, at least to our knowledge), these restrictions will take a different form.

3.12 Application of ambiguity to combat parasitic echoes

If we have a useful target with a radar cross-section σ_u and a parasitic target with a radar cross-section σ_p located at (θ, F) with respect to the useful target, the power of the parasitic signal at the position of the useful signal is given by

$$\sigma_p|\mathscr{A}(\theta, F)|^2$$

relative to the power σ_u of the useful signal. Now, if we have a large number of parasitic targets with radar cross-sections σ_{pi} located at (θ_i, F_i) with respect to the useful target and returning signals which are not in phase (which is usually the case), the power of the parasitic signal at the position of the useful signal is given by

$$\sum \sigma_{pi} |\mathscr{A}(\theta_i, F_i)|^2$$

Similarly, if we have diffuse parasitic echoes (clutter) characterized by an equivalent surface density $V(\theta, F)$ (expressed in $\mathrm{m^2\,Hz^{-1}\,s^{-1}}$), their power at the position of the useful signal is

$$\iint |\mathscr{A}^2(\theta, F)|\, V(\theta, F)\, \mathrm{d}\theta\, \mathrm{d}F$$

relative to σ_u. If this integral is larger than σ_u, there is little hope of obtaining correct information about the useful target. Conversely, if the integral is smaller than σ_u there is a good chance that correct information can be obtained. If $V(\theta, F)$ is constant, the integral will not depend on the transmitted signal. Fortunately, this is never the case, and therefore engineers are able to choose an appropriate transmitted signal.

For example, assume that we are dealing with a useful target with a radar cross-section of $1\ \mathrm{m^2}$ in the presence of diffuse parasitic echoes such that $V(\theta, F) = 10^4\ \mathrm{m^2\,Hz^{-1}\,s^{-1}}$ for $5000\ \mathrm{Hz} < F < 5100\ \mathrm{Hz}$ and $-200\ \mu\mathrm{s} < \theta < 200\ \mu\mathrm{s}$ and is zero outside these intervals (Fig. 3.28). This example corresponds to a clutter occupying a large area of space around the useful target with a density of $10^{-8}\ \mathrm{m^2}$ of radar cross-section per cubic meter when a radar strobe of azimuth 10^{-2} rad and elevation 6×10^{-3} rad is used. The useful target is 100 km from the radar.

If the transmitted signal is a non-frequency-modulated gaussian signal of duration 10 μs,

$$\iint |\mathscr{A}^2(\theta, F)|\, V(\theta, F)\, \mathrm{d}\theta\, \mathrm{d}F \approx 2 \times 10^{-5} \times 100 \times 10^4 = 20$$

and the useful target cannot be located correctly (Fig. 3.29). If the transmitted signal is a non-frequency-modulated gaussian signal of duration 400 μs (Fig. 3.30),

$$\iint |\mathscr{A}^2(\theta, F)|\, V(\theta, F)\, \mathrm{d}\theta\, \mathrm{d}F \approx 4 \times 10^{-4} \times 10^2 \times 10^4 \times 10^{-6} \approx 4 \times 10^{-4}$$

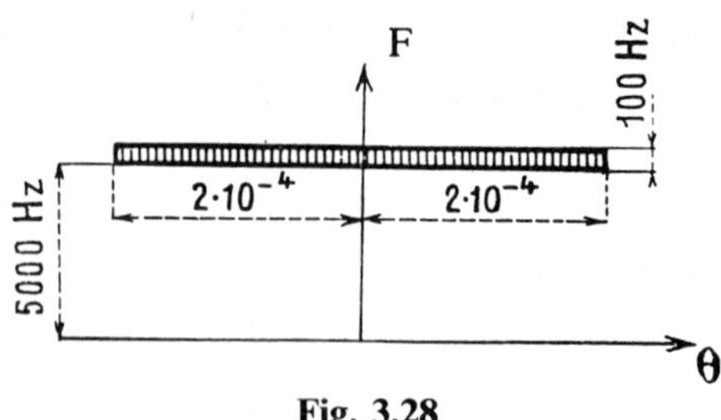

Fig. 3.28

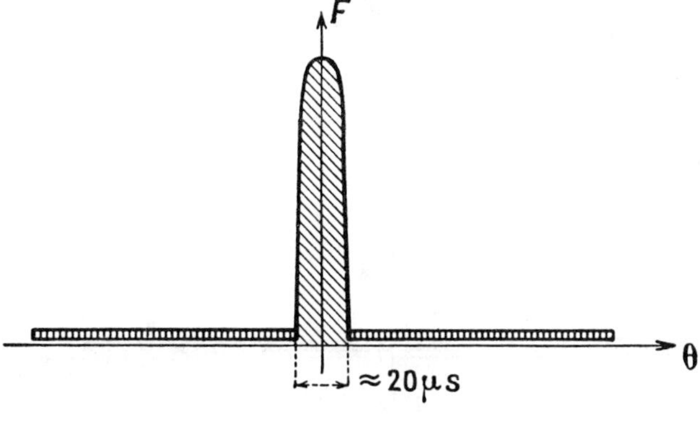

Fig. 3.29

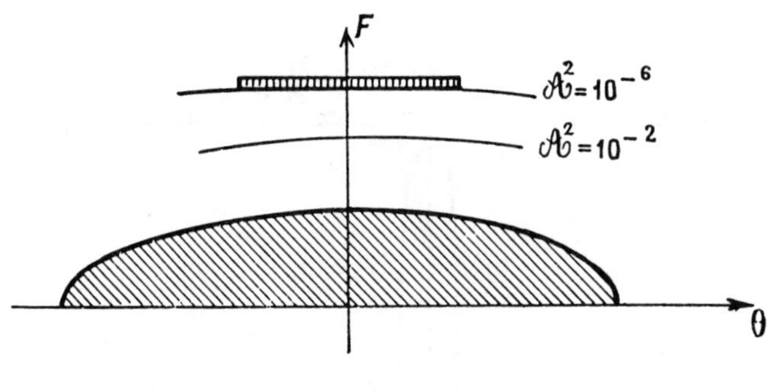

Fig. 3.30

which is negligible. However, the radar can only distinguish two targets with the same radial velocity if their ranges differ by more than 75 km (500 μs), and the accuracy of measuring ranges with the radar will be poor.

Another possibility involves using a gaussian signal of duration $T = 10\,\mu\text{s}$ with linear frequency modulation with $k = 0.01T^2$ (compression ratio of the order of 100). The result shown in Fig. 3.31 is obtained, and it can be seen that

$$\iint |\mathscr{A}^2(\theta, F)|\, V(\theta, F)\, \mathrm{d}\theta\, \mathrm{d}F \approx 2 \times 10^{-7} \times 100 \times 10^4 = 0.2$$

which is good. The same result would have been obtained with a non-frequency-modulated gaussian signal of duration $T = 0.1\,\mu\text{s}$ (i.e. all other things being equal, with a peak power 100 times higher). It is also said that the clutter has been reduced in the compression ratio.

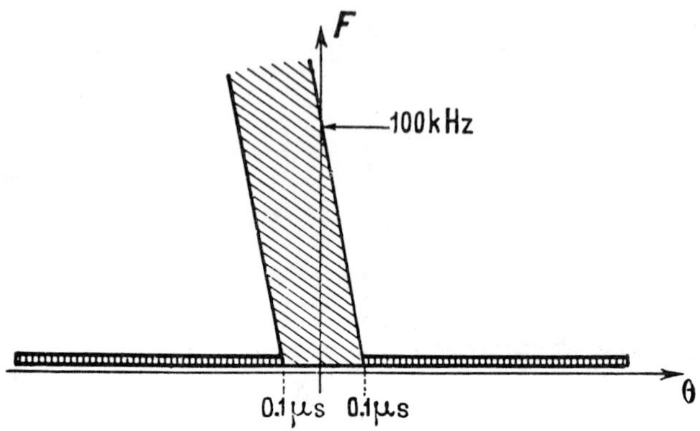

Fig. 3.31

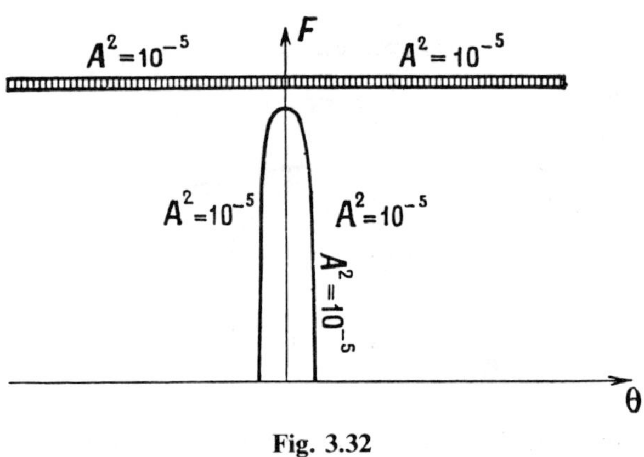

Fig. 3.32

Finally, it is possible to use a transmission signal with random phase modulation, as described in Section 3.11.5, with $\Delta f = 10\,\text{MHz}$ and $T = 0.01\,\text{s}$ ($T\Delta f = 10^5$). This gives a value of

$$4 \times 10^{-4} \times 10^2 \times 10^4 \times 10^{-5} \approx 4 \times 10^{-3}$$

for the double integral, which is completely negligible (Fig. 3.32). However, the range resolving power is of the order of 20 m and the radial velocity resolving power is of the order of 100 Hz (these values are meaningful for two similar targets), and the peak power to be transmitted is a factor of 1000 lower than in the case of the compression radar considered above, all other things being equal. It should be noted for completeness, even in an example, that if the parasitic targets are point targets, are few in number and are very powerful, their influence is constant in a radar system using such a signal irrespective of the position of the useful target. This is not the case for a radar system using the signal shown in Fig. 3.29 (classical radar) for which the parasitic targets only have an adverse effect in practice if they are less than 2000 m from the useful target. In the presence of a single parasitic point target with a radar cross-section of 10^6 m^2 the random phase transmission radar is incapable of correctly locating a useful target with a radar cross-section of less than 10 m^2, whereas classical radar does not have this drawback. This also draws attention to the inadequacy of the ambiguity concept. This is that, in such a case, we can very rapidly determine the location of a single parasitic target in terms of θ and F and estimate its level and hence subtract the corresponding signal from the received signal $y(t)$, thus eliminating, at least partially, its adverse effects.

3.13 Conclusions

In this section we summarize the results obtained in this chapter, so that the reader, even if he has not understood the theoretical discussion (which is usually the fault of the author) need only remember the following conclusions. A knowledge of these conclusions is essential for any radar engineer.

It should be remembered that what follows applies to ideal radar and that in practice a radar system cannot be better than ideal; however, practical radar systems often almost reach the theoretical limits of the ideal radar.

(a) In order for the probability of detecting a target to be acceptable, it is necessary and sufficient for the ratio R of the energy received during a measurement to the power density of the noise (or jamming) per hertz to be acceptable. The range of a radar system (with a given antenna, wavelength and target) therefore depends only on the ratio of the energy transmitted during a measurement to the power density of the noise (or jamming) per hertz. In particular, the range of a radar system does not depend on the form of a transmitted signal or on the shape or width of its spectrum. For example, two good identical radar systems transmitting at the same rate a signal of 100 MW peak power for 4 μs and a signal of 100 kW peak power for 4 ms respectively have the same range.

(b) For a given quality of detection, the range measurement accuracy which a radar can achieve depends only on the shape of the spectrum of the transmitted signal (only the amplitude of the spectrum is involved so that the phase can be modified without any effect on the accuracy or resolving power). For a given range, the range error is approximately inversely proportional to the width of the transmitted spectrum. As the target approaches the radar (R increasing), the accuracy improves.

(c) For a given detection quality and carrier transmission frequency, the accuracy of measuring the radial velocity which can be achieved by a radar system depends only on the envelope of the useful transmitted signal. The error in measuring the radial velocity is approximately inversely proportional to the duration of the useful transmitted signal (or the duration of the measurement).

(d) The probability of resolving two targets does not depend on the transmitted energy. Two targets with the same velocity and a given separation can only be distinguished if the shape of the transmitted spectrum is correct and in practice if the transmitted spectrum is sufficiently wide (only the amplitude of the spectrum is involved and hence the phase can be modified without any effect on the accuracy or resolving power). Similarly, it is only possible to distinguish two targets at the same range with different radial velocities if the shape of the transmitted signal is correct and in practice if the transmitted signal has sufficient duration (or if the duration of the measurement is sufficient).

It immediately follows that the radar system with the best qualities is that system which transmits the longest pulses and the widest spectrum. Traditional radar engineers may regard these properties as contradictory. In fact, they are only contradictory if the transmitted signal is amplitude modulated. If frequency (or phase) modulation is used, it is very easy to obtain a long signal occupying a wide spectrum. However, the main advantage of transmitting a long signal occupying a wide spectrum rather than a short signal (which, unfortunately, is what most existing radar systems do) lies in the fact that, because the range is determined by the energy transmitted in the signal, it is necessary to transmit enormous peak power with short signals whereas the peak power can be greatly reduced with long signals.

3.13.1 Choice of signal to be transmitted

Three requirements must be taken into account in the choice of signal.

(1) The first relates to the range required, which does not determine the power but the energy to be transmitted for the duration of the measurement. This energy is often easier to achieve by transmitting for a long time at low power than at high power for a short time. However, the measurement time

is limited to a maximum which depends on the maximum acceptable time for obtaining the measurement result, or the time during which the parameters to be measured are sufficiently constant, depending on the measurement accuracy required, or the time during which it can be assumed that the fluctuations in the received signal caused by the behavior of the target and the movement of the radar antenna (see Chapter 5) are small.

(2) The second consideration relates to the accuracy required for the range and radial velocity measurements. This requirement eliminates signals with too narrow a spectrum (inaccurate range) or signals which are too short (inaccurate radial velocity), and also eliminates signals which produce ambiguous measurements.

(3) The third consideration takes into account the amplitude, the range and the radial velocity of targets other than the primary target with respect to the same variables of the primary target.

Let us assume that we have chosen a transmission signal $S(t)$. Then, corresponding to this signal through a particular mathematical operation we have the ambiguity function

$$|\mathscr{A}^2(\theta, F)|$$

which is a function of the difference θ between the range of a parasitic target and that of the primary target and the difference F between the radial velocity of the parasitic target and that of the primary target. The function $|\mathscr{A}^2(\theta, F)|$ can be loosely defined as representing the degree of interference produced by the parasitic target in the detection and location of the primary target. The configuration of the parasitic targets (clutter) is represented by a function $V(\theta, F)$ which gives the amplitude of the clutter as a function of θ and F, and hence the interference produced by these parasitic targets is given by

$$\iint |\mathscr{A}^2(\theta, F)|\, V(\theta, F)\, \mathrm{d}\theta\, \mathrm{d}F$$

It is clear that it is necessary to choose $\mathscr{A}^2(\theta, F)$, and therefore the transmitted signal $S(t)$, so that this integral is as small as possible. Thus we are now able to select the optimum modulation for the transmitted signal, after which we can choose the type of receiver to be used.

It should be noted that this work is not always carried out completely because the transmitted signal in the direction of the target is the result of the modulation by the antenna of the signal at the output of the transmitter. This is why in a conical scanning radar the movement of the antenna introduces an amplitude modulation of the transmitted signal by the equipment as a whole and the receiver has to be matched not to the signal at the output of the transmitter but to the signal received by the target, which is very different.

Remark 3.12

It is not sufficient to choose the transmitted signal $S(t)$ in order to minimize the interference caused by the clutter, i.e. to minimize the integral

$$\iint |\mathscr{A}^2(\theta, F)|\, V(\theta, F)\, \mathrm{d}\theta\, \mathrm{d}F$$

but it is necessary to choose $S(t)$ in order to maximize the response of the useful signal, i.e. the expression in which $V(\theta, F)$ represents the geography of the useful target. This consideration does not apply when the dimensions of the useful target are small with respect to the range resolving power of the radar or when the Doppler spectrum of the useful target is narrow compared with the Doppler frequency resolving power which has often been the case to date. However, this consideration must be taken into account when the target is at a large range or has a high velocity (e.g. the Earth or the Moon) or when the range resolving power becomes very good, leading to new developments in radar systems.

So far we have assumed that the signal reflected by the target has the same form as the transmitted signal. In practice, however, although the signal from the target is derived from the transmission signal, it is modulated by the target, and it is necessary to use a receiver matched not to the transmitted signal but to the signal received when $S(t)$ has been transmitted to a particular target. The next stage in the development of radar will no doubt involve the use of a receiver matched to a given target when it is illuminated by a given signal transmitted using a given antenna, and it will thus be possible to obtain a substantially improved response, i.e. less noise, and to obtain a significant response for a given target only. The end product could be, for example, a radar system equipped with several outputs, each corresponding to a given type of target.

3.13.2 Choice of receiver

As a result of the convolution theorem, a radar receiver can be produced using either correlators or a matched filter (see Chapter 4, Section 4.3). When correlators are used, we have a correlation radar, the earliest example of which is the pulse Doppler radar. When a matched filter is used, we are dealing with a matched filter radar, of which the earliest example is the classical radar and a more recent example is the chirp pulse compression radar.

The choice between the two types of radar systems is based on technological and economic factors.

Chapter 4

Analysis of the Operating Principles of some Types of Radar

Our aim in this chapter is not to give a complete description of existing radar systems but to explain their underlying philosophy.

4.1 Noise radar

It has been shown in the preceding chapters that the ideal radar should integrate the product

(received signal) × (transmitted signal delayed by t_0)

over the measurement time and should find the value of t_0 giving the maximum. In other words, we have an infinite number of transmitted signal samples delayed by all possible and imaginable values of t_0. Each of these samples is multiplied by the received signal, the result is integrated over a time T and the value of t_0 for which the integral gives the largest result is found. In practice, only a finite number of samples of the delayed transmitted signal are used.

(a) It is possible to use just one sample, in which case the integrator will give the maximum signal when the target passes through the distance corresponding to the sample delay.

(b) A large number of samples separated from each other by the error in the location of the target can be used.

(c) Finally, a small but sufficient number of samples separated from each other by the sampling interval $1/\Delta f$ can be used.

Cases (b) and (c) differ from case (a) only in the number of correlators (the complexity of the equipment). It is therefore possible to make do with a complete analysis of case (a), i.e. an examination of the behavior of the radar which looks for a target at a specific range at a given time or which measures, to a specific accuracy, the times when a target is at a specific given range.

4.1.1 Target with a known velocity

For example, assume that we are dealing with a target with radial velocity lying between $V_r - v_r$ and $V_r + v_r$ with $V_r = 300\,\mathrm{m\,s^{-1}}$ and $v_r = 5\,\mathrm{m\,s^{-1}}$. Assume further that the transmitted signal has a square-wave spectrum of width Δf centered on 150 MHz ($\lambda = 2$ m) and that we wish to know the time when the target is at a range $D = 500$ km from the radar to an accuracy of ± 15 m. Assume also that we are satisfied with a false alarm probability of 10^{-3} and a detection probability of 0.9. This means that at the limit of the range (see Chapter 3, Fig. 3.4) we will have $R = 20$ (13 dB). The value of B is therefore determined by

$$15\,\mathrm{m} \rightarrow 10^{-7}\,\mathrm{s} = \frac{1}{2\pi B\sqrt{(20)}}$$

$$2B = 700\,\mathrm{kHz}$$

and, according to Chapter 3, Remark 3.7,

$$\Delta f = 1.2\,\mathrm{MHz}$$

If we adopt this value for Δf, we also know that since the target has a radial velocity V_r of $300\,\mathrm{m\,s^{-1}}$ there is no point in making a measurement over a time that is longer than the time that the target takes to travel 15 m, i.e. for a time longer than

$$15/300 = 0.05\,\mathrm{s} = T$$

A radial velocity v_r corresponds to a Doppler deviation

$$f_D = \frac{2v_r}{\lambda} = 5\,\mathrm{Hz}$$

This value is less than $1/T$: therefore the velocity is known with sufficient accuracy.

The mean radial velocity corresponds to a Doppler frequency of 300 Hz. The signal to be transmitted will therefore be determined when the energy to be transmitted for the time T is known. To find this, we assume that the transmitter antenna has a circular aperture of diameter 20 m and therefore an area A_t of $300\,\mathrm{m}^2$, i.e. a maximum theoretical gain

$$G_t = \frac{4\pi A_t}{\lambda^2} = \frac{4\pi \times 300}{4} = 900$$

The actual gain will be smaller, of the order of 600 (28 dB). The power density at the range D is therefore given by

$$\frac{600P}{4\pi D^2} = p$$

The target is assumed to reradiate isotropically a certain proportion $p\sigma_t$ of the transmitted power, by virtue of the definition of the radar cross-section σ_t of the target.

It should be noted in passing that the radar cross-section of a target is only distantly related to its actual area, i.e. to its dimensions; it depends on the wavelength and the presentation of the target, and it fluctuates with time (see Chapter 5) which also makes it necessary, as will be seen, to take R larger than 13 dB in many cases. However, we can assume that, for a given wavelength, a given target in a particular presentation has a corresponding order of magnitude for σ_t which we shall take equal to 1 m^2 (this is the normal value for a small plane). The target therefore reradiates a power density equal to $G_t P\sigma_t/4\pi D^2$, and the power finally received at the receiver antenna, which is assumed to be identical with the transmitter antenna but displaced to one side and has an effective reception area $A_R = \lambda^2 G_t/4\pi$, is given by

$$\frac{G_t P\sigma_t}{(4\pi D^2)^2} A_R = \frac{12 \times 10^4 P}{(4\pi D^2)^2}$$

The energy E received from the target as a function of the transmitted energy E_t can then be written as

$$E = \frac{G_t A_R \sigma_t}{(4\pi^2)^2} E_t = \frac{12 \times 10^4 E_t}{(4\pi^2)^2}$$

It is now necessary to know the power density N_0 of the noise accompanying the useful signal. If there is no jamming, the value of N_0 is given by

$$N_0 = kT_N$$

where k is the Boltzmann constant which is equal to 1.4×10^{-23} W Hz^{-1} K^{-1} and T_N is the effective noise temperature of the reception system (in Kelvin). T_N is the sum of the noise temperatures of each of the components of the reception system (taking into account, if necessary, the attenuation produced by these components). For simplicity we can write

$$T_N = T_M + T_R$$

where T_M is the noise temperature of the antenna and the microwave guides and T_R is the noise temperature of the receiver itself (which may have a microwave component). We then define the noise factor of the reception system by the relationship

$$N_{RS} = \frac{T_N}{T_0} = \frac{T_M + T_R}{T_0}$$

where T_0 is the reference temperature which is equal to 290 K. The noise factor of the receiver is defined as that which would be found for the complete reception system if $T_M = T_0 = 290$ K, i.e.

$$N_R = \frac{T_0 + T_R}{T_0} = 1 + \frac{T_R}{T_0}$$

and therefore

$$N_{RS} = \frac{T_M}{T_0} + N_R - 1$$

Two numerical examples are given below:

$$(1)\ T_R = 2600\,\text{K} \quad N_R = \frac{2890}{290} = 10\,\text{dB}$$

$$T_M = 100\,\text{K} \quad N_{RS} = \frac{2700}{290} = 9.7\,\text{dB} \approx N_R$$

$$(2)\ T_R = 290\,\text{K} \quad N_R = \frac{580}{290} = 3\,\text{dB}$$

$$T_M = 100\,\text{K} \quad N_{RS} = \frac{390}{290} = 1.3\,\text{dB} \neq N_R$$

We finally find that, at the range D

$$R = \frac{2G_t A_R E_t \sigma_t}{(4\pi D^2)^2 k T_N} = \frac{2 \times 1.2 \times 10^5 E_t}{(4\pi \times 25 \times 10^{10})^2 \times 1.4 \times 10^{-20}} \quad \left(R = \frac{2E}{N_0}\right)$$

where we have taken $T_N = 1000$ K which is a reasonable value and corresponds to a noise factor of approximately 5 dB. Since $R = 20$ (no allowance is made for losses due to imperfections in the receiver), the transmitted energy E_t is given by

$$E_t = \frac{RkT_N(4\pi)^2 D^4}{2G_t A_R \sigma_t} = \frac{20 \times 1.4 \times 10^{-20} \times 160 \times 625 \times 10^{20}}{2 \times 1.2 \times 10^5} \tag{4.1}$$

where E_t is in joules. In this case,

$$E_t \approx 15\,\text{J}$$

If we assume that the transmitted signal has a constant power throughout its duration (peak power), this power must be

$$\frac{E_t}{T} = \frac{15}{0.05} = 300\,\text{W}$$

In practice, in order to ensure that the target receives the transmitted signal when it is at the required range and if it is not known *a priori* when it will be there, it is necessary to transmit an average power of 300 W continuously. The transmission signal will be, for example, a noise uniformly occupying the frequency band between 149.4 and 150.6 MHz with an average power of 300 W.

Let us now assume that the target is transmitting jamming in the form of gaussian noise with a density p of $10^{-6}\,\mathrm{W\,Hz^{-1}}$ (a 50 W transmitter occupying a band of 50 MHz and using an isotropic antenna). Now, the receiver noise will be negligible compared with the jamming and the ratio R will be given by

$$R = \frac{2G_t E_t \sigma_t}{4\pi D^2} \frac{1}{p}$$

which leads to

$$E_t = \frac{2\pi D^2 R}{G_t \sigma_t} p \approx 5 \times 10^4\,\mathrm{J} \tag{4.2}$$

and a power (peak or mean) of 1 MW.

These results can be summarized as follows. In the absence of jamming the transmitted energy is proportional to the fourth power of the range, is inversely proportional to the gain of the transmission antenna and to the surface gain of the reception antenna, is proportional to the square of the wavelength if the antenna area is given and is inversely proportional to the square of the wavelength if the antenna gain is given. In the presence of a target producing jamming noise the transmitted energy is proportional to the square of the range and is inversely proportional to the gain of the transmission antenna (the reception gain is negligible).

Figure 4.1 shows the block diagram of a noise radar of this type (to our knowledge this diagram does not correspond to any practical implementation). Unit 1 creates continuous noise uniformly occupying a band 1.2 MHz wide centered on 5 MHz. Unit 2 (mixer and filter) shifts this noise spectrum around 150 MHz. Unit 4 is a local oscillator operating at 145 MHz. Unit 3 is a power amplifier which amplifies the level of the signal from 2 and transmits to antenna 5 a microwave signal formed by a gaussian noise uniformly occupying a band of 1.2 MHz centered on 150 MHz. The signal received by antenna 8 is amplified in a low noise preamplifier and its frequency is changed in the mixer 10 so that the output is a signal centered on 5 MHz. This mixer is followed by the bandpass filter 11 which is centered on 5 MHz and has a pass band of 1.2 MHz. A small part of the transmitted signal is frequency shifted to about 5 MHz in 6 and is then delayed by τ_0 (corresponding to D) in the delay line 7. The output of 7 is multiplied by the output of 11 in a multiplier and the product is amplified in 12. The output of 12 is filtered in filter 13,

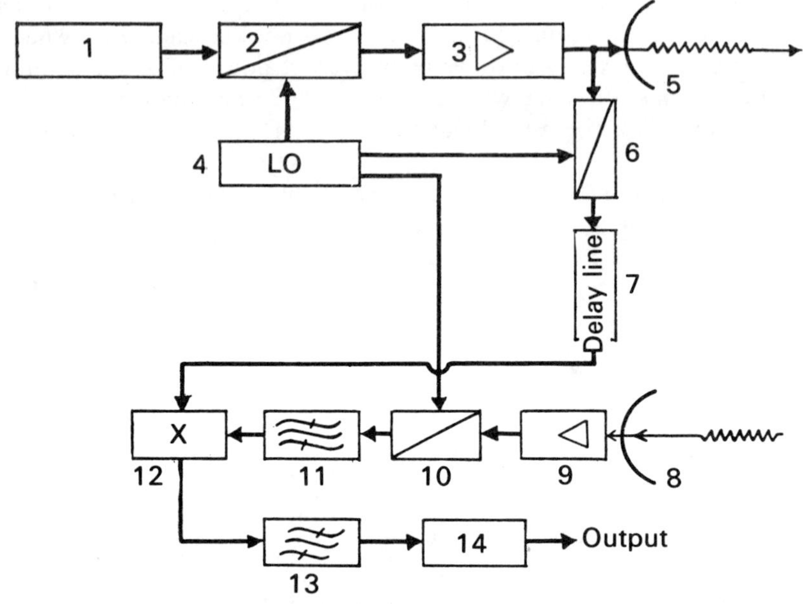

Fig. 4.1 Block diagram of a noise radar (LO, local oscillator).

which is centered on 300 Hz and has a pass band of 20 Hz (at 3 dB; see Chapter 1, Section 1.8.1), and is then used in 14 (converted to base band and then displayed for the operator or the computer).

The operation of the system is relatively simple. When the target is at a range τ close to τ_0, unit 9 receives the signal detected by the receiver antenna. The main function of the low noise preamplifier, which can never output a power greater than a certain maximum, is to protect the extremely fragile mixer 10. As this preamplifier has a significant gain, only its noise need be taken into account; the mixer noise, however large it may be, is negligible. The low noise preamplifier transmits a useful signal centered on 150 MHz and a noise covering a very broad band, particularly the frequencies close to 140 MHz. If no precautions are taken, the noise around 140 MHz (not accompanied by the useful signal) will give a parasitic noise around 5 MHz which is synchronous with the 145 MHz of the local oscillator, and this effect will double the receiver noise. This is prevented by inserting a filter with pass band 5 MHz, for example, centered on 150 MHz after the preamplifier 9.

Finally, we obtain at the output of 10 the useful signal with mean power E_R/T and a parasitic noise with mean power $kT_N \times 5 \times 10^6$. Thus the signal-to-noise power ratio is (laboratory technicians' definition; see Section 3.4)

$$\frac{E_{\mathrm{R}}}{T}\,\frac{1}{kT_{\mathrm{N}} \times 5 \times 10^6} = \frac{E_{\mathrm{R}}}{kT_{\mathrm{N}}}\,\frac{1}{T \times 5 \times 10^6}$$

$$\frac{S_1}{N_1} = \frac{R/2}{0.05 \times 5 \times 10^6} = \frac{10}{0.25 \times 10^6} = 4 \times 10^{-5} = -44\,\mathrm{dB}$$

where the signal-to-noise power ratio is defined as

$$\frac{S}{N} = \frac{\text{mean useful power}}{\text{mean parasitic power}}$$

At the output of 11, the noise power is only equal to

$$kT_{\mathrm{N}}\,\Delta f = kT_{\mathrm{N}} \times 1.2 \times 10^6$$

and the signal-to-noise power ratio is equal to (still with the same definition)

$$1.8 \times 10^{-4} = -38\,\mathrm{dB}$$

Unit 11 multiplies the reference useful signal $S(t - \tau_0)$ centered on 5 MHz by the received signal (accompanied by noise), i.e. the signal $S(t - \tau)$ centered on 5 MHz + 300 Hz (to the nearest 5 Hz). The product of these two signals gives a component at 300 ± 5 Hz and a component at 10 MHz + 300 Hz which is useless. Finally, the bandpass filter with width 20 Hz centered on 300 Hz integrates (averages) the useful component over a time equal to $1/20\,\mathrm{s} = 0.05\,\mathrm{s} = T$. When τ varies, the output of 13 represents the autocorrelation function $\varrho(\tau - \tau_0)$ of the transmission signal which is assumed to be centered on 5 MHz, and the demodulation (or, more simply, the detection) of this output gives the autocorrelation function of the transmission centered on zero frequency. We therefore have a signal-to-noise ratio of $R/2$, i.e. 10 dB, at the output (laboratory technicians' definition). The correlation has therefore improved the signal-to-noise ratio by 10 dB + 38 dB = 48 dB, which represents the product $T\Delta f$. Thus if we use an oscilloscope to examine the output signal of 11 (or 10), only the parasite noise will be seen as the useful signal is extremely weak, whereas the output of 13 will show a signal clearly differentiated from the noise. The multiplier 12 must have a dynamic range of at least 48 dB, i.e. it must be able to process, in the same manner, signals with levels which differ by 48 dB.

4.1.2 Target with unknown velocity

If the radial velocity of the target is unknown (it is assumed that the sign is known, but if this is not the case only a small modification of the procedure described below is required), the single chain 13–14 is replaced by a number of analog chains with filters of width 20 Hz (at 3 dB) centered on 240 Hz, 260 Hz, 280 Hz, 300 Hz, 320 Hz etc. Interpolation between two adjacent filters gives the radial velocity with an error of $2.5\,\mathrm{m\,s^{-1}}$ (Chapter 3, eqn (3.23)). These filters are called Doppler filters. This procedure enables us to distinguish two targets whose velocities differ by more than $20\,\mathrm{m\,s^{-1}}$ (see Chapter 3, Section 3.9). If we want to eliminate slow targets and, in particular, fixed targets, it is sufficient to remove the filters corresponding to the low velocities (in practice, this operation may be hampered by transmitter leakage directly in the receiver antenna).

Remark 4.1

Figure 4.1 is the standard diagram for a correlation proximity fuse in which it is sufficient to determine if and when the target is situated at a certain defined range. If continuous surveillance of an entire range zone is required, it is necessary, as already mentioned, to have a number of receivers, each of which corresponds to a different value of τ_0. However, in this case it is impossible to distinguish two targets with similar velocities which are separated by less than 120 m because

$$\theta_{\mathrm{min}} = \frac{1}{1.2 \times 10^6} = 0.8\,\mu\mathrm{s}$$

Remark 4.2

The ambiguity zone is an ellipse whose axes are formed by the coordinate axes and have a length of $d = 120\,\mathrm{m}$ and $v_{\mathrm{R}} = 20\,\mathrm{m\,s^{-1}}$.

Remark 4.3

In practice we often use a random signal of the form $S(t) = A \sin(2\pi f_0 t) F(t)$, where $F(t)$ is either 1 or -1 depending on an equitable random sample taken at intervals of $1/\Delta f$. We can then replace the expression $\int_T S(t - t_0)\, y(t)\, \mathrm{d}t$ by

$$\frac{1}{A}\int_T S(t - t_0)\, y(t)\mathrm{d}t = \int \sin[2\pi f_0(t - t_0)]\, F(t - t_0)\, y(t)\, \mathrm{d}t$$

However, since replacing $\sin[2\pi f_0(t - t_0)]$ by $\sin(2\pi f_0 t)$ amounts to making an error on t_0 of less than $1/2f_0$, which is always negligible (see Chapter 3, Section 3.4),

$$\int_T [\,y(t)\, \sin(2\pi f_0 t)]\, F(t - t_0)\, \mathrm{d}t$$

is usually calculated in practice, i.e. the received signal $y(t)$ will be demodulated by the transmission carrier f_0, then multiplied by 1 or -1 depending on $F(t - t_0)$ and finally passed through a bank of Doppler filters.

Remark 4.4

In practice, the function $F(t)$ will often be pseudorandom (for example, following the sequence of 1s and 0s of a pseudorandom sequence, which is a binary code with maximum length easily obtained from a loop shift register).

4.2 Pulse Doppler radar

Pulse Doppler radar transmits a signal of the form $A \sin(2\pi f_0 t)$ modulated by zero or unity, i.e. the transmitted signal can be represented by the mathematical expression

$$S(t) = A \sin(2\pi f_0 t) F(t)$$

where $F(t)$ is a function which can only be equal to zero or unity:

$$\begin{array}{ll} \text{equal to 1} & \text{for } 0 \leqslant t \leqslant T_0 \\ & \text{for } t_1 \leqslant t \leqslant t_1 + T_0 \\ & \quad \vdots \qquad\qquad \vdots \\ & \text{for } t_k \leqslant t \leqslant t_k + T_0 \\ \text{equal to 0} & \text{otherwise} \end{array}$$

In practice, $t_1, t_2, \ldots, t_k$ are often divided into arithmetic progressions in time with repetition period T_R (but this is not general).

If noncoherent transmitter tubes are used the phase of the carrier $A \sin(2\pi f_0 t)$ may vary from one pulse to the next. If this is the case a superheterodyne receiver is used in which the phase of the local oscillator (defined as coherent) follows that of the transmitter in order to return to the situation when the transmitter is coherent (constant phase of the carrier). We will therefore only consider this case.

Although pulse Doppler radar can be received in a matched filter (see Section 3.1), correlation reception, which is very similar to that which has been described for noise radar, is usually used. We therefore calculate the expression

$$\int_T S(t - t_0) y(t) \, dt$$

for a finite number of values of t_0, i.e. $t_{01}, t_{02}, \ldots, t_{0k}$. This expression can be simplified to

$$\int_T [y(t) \sin(2\pi f_0 t)] F(t - t_0) \, dt$$

and thus reception will consist of demodulating the received signal $y(t)$ by the transmission carrier (demodulation by the frequency of the local oscillator followed, after amplifying and minimum filtering with intermediate frequency, by repeated demodulation by the intermediate frequency with the precautions specified in Section 4.7 in order to avoid "blind phases") and then obtaining the product of

$$[y(t)\sin(2\pi f_0 t)] \quad \text{and} \quad F(t - t_0)$$

Then

$$[y(t)\sin(2\pi f_0 t)]$$

is allowed to enter receiver 1 during the intervals $[t_{01}, t_{01} + T_0]$, $[t_{01} - t_1, t_{01} - t_1 + T_0]$ etc., to enter receiver 2 during the intervals $[t_{02}, t_{02} + T_0]$, $[t_{02} - t_1, t_{02} - t_1 + T_0]$ etc. Since the process requires a certain number of receivers of this type (range-gated receivers), each of them is equipped with a certain number of Doppler filters as in Fig. 4.1.

Remark 4.5

If the radar is a two-dimensional search radar, the duration T of the coherent measurement (reciprocal of the width of a Doppler filter) will normally be the transit time of the radar beam over a target.

Remark 4.6

The Doppler filters must have the following two properties: they must be able to perform a coherent integration over time (this is not a severe constraint), and they must have a very small gain in the neighborhood of zero frequency (and frequencies k/T_R) so that parasitic targets are effectively rejected at low radial velocities (this requires great care). Thus, in practice, each range-gated receiver consists of an expensive rejector filter whose function is to give a very small gain at frequencies near zero (modulo $1/T_R$) followed by Doppler filters which are much less expensive (Fig. 4.2).

Remark 4.7

As will be seen in Chapter 7, for economic reasons coherent integration is not strictly enforced. Therefore it is possible to reduce the number of Doppler filters, even if the integration (after detection) is completed in the noncoherent mode, hence reducing the cost–performance ratio.

4.3 New algorithm for a radar receiver

In order to show that it is possible to develop radar systems which are different from known types (correlation or matched filter radars), a new type of radar receiver [5, 6] is described briefly in this section. The following process takes place in the receiver.

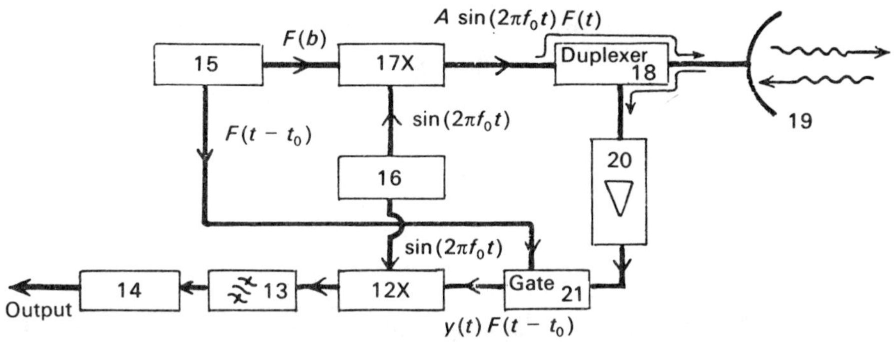

Fig. 4.2 Range-gated receiver.

(1) The received signal $y(t)$ is added to a replica of the transmitted signal $S(t)$ (in practice, the transmitter leakage) to give

$$A(t) = y(t) + S(t)$$

(2) The square of the amplitude of the Fourier transform of $A(t)$ is calculated (usually using a spectrum analyzer), i.e.

$$|\mathscr{A}(f)|^2 = |\mathscr{Y}(f) + \phi(f)|^2$$

which gives

$$|\mathscr{Y}(f)|^2 + |\phi(f)|^2 + \mathscr{Y}(f)\phi^*(f) + \mathscr{Y}^*(f)\phi(f)$$

(3) The Fourier transform of this expression is taken, producing a sum of four terms which are functions of t_0. The first two terms are the autocorrelation functions of $y(t)$ and $S(t)$, which become very small as soon as t_0 becomes large and which can be neglected in certain applications, the third term is $C(t_0)$ and the last term is $C(-t_0)$. In practice it is possible to use a second spectrum analyzer instead of a Fourier transformer.

4.4 Classical pulse radar

4.4.1 Classical concept

Classical pulse radar transmits square-wave pulses with duration T_p and repetition frequency F_R which are not intentionally frequency modulated. The pulses have a peak power P_p, and the mean transmitted power P_m is obviously given by

$$P_p T_p F_R = P_m$$

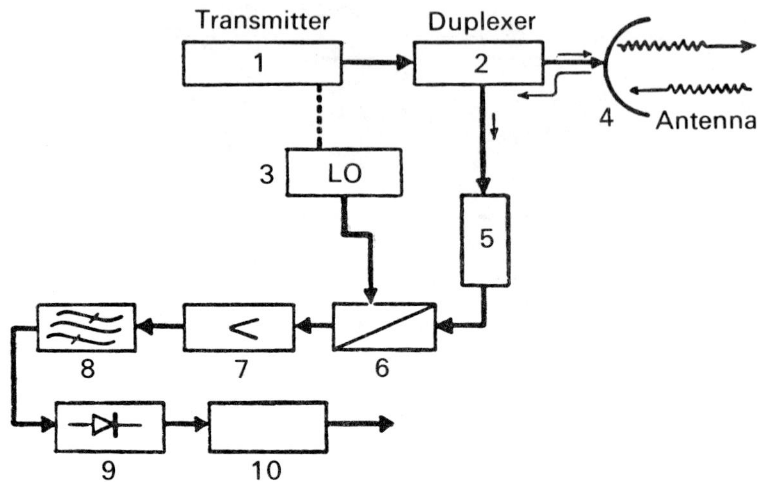

Fig. 4.3 Classical pulse radar (LO, local oscillator).

where $T_p F_R$ is the duty cycle of the radar. For example, if $P_p = 20\,\text{MW}$, $T_p = 4\,\mu\text{s}$ and $F_R = 250\,\text{Hz}$, the duty cycle is 0.001 and $P_m = 20\,\text{kW}$. The energy transmitted during the pulse is radiated from a transmission antenna and the energy radiated by the target is received in a receiver antenna. The receiver antenna and the transmitter antenna are normally combined, which makes it necessary to install a device called a duplexer to prevent any of the transmitted energy from perturbing the receiver during transmission.

The operation of the radar is illustrated in the block diagram in Fig. 4.3. The received signal is amplified in a low noise preamplifier followed by a filter to eliminate the image frequency (see Section 4.1). The signal frequency is then changed so that its spectrum is centered on the intermediate frequency (e.g. 30 MHz). This is done by using a local oscillator to transmit a frequency differing by 30 MHz from the central transmission frequency (by means of automatic frequency control (AFC) or manual frequency control (MFC)). The output signal from mixer 6 therefore consists of one or more square-wave pulses from the target accompanied by a gaussian noise occupying a very wide spectrum (equal to the width of the pass band of the low noise preamplifier filter). It is amplified and filtered in the receiver with an intermediate frequency (in practice, the filter forms an integral part of the amplifier) which means that only frequencies between 30 MHz $- \Delta f/2$ and 30 MHz $+ \Delta f/2$ are output from the receiver. Therefore, the system behaves as if a signal with a useful spectrum of width Δf had been transmitted.

The Doppler frequency due to the velocity of the targets is generally very small compared with both Δf and $1/T_p$ and we will ignore it at this stage.

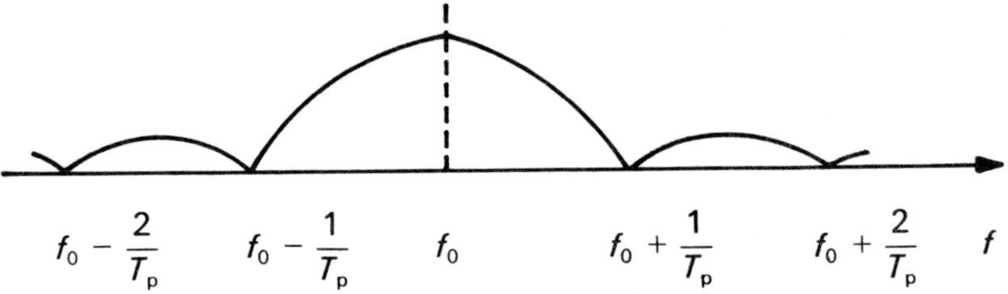

Fig. 4.4

The overall problem is therefore to define the value of Δf. The transmitted signal has a Fourier transform whose amplitude is shown in Fig. 4.4, where f_0 is the central transmission frequency. The received signal therefore occupies a spectrum whose Fourier transform is the same except for the replacement of f_0 by the intermediate frequency (in this case 30 MHz). If Δf is very large compared with $1/T_p$, apart from the central lobe, a large number of sidelobes are retained which only contain a small amount of useful energy although the noise power will be significant. Therefore at the output of 8 we have square-wave pulses in a noisy background. If Δf is very small compared with $1/T_p$, the signal is completely destroyed.

We can understand what is happening by examining the filtering of a square-wave pulse of length T_p and unit height accompanied by white noise by a square-wave filter of width $2B$. The Fourier transform of a square-wave pulse $S(t)$ of duration T_p and height $1/T_i$ is

$$\Phi(f) = \frac{\sin(\pi f T_p)}{\pi f T_p}$$

When this pulse is filtered by a square-wave filter of width $2B$, the resulting signal is

$$\sigma_N(t) = \int_{-B}^{+B} \frac{\sin(\pi f T_p)}{\pi f T_p} \exp(2\pi j f t)\,\mathrm{d}f$$

whose central value (for $t = 0$) is given by

$$\sigma_N(0) = \int_{-B}^{+B} \frac{\sin(\pi f T_p)}{\pi f T_p}\,\mathrm{d}f = 2\int_0^B \frac{\sin(\pi f T_p)}{\pi f T_p}\,\mathrm{d}f$$

Then, if we set $\pi f T_p = u$,

$$\sigma_N(0) = \frac{2}{\pi T_p}\int_0^{\pi B T_p} \frac{\sin u}{u}\,\mathrm{d}u$$

and

$$\mathrm{d}f = \frac{\mathrm{d}u}{\pi T_p}$$

Table 4.1

πBT_p	0.5	1	1.5	1.7	1.9	2.1	2.3	2.5	3
$\mathrm{Si}(\pi BT_p)$	0.49	0.95	1.32	1.45	1.56	1.65	1.72	1.79	1.85
$(\pi BT_p)^{1/2}$	0.71	1	1.22	1.30	1.38	1.45	1.52	1.58	1.73
$\dfrac{\mathrm{Si}(\pi BT_p)}{(\pi BT_p)^{1/2}}$	0.70	0.95	1.08	1.11	1.13	1.14	1.13	1.12	1.07

Hence

$$\sigma_N(0) = \frac{2}{\pi T_p} \mathrm{Si}(\pi BT_p)$$

Since the power of the output noise is proportional to B and therefore its standard deviation to $B^{1/2}$, the search for the optimum value of B is equivalent to the search for the value of B which makes the expression

$$\frac{2}{\pi T_p B^{1/2}} \mathrm{Si}(\pi BT_p) = \frac{2}{(\pi T_p)^{1/2}} \frac{\mathrm{Si}(\pi BT_p)}{(\pi BT_p)^{1/2}}$$

a maximum. For $\pi BT_p \to 0$,

$$\mathrm{Si}(\pi BT_p) \sim \pi BT_p$$

and

$$\frac{\mathrm{Si}(\pi BT_p)}{(\pi BT_p)^{1/2}} \sim (\pi BT_p)^{1/2} \to 0$$

For $\pi BT_p \to \infty$,

$$\mathrm{Si}(\pi BT_p) \to \pi/2$$

$$\frac{\mathrm{Si}(\pi BT_p)}{(\pi BT_p)^{1/2}} \to 0$$

Values of functions of πBT_p are given in Table 4.1. The maximum is obtained for

$$\pi BT_p \approx 2.15$$

$$BT_p = 0.68$$

$$2B = 1.37/T_p$$

In practice, as the rising and falling edges of the pulse are not vertical, we take $2B$ to be smaller, of the order of $1.2/T_p$. Under these conditions, the square-wave pulse which enters the filter no longer has a square waveform when it leaves the filter, as shown in Fig. 4.5. Figure 4.6 shows the variations in $\Phi^2(f)$ (square of the Fourier transform of the signal) as a function of f. It can be

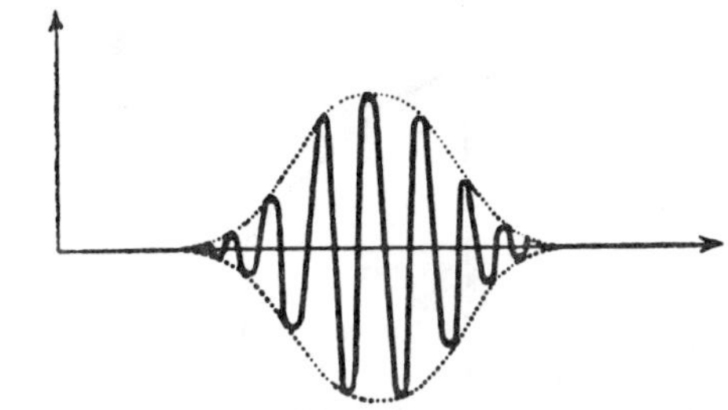

Fig. 4.5

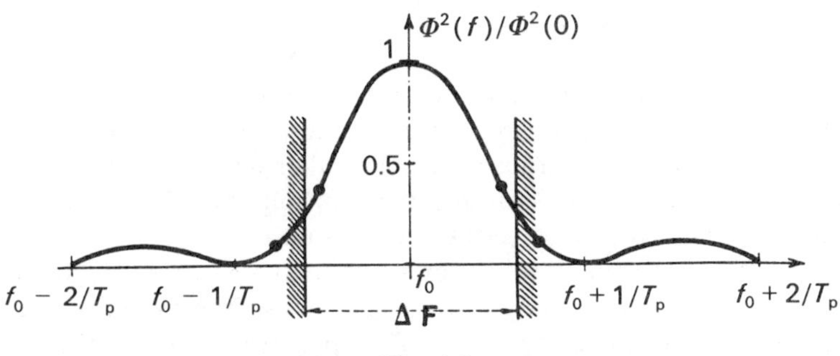

Fig. 4.6

seen that almost none of the transmitted energy is lost by only sampling the part between $f_0 - 0.6/T_p$ and $f_0 + 0.6/T_p$ (the loss is approximately 1 dB). The output signal of filter 8 is finally detected in unit 9 and applied in unit 10 (bringing to base level etc).

4.4.2 A new concept

It was shown in Section 3.1 that in order to obtain an ideal receiver it is necessary to have a receiver with a transmission characteristic such that the modulus of the gain of the filter is equal to $|\Phi(f)|$ (see Fig. 4.4) and the phase shift produced by the filter is equal to the opposite of the argument of $\Phi(f)$ (i.e. it is zero for $f_0 - 1/T_p < f < f_0 + 1/T_p$, 180° for $f_0 - 2/T_p < f < f_0 - 1/T_p$ etc.), where $\Phi(f)$ is the Fourier transform of the transmitted pulse. Such a filter, which is known as matched filter, is difficult to implement.

In practice, if the loss of the energy contained at frequencies below $f_0 - 0.6/T_p$ and above $f_0 + 0.6/T_p$ is accepted (approximately 20% of the transmitted energy) and only frequencies between $f_0 - 0.6/T_p$ and

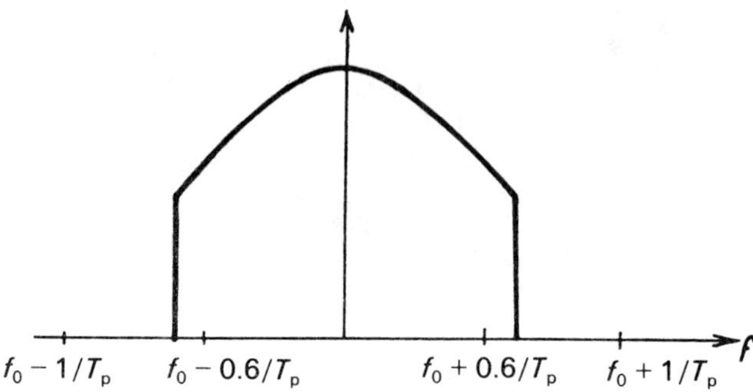

Fig. 4.7

$f_0 + 0.6/T_p$ are retained at the receiver, the system behaves as if a signal $S(t)$, whose Fourier transform has zero phase (zero argument) and the amplitude shown in Fig. 4.7, had been transmitted. The filter matched to such a signal should then have zero phase shift and the transmittance shown in Fig. 4.7. The bandpass filter used gives very similar results.

Thus, compared with an ideal radar, the unavoidable loss is due to the deliberate loss (of approximately 1 dB) of the energy transmitted outside the band between $f_0 - 0.6/T_p$ and $f_0 + 0.6/T_p$. Apart from this loss, the receiver (at $1.2/T_p$ of the pass band) of classical radar appears to be very similar to the matched filter.

Equations (4.1) and (4.2), which were calculated for correlation radar, are therefore also valid in this case, and it is of interest to compare eqn (4.1) with the well-known radar equation which is obtained as follows. The transmitted power density at range D is given by

$$p = \frac{G_t P_p}{4\pi D^2}$$

The power radiated by the target (which is assumed to be isotropic) is therefore

$$\frac{G_t P_p \sigma_t}{4\pi D^2}$$

and the reception antenna captures

$$\frac{G_t P_p \sigma_t}{4\pi D^2}\frac{A_R}{4\pi D^2} = \frac{G_R G_t P_p \sigma_t \lambda^2}{(4\pi)^3 D^4}$$

which constitutes the received peak power (G_R is the antenna gain at reception). Since the power of the parasitic noise of the receiver is equal to $kT_N \Delta f$, the signal-to-noise ratio can be written

$$\frac{S}{N} = \frac{G_R G_t P_p \sigma_t \lambda^2}{(4\pi)^3 D^4 k T_N \Delta f}$$

which is the radar equation. If we replace Δf by $1.2/T_p$, we also find

$$1.2\frac{S}{N} = \frac{G_R G_t P_p T_p \sigma_t \lambda^2}{(4\pi)^3 D^4 k T_N} = \frac{A_R G_t \sigma_t (P_p T_p)}{k T_N (4\pi D^2)^2}$$

If we had used eqn (4.1), taking into account the loss of approximately 1 dB (ratio of 1.25) and the fact that $P_p T_p$ is exactly the energy E_t transmitted during the measurement time T_p, we would have obtained

$$\frac{G_R G_t \sigma_t \lambda^2 (P_p T_p)}{(4\pi)^3 D^4 k T_N} = \frac{A_R G_t \sigma_t (P_p T_p)}{k T_N (4\pi D^2)^2} = \frac{1.25}{2} R$$

Therefore, provided that the received signal is filtered into a bandpass filter with passband $\Delta f = 1.2/T_p$, the curves giving the appropriate values of R that ensure a given false alarm probability (for a given false alarm) are the same as those which give the appropriate values for twice the signal-to-noise power ratio of a perfect classical radar. Thus, if only one pulse is received per target, the energy that must be transmitted in one pulse in order to detect a target with a radar cross-section of 1 m^2 under the conditions assumed in Section 4.1 is (in the absence of jamming) $1.25 \times 15\,\text{J} \approx 20\,\text{J}$. If we require the same accuracy of ± 15 m for the range measurement, Δf must be at least 1.2 MHz (a practical radar cannot be better than the ideal radar), i.e. T_p cannot exceed 1 µs. Then the peak transmitted power will be at least of the order of $20/10^{-6} = 20$ MW (in the correlation radar 300 W was sufficient), and a minimum peak power of 65 000 MW, which cannot be achieved, is required in the presence of the jamming target assumed in Section 4.1.

The maximum repetition frequency is often based on the requirement that the receiver should not receive a signal corresponding to the first pulse after having sent a second pulse (this would make range measurement ambiguous because the only procedure available for determining target ranges is to measure the time between the beginning of the transmitted pulse and a characteristic point (e.g. the middle) of the received pulse). Thus the maximum value of F_R for a range D of 500 km (3.3 ms) is 300 Hz. If we take $T_p = 1$ µs and $F_R = 300$ Hz, we obtain a duty cycle of the order of 0.0003. In the above, we have ignored the various losses (in particular, in the microwave guide), atmospheric attenuation, target fluctuation and the fact that the radar generally carries out the measurement on N consecutive pulses. The last two points will be treated in subsequent chapters.

Remark 4.8

Since the spectrum that is not located in the band between $f_0 - 0.6/T_p$ and $f_0 + 0.6/T_p$ is eliminated on reception, it makes no sense to incur large costs in efforts to produce it. Therefore it is unnecessary to expend time and money as has frequently been done, on attempts to obtain very steep rising and falling edges for the transmission pulse which is only required to be symmetric.

4.5 Pulse compression radar

4.5.1 Introduction

The transmitter used in the classical radar system shown in Fig. 4.3 has not been described. In practice, there are two types: magnetron transmitters, which consist of an oscillator tube which is only turned on during the pulses T_p (keyed by a modulator), and amplifier transmitters, which contain a number of tubes in cascade.

Figure 4.8 shows a block diagram of such an amplifier transmitter, which can be used to replace unit 1 in Fig. 4.3. Unit 11 transmits square-wave pulses of duration T_p with a repetition frequency F_R and a carrier frequency of 30 MHz. Unit 13 converts the spectrum of the 30 MHz signal from 12 to the transmission frequency f_0 by heterodyning with the local oscillator 3. Unit 14 consists of a cascade of amplifier tubes, each of which transmits a power much higher than the preceding one (for example, the chain consists of a traveling wave tube providing a peak power of 10 W, an amplifier klystron providing a peak power of 30 kW and a final amplifier klystron providing a peak power of 30 MW).

Since the energy transmitted at frequencies outside the interval $[f_0 - 0.6/T_p, f_0 + 0.6/T_p]$ is completely ignored at reception, it is not worth transmitting them and the operation of the system is not affected if a filter is inserted between unit 11 and unit 13 (similar to that used at reception). This filter has the following characteristics: it does not modify the amplitudes

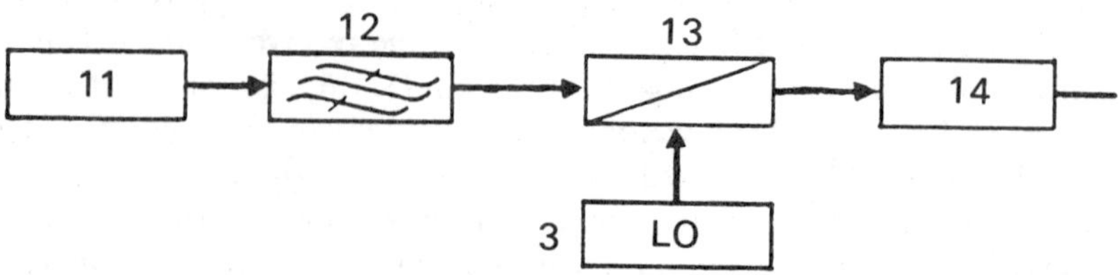

Fig. 4.8 Block diagram of an amplifier transmitter (LO, local oscillator).

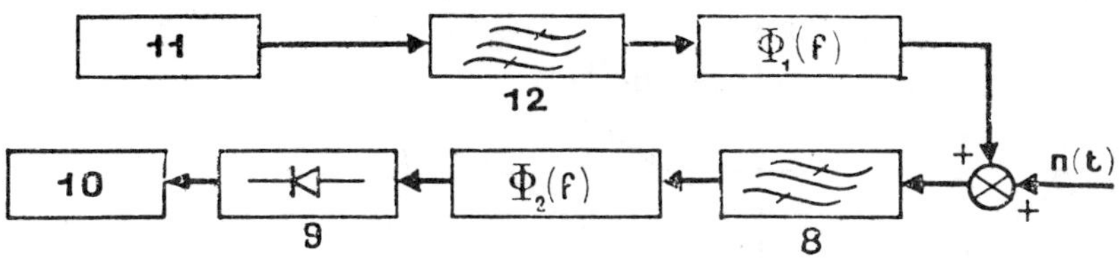

Fig. 4.9

or phases of the spectral components between 30 MHz $- 0.6/T_p$ and 30 MHz $+ 0.6/T_p$, and it suppresses the other components. However, if we insert a filter which passes everything between 12 and 13 with a characteristic $\Phi_1(f)$ such that $|\Phi_1(f)| = 1$ and the argument of $\Phi_1(f)$ is not equal to zero (the filter only modifies the phases of the spectrum which pass through it) and if we also insert a complementary filter which passes everything between 8 and 9 (Fig. 4.3) with characteristic $\Phi_2(f)$ such that

$$\text{phase } \Phi_2(f) = -\text{phase } \Phi_1(f)$$

the radar receiver remains a matched filter and the amplitudes (moduli) of the transmitted spectral components do not change. In other words, the detection range and the quality of the range measurement are not affected by the insertion of the filters Φ_1 and Φ_2. In practice, the system behaves as if it were the simplified chain shown in Fig. 4.9.

The output of 11 is a square-wave pulse at 30 MHz of duration T_p. The output of 12 is a pulse at 30 MHz with duration close to T_p and the shape shown in Fig. 4.5. The output of Φ_1 is a pulse whose spectrum always has a width of $1.2/T_p$ but which is phase (or frequency) modulated. The output of 8 is formed by the output of Φ_1 with the addition of a gaussian noise which uniformly occupies a spectrum of width $1.2/T_p$. The output of Φ_2 is a pulse of the same shape as the output of 12 (which is not frequency modulated) and which is strictly that which would have been obtained if Φ_1 and Φ_2 did not exist. Compared with classical radar, the only difference is that the length T_t of the output signal of Φ_1 may be much larger than T_p because it is phase modulated and that the transmitted signal is much longer than T_p. T_t/T_p is called the compression ratio of the radar.

The following simple arguments help to illustrate the operation of a pulse compression radar in a physically meaningful way. Let us compare the spectrum of a continuous wave with carrier frequency f_0, which is frequency modulated by a sawtooth wave with modulation frequency 1 kHz and a peak-to-peak frequency deviation of 20 kHz (Fig. 4.10), with the spectrum carrier at f_0 modulated by a series of quasi-gaussian pulses of length 100 μs which repeat every millisecond (Fig. 4.11). The spectrum of the wave illustrated

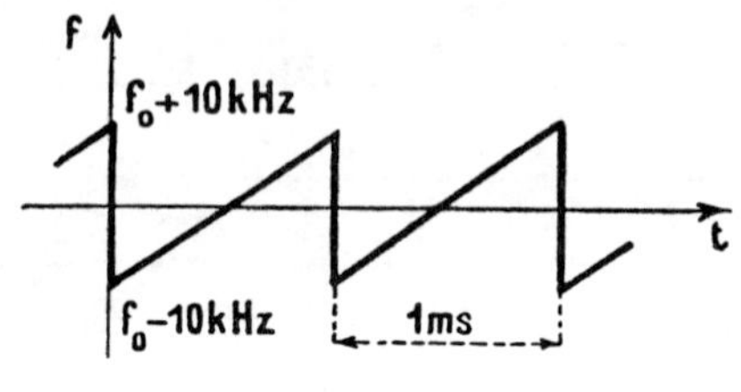

Fig. 4.10

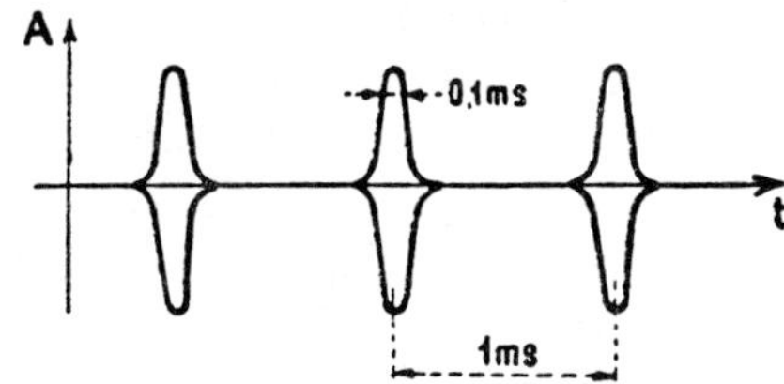

Fig. 4.11

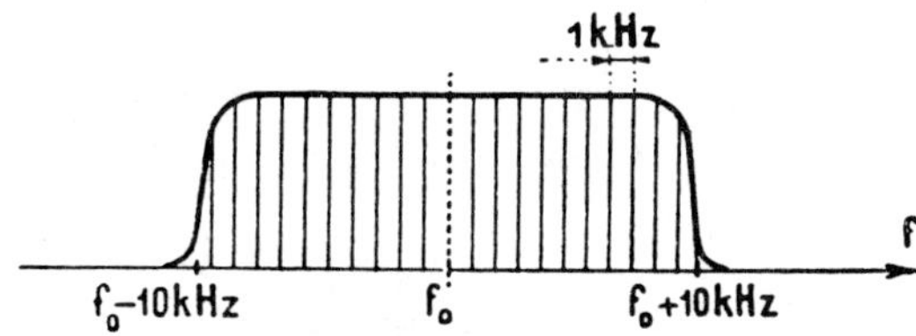

Fig. 4.12

in Fig. 4.10 is shown in Fig. 4.12. It consists of lines separated by 1 kHz whose envelope is approximately a rectangle of width 20 kHz centered on f_0. However, given the accuracy of the measured spectrum, the spectrum of the wave shown in Fig. 4.11 can also be represented by Fig. 4.12 and consists of lines separated by 1 kHz whose spectrum centered on f_0 has a width of 20 kHz. The only difference between the spectra of the waves shown in Figs 4.10 and 4.11 is that the phases of the lines are not the same. Thus, by allowing the wave shown in Fig. 4.11 to pass through a filter which changes the phase but does not modify the amplitude of the components it is possible to transform a series of 0.1 ms signals into a continuous wave. Conversely, a complementary filter can be used to transform the continuous wave shown in Fig. 4.10 into a series of 0.1 ms signals.

Let us assume that we have produced the complementary filters Φ_1 and Φ_2 (or that we have been able to produce a device which gives the same result as the output of $\Phi_1(f)$). Compared with the classical radar described in Section 4.4, we have to transmit pulses of duration T_t much longer than T_p but containing the same energy E_t and occupying the same spectrum as the pulses transmitted by classical radar. The transmitted peak power is a factor of T_p/T_t smaller than that of classical radar ($T_p/T_t \ll 1$) and the signal-to-noise ratio at the receiver input is smaller by the same factor. When the long pulse received in the filter Φ_2 is compressed, the amplitude of the useful signal is multiplied by $(T_t/T_p)^{1/2}$ and hence its power is multiplied by T_t/T_p. The noise entering Φ_2 has random phases, which remain random if they are modified in Φ_2 in a known manner:

$$\text{random quantity} + \text{known quantity} = \text{random quantity}$$

The power of the noise output from Φ_2 is therefore equal to the input power (the noise is not compressed). Finally, the signal-to-noise ratio at the output of Φ_2 is multiplied by T_t/T_p with respect to the input. Thus we have recovered the signal-to-noise ratio of classical radar, having gained in the compression the decibels lost in transmission.

Let us now return to the radar problem described above. We have seen that, in the absence of jamming, classical radar should transmit pulses of length 1 μs and peak power 20 MW. If we use a compression ratio of 100, we obtain the same performance with a radar (with the same average power) transmitting 100 μs pulses occupying a spectrum of 1.2 MHz and a peak power of the order of 200 kW.

4.5.2 Pulse compression procedures

The signal at the output of the matched receiver has a duration which is essentially equal to the reciprocal of the width Δf of the transmitted spectrum. If the transmitted signal has a duration T, it is then said that the matched receiver has compressed the signals by the compression ratio

$$T\Delta f = \varrho$$

There are two major families of pulse compression systems: the active generation family and the passive generation family.

4.5.2.1 Active generation

A signal whose spectrum $\Phi(f)$ is such that the filter at the reception is matched is created artificially and directly. For example, if we use at reception a delay line whose group propagation time T_R decreases linearly with the frequency f (with a gradient $dT_R/df = K$), the transmission signal corresponding to it is linearly frequency modulated (the frequency increasing linearly with a gradient $\mathrm{d}f/\mathrm{d}t = 1/K$) with a duration T such that $T^2K \gg 1$ (K has the dimensions of time squared). This can be shown mathematically and physically.

Mathematically, $\Phi(f)$ has a quasi-constant modulus for

$$f_0 - \frac{T}{2K} < f < f_0 + \frac{T}{2K}$$

and a quasi-zero modulus for

$$f < f_0 - \frac{T}{2K} \text{ and } f > f_0 + \frac{T}{2K}$$

although the argument of $\Phi(f)$ is very close to

$$-\pi K(f - f_0)^2$$

The complete calculation of $\Phi(f)$ has been carried out by a large number of workers [7], and there is no point in reproducing it here. Figure 4.13 shows the shape of $|\Phi(f)|$ when $T\Delta f$ is not infinitely large. The argument of $\Phi(f)$

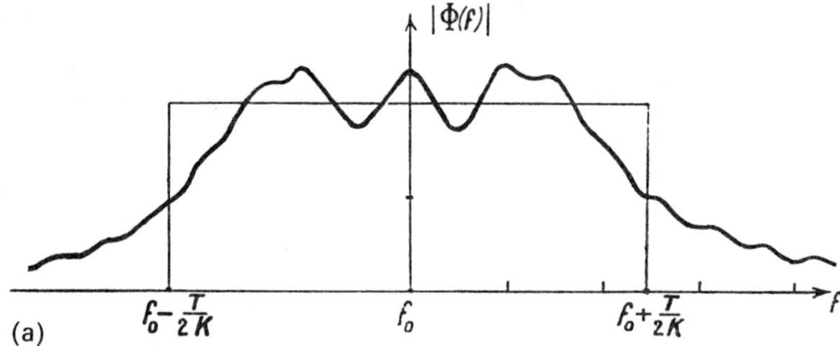

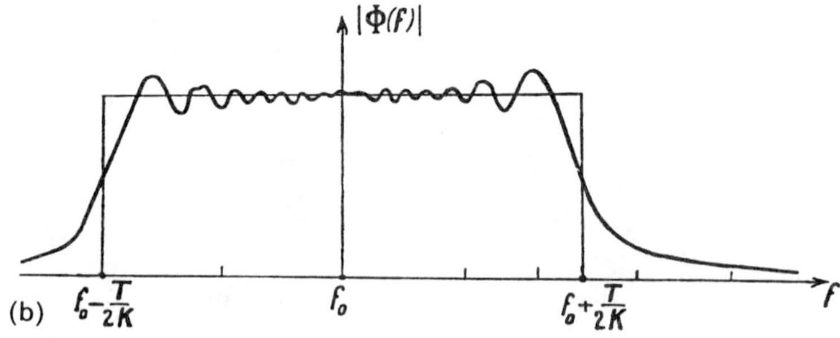

Fig. 4.13 $\Phi(f)$ for (a) $T^2/K = T\Delta f = 10$ and (b) $T^2/K = T\Delta f = 60$.

can also be written

$$-\frac{K(\omega - \omega_0)^2}{4\pi} = -\frac{K\omega^2}{4\pi} + \frac{K\omega\omega_0}{2\pi} - \frac{K\omega_0^2}{4\pi}$$

Since the delay line at the reception has a propagation time of the form

$$T_R = T_0 - Kf = T_0 - K\frac{\omega}{2\pi}$$

it acts as a filter which does not modify the amplitudes of the signal components passing through it and adds to their phase

$$\Phi = -T_0\omega + \frac{K\omega^2}{4\pi} + \varphi_0$$

i.e. the opposite of the phase of $\Phi(f)$ if we take into account the fact that terms of the form $T_0\omega$ delay the entire signal and the constant terms shift the carrier phase. Therefore, provided that a bandpass filter which passes the band $[f_0 - T/2K, f_0 + T/2K]$ is inserted in the receiver, the receiver is matched to the transmitted signal. It should be noted that after reception we receive a signal which is not frequency modulated (zero phase) and whose spectrum has a constant amplitude between $f_0 - T/2K$ and $f_0 + T/2K$, i.e. a signal of the form (Fig. 4.14)

$$S_u(t) = K_1 \cos(2\pi f_0 t)\frac{\sin(\pi Tt/K)}{\pi Tt/K}$$

containing a principal lobe of total width $2K/T = 2/\Delta f$, where Δf is the width of the spectrum, and a 3 dB width of $K/T \approx 1/\Delta f$, and a number of secondary lobes of which the first is located 13 dB below the principal lobe.

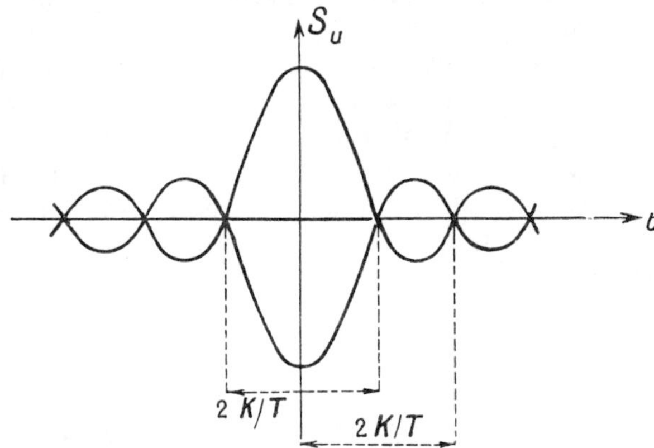

Fig. 4.14

Physically, the frequency f, which is transmitted a certain time before the last frequency $f_0 + T/2K$, is delayed at reception by an equal time. Therefore all the frequencies leave the receiver (delay line) together i.e. in phase, giving a signal which is not frequency modulated and has a high power.

Remark 4.9

The phase variation inside the signal is given by

$$\pi K \left(\frac{T^2}{4K^2} \right) = \frac{\pi T^2}{4K} = \frac{\pi T \Delta f}{4} = \frac{\pi \varrho}{4}$$

where Δf is the width of the spectrum and ϱ is the compression ratio. The higher is ϱ, the larger is the total phase variation and the better must be the relative accuracy of the phase (frequency) curve of the reception filter for a given absolute phase error (and a given compression quality). In other words, as ϱ increases the correct implementation of the response of the transmission or reception filter becomes more critical.

4.5.2.2 Passive generation

The simplest type of passive generation uses a short (input) signal which is not frequency modulated and whose spectrum, centered on the frequency f_0, has a width Δf, i.e. all the component frequencies are in phase ("simultaneous") and pass through a filter (filter 1) which introduces a delay $T_T = F_1(f)$ without affecting the amplitudes of the components. If a second filter (filter 2) whose transmission time is of the form

$$T_R = F_2(f) = T_0 - F_1(f) = T_0 - T_T$$

is used at reception, all the frequencies will finally be in phase at the output of the filter and short signal identical with the first signal (except for a delay T_0) will be produced, although the transmitted signal could be very long.

Several procedures can be used to obtain this result. The first is as follows. The output signal of filter 1 is recorded in the form of a signal corresponding to the relationship $T_T' = F_1(f)$. The recording is time reversed before being read and thus supplies a signal corresponding to the relationship $T_T = T_1 - F_1(f)$ which is transmitted. Filter 1 is then used at reception instead of filter 2, and this introduces a delay

$$T_R = T_T' = F_1(f) = T_1 - T_T$$

This procedure, which has been patented since 1945 in England and the USA, has probably never been used. In radar applications it requires a very high recording quality and a very expensive recorder, although it would probably be more feasible for signals with a low pass band.

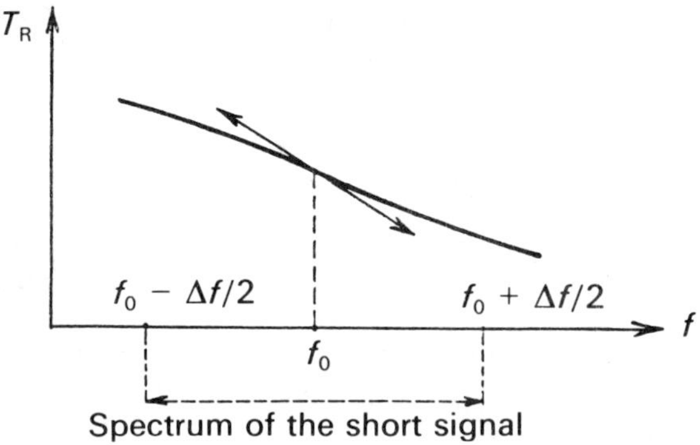

Fig. 4.15

In the second process [8] filter 1 is used in such a way that the curve $T_T' = F_1(f)$ has an inflection point for the central frequency f_0 of the spectrum of the short incident signal (Fig. 4.15). In practice, the incident signal may be a short signal modulated onto a carrier frequency f_0 so that its spectrum is centered on f_0 and is symmetric about f_0. In this case, it is usually valid to assume that the useful part of the curve $T_R = F_1(f)$, i.e. between $f_0 - \Delta f/2$ and $f_0 + \Delta f/2$, is symmetric about f_0, so that symmetry with respect to a vertical line passing through f_0 is equivalent to symmetry with respect to a horizontal line. Therefore if we reverse the spectrum of the outgoing signal of filter 1 with respect to f_0, we obtain a signal corresponding to

$$T_T = T_2 - T_T' = T_2 - F_1(f)$$

which is transmitted. The spectrum is reversed by making the output signal of filter 1 oscillate with a frequency $2f_0$ (in a balanced modulator). An incident frequency f gives two frequencies $2f_0 + f$ and $2f_0 - f$. The frequencies $2f_0 + f$ are eliminated and the frequencies $2f_0 - f$, i.e. $f_0 + (f_0 - f) = f_0 - (f - f_0)$ give the spectrum reversed with respect to f_0. Therefore at reception it is sufficient to let the signal pass through a filter identical with filter 1 which introduces a delay:

$$T_R = F_1(f) = T_2 - T_T$$

This procedure is very common.

Figure 4.16 shows an example of the third procedure. A short signal whose spectrum is almost constant in the band $[f_0 - \Delta f/2, f_0 + \Delta f/2]$ and zero outside it, and which therefore has a duration of $1/\Delta f$ at 3 dB, enters the network shown in Fig. 4.16 at A. We therefore find this signal at A, the same signal shifted by a delay $4/\Delta f$ due to the delay line (DL) at D, the same signal

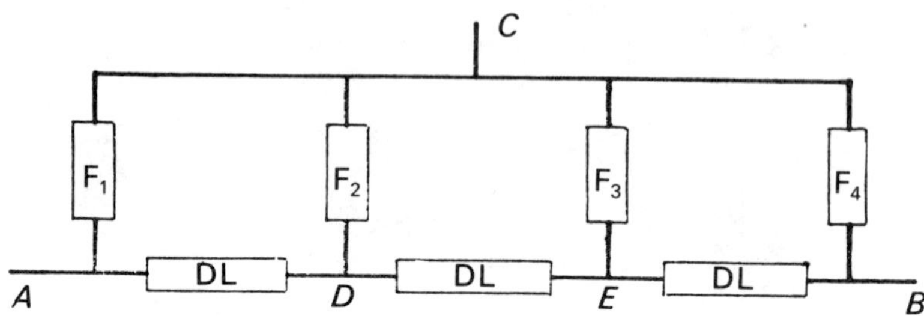

Fig. 4.16 Parallel procedure (DL, delay line).

delayed by $8/\Delta f$ at E and the same signal delayed by $12/\Delta f$ at B. F_1 is a bandpass filter which transmits the band $[f_0 - \Delta f/2, f_0 - \Delta f/4]$, F_2 is a bandpass filter which transmits the band $[f_0 - \Delta f/4, f_0]$ and so on. The output of F_1 is therefore a signal with carrier frequency $f_0 - 3\Delta f/8$ and duration $4/\Delta f$, the output of F_2 is a signal with carrier frequency $f_0 - \Delta f/8$ and duration $4/\Delta f$ which is produced $4/\Delta f$ after the preceding signal and so on. Finally the signal emerging at C, which is the transmission signal, has a total duration of $4 \times 4/\Delta f = 16/\Delta f$ and is frequency modulated by increasing frequencies. If this signal enters the network at B, instead of A, it leaves again at C identical to the input signal. It is clear that the frequencies of the band $[f_0 - \Delta f/2, f_0 - \Delta f/4]$ will be delayed by $0 + 12/\Delta f$, the frequencies of the band $[f_0 - \Delta f/4, f_0]$ will be delayed by $4/\Delta f + 8/\Delta f$ and so on, i.e. all the frequencies will be delayed by the same amount $12/\Delta f$. It should also be noted that if the incident signal had entered at B it would have left at C frequency modulated by decreasing frequencies, and that the systems A → C and C → A are reciprocal as are the systems B → C and C → B. A similar system containing n filters $F_1, \ldots, F_n$ and $n - 1$ delay lines would provide an expansion and compression ratio of n^2.

This third procedure, which is also known as the parallel procedure, can be generalized to the Turin procedure (Fig. 4.17) in which a delay line with n connection points (separated by a delay τ) and n amplifiers is used. Let $G_1(p), G_2(p), \ldots, G_n(p)$ be the transfer functions of the amplifiers $G_1, G_2, \ldots, G_n$. The network traversed between B and C is the filter matched to the signal obtained at C when a Dirac delta function $\delta(t)$ is input at A provided that

$$G_1(\mathrm{j}\omega) + G_2(\mathrm{j}\omega)\exp(-\tau\mathrm{j}\omega) + \cdots + G_n(\mathrm{j}\omega)\exp[-(n-1)\tau\mathrm{j}\omega]$$

is the imaginary conjugate of

$$\{G_n(\mathrm{j}\omega) + G_{n-1}(\mathrm{j}\omega)\exp(-\tau\mathrm{j}\omega) + \cdots$$
$$+ G_1(\mathrm{j}\omega)\exp[-(n-1)\tau\mathrm{j}\omega]\} K_2 \exp(T_1\mathrm{j}\omega)$$

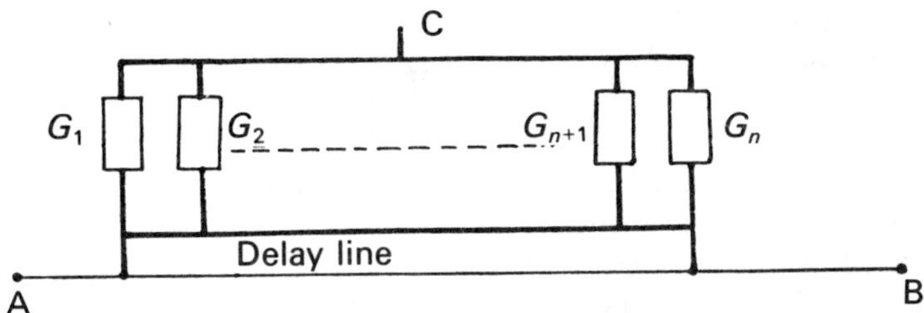

Fig. 4.17 Turin network.

where K_2 and T_1 are two real constants (independent of ω). It is obvious that if the network A $\rightarrow$ C is preceded or followed by a bandpass filter which does not modify the phases in the passband, the system remains correct provided that an identical bandpass filter is inserted at reception. The third parallel procedure is therefore only a particular case of the Turin procedure.

The above relationship is satisfied if all the $G_i(\mathrm{j}\omega)$ are real numbers with $K_2 = 1$ and $T_1 = (n - 1)\tau$, i.e. if the amplifiers G_i do not change the phase of the signals passing through them (except by 180°). This is why the system shown in Fig. 4.16 remains correct if the filters F_i are combined with amplifiers of this type with different gains G_i depending on the number of the filter (however, as the amplifiers are not reciprocal in this case the entire system must be used exclusively from A to C or from B to C).

In the case when the amplifiers G_i have real gains and a wide pass band, the envelope of the output signal of the Turin network at C when a Dirac delta function is input at A is given by (Fig. 4.18)

$$S_1(t) = G_1\delta(t) + G_2\delta(t - \tau) + G_3\delta(t - 2\tau) + \ldots$$

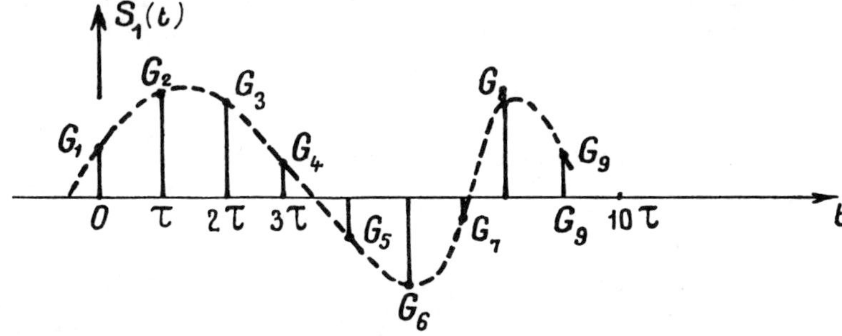

Fig. 4.18 Envelope of the output signal of the Turin network at C when a Dirac delta function is input at A.

The Fourier transform $\Phi_1(f)$ of $S_1(t)$ is periodic with period $1/\tau$ and, apart from a constant factor, is given by

$$\ldots \Phi(f + 2/\tau) + \Phi(f + 1/\tau) + \Phi(f) + \Phi(f - 1/\tau) + \Phi(f - 2/\tau) + \ldots$$

where $\Phi(f)$ is the Fourier transform of the envelope $S(t)$ of the signal $S_1(t)$; in any case $\Phi_1(f)$ is zero outside the frequency interval $[-1/2\tau, +1/2\tau]$. Therefore if we let $S_1(t)$ pass through a lowpass filter which eliminates frequencies above $1/2\tau$ (and does not modify componentes with frequencies below $1/2\tau$), we obtain the envelope $S(t)$ of $S_1(t)$ at the output of this filter, and the filter matched to $S(t)$ will be the cascade of the network in the direction B → C preceded or followed by a lowpass filter with a cut-off frequency $1/2\tau$. If, for example, we wish to transmit a signal

$$S(t) = \sin\left[2\pi\left(F_1 t + \frac{t^2}{2K}\right)\right] \quad \text{for} \quad 0 < t < T$$

$$S(t) = 0 \quad \text{for} \quad t > T \quad \text{and} \quad t < 0$$

i.e. a signal with a duration T which is linearly frequency modulated between F_1 and $T/K + F_1$ ($T^2/K \gg 1$) and whose useful spectrum lies between F_1 and $T/K + F_1$, we can take

$$\tau = \frac{1}{2(F_1 + T/K)}$$

and

$$G_i = \sin\left\{2\pi\left[F_1(i-1)\tau + \frac{(i-1)^2\tau^2}{2K}\right]\right\}$$

and let the network A → C (a Dirac delta function is input at A) be followed by a lowpass filter with cut-off frequency $T/K + F_1$. (In practice and for simplicity, it is sufficient to take G_i as 1 or -1 depending on its sign). At reception, we use the network B → C cascaded with a lowpass filter with cut-off frequency $T/K + F_1$ (Fig. 4.19). This procedure is very versatile because it allows, at least in theory, any signal $S(t)$ and the associated matched filter to be implemented at will.

4.5.3 General properties of pulse compression

4.5.3.1 Improvement of the signal-to-noise ratio

The long signal (like the short signal) occupies a spectrum of width Δf and (for an assumed constant level throughout its duration T) has a power $P = E/T$, where E is its energy. Therefore the signal-to-noise power ratio before compression is equal to $P/N_0\Delta f$, where N_0 is the spectral density (assumed uniform) of the noise, or $E/N_0 T\Delta f$. After compression, the

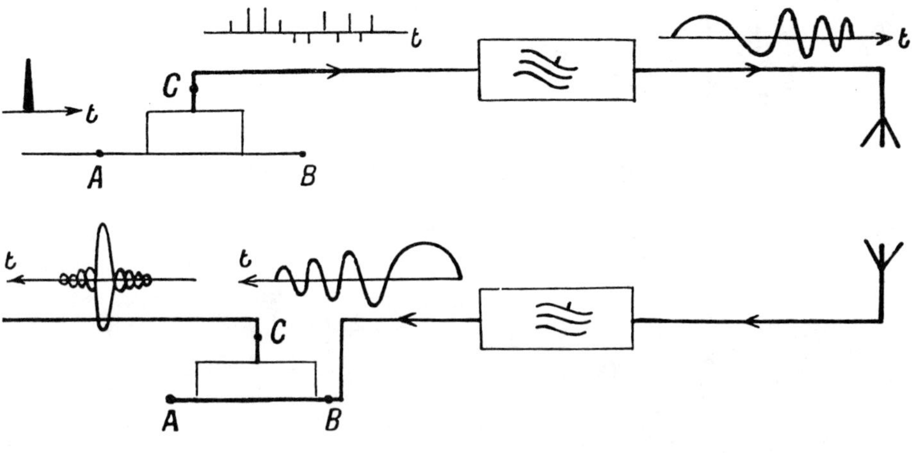

Fig. 4.19

signal-to-noise power ratio (laboratory technicians' definition) is equal to E/N_0 (see Chapter 3, Remark 3.4). Thus the compression operation has multiplied the signal-to-noise power ratio by $T\Delta f$ (obviously, the same result is obtained as in correlation reception).

4.5.3.2 Sidelobes

In the example used to illustrate active generation it was seen that the compressed signal consisted of a principal lobe surrounded by sidelobes 13–14 dB below the principal lobe. This occurs because the transmitted spectrum is rectangular. The third procedure for passive generation gives a similar result since it has been shown that the compressed short signal is identical with the input signal which is assumed *a priori* to be of this form. This behavior, which can be a problem, is not obtained in the second procedure for passive generation because the compressed signal is identical with the input signal for which no assumption of this type has been made. However, for technological reasons, the transmitter providing the long signal will generally operate as a saturated amplifier in order to supply a long signal with constant level throughout its duration, and when this long transmitted signal has a constant level its spectrum becomes quasi-rectangular, therefore producing a compressed signal with sidelobes.

The filters $F_1, \ldots, F_n$ used in the third procedure for passive generation can be assigned different gains G_i without altering the quality of the matching to the reception. For example, if there are 10 filters with gains $G_1 = -7$ dB, $G_2 = -4$ dB, $G_3 = -2$ dB, $G_4 = -1$ dB, $G_5 = 0$, $G_6 = 0$, $G_7 = -1$ dB, $G_8 = -2$ dB, $G_9 = -4$ dB and $G_{10} = -7$ dB, the (power) spectrum of the

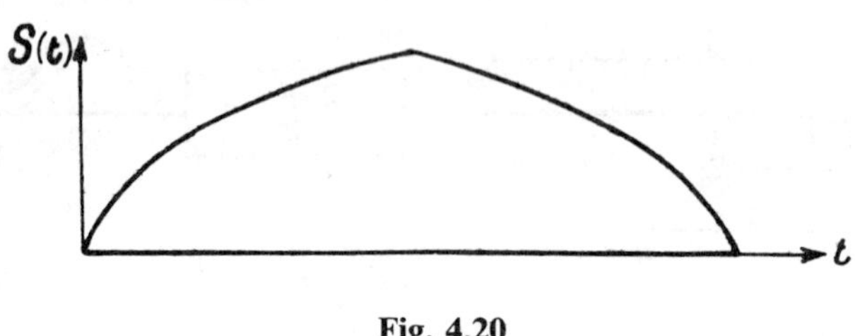

Fig. 4.20

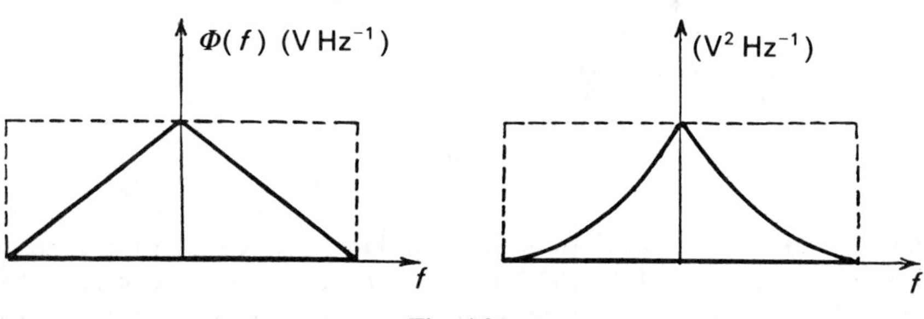

Fig. 4.21

transmitted signal will be triangular and the compressed signal will consist of a principal lobe with total width equal to $4/\Delta f$ (double width) and width at 3 dB equal to $1.3/\Delta f$ (increased by 30%) and sidelobes of which the first is 27 dB below the principal lobe.

Then the transmitted signal is also triangular in terms of power, i.e. it is flattened out in terms of voltage (Fig. 4.20). If we now assume that the amplifiers at reception have gains $G_1 = -14$ dB, $G_2 = -8$ dB, $G_3 = -4$ dB, $G_4 = -2$ dB, $G_5 = 0$, $G_6 = 0$, $G_7 = -2$ dB, $G_8 = -4$ dB, $G_9 = -8$ dB and $G_{10} = -14$ dB, the compressed signal will have the same form but the reception filter will no longer be matched to the transmitted signal. The system behaves as if we had used a matched filter followed by a bandpass filter with a triangular voltage characteristic. This is known as reducing the sidelobes by spectral weighting. Thus the level of the useful signal is divided by 2 and the noise power is divided by 3, so that the signal-to-noise ratio is divided by 1.33 with a consequent loss of 1.3 dB (Fig. 4.21).

The reduction of the sidelobes in a weighting filter is accompanied by broadening of the compressed signal and a loss in the signal-to-noise ratio owing to the mismatch in amplitude. This loss is often acceptable. When it is not, but none the less small sidelobes are required, it is necessary to

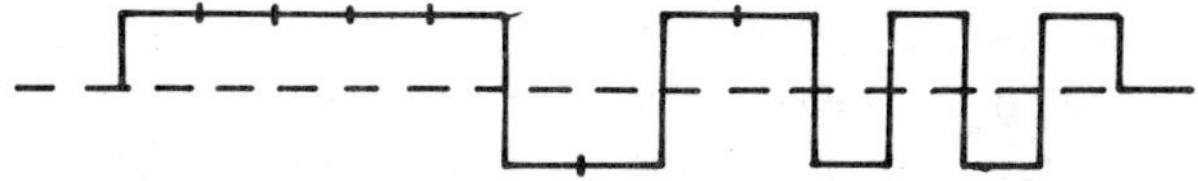

Fig. 4.22 Barker signal (transposed to video).

transmit a signal $S(t)$ whose autocorrelation function has small sidelobes. Two types of signal are frequently used for this purpose.

The first type is the family of binary phase-modulated signals using the Barker code. Such signals enable an odd compression ratio ϱ with a maximum value of 13 to be obtained, with the sidelobes having a maximum level (in amplitude) equal to the compression ratio (for $\varrho = 13$, this level is 22 dB). Figure 4.22 shows a Barker signal (transposed to video) for $\varrho = 13$. Polyphase signals, which belong to the same family of phase-modulated signals, can also be used (the phase takes a discrete number of values). For example, by using eight discrete values of the phase ($0, \pi/4, \pi/2, 3\pi/4, \pi, 5\pi/4, 3\pi/2, 7\pi/4$) it is possible to obtain a compression ratio of 64 with the level of the worst sidelobe being at -32 dB.

The second type is the family of signals with a rectangular envelope in which the frequency varies quasi-linearly with time. It can be shown [9] that if the instantaneous frequency f of the transmitted signal $S(t)$ varies as a function of the time t such that $\mathrm{d}f/\mathrm{d}t$ is a minimum at the center of the signal, the spectrum $|\Phi(f)|^2$ of the signal $S(t)$ has a maximum at the central frequency (and a minimum at the "edges of the spectrum") because, to a first approximation, $|\Phi(f)|^2$ varies as $\mathrm{d}t/\mathrm{d}f$. In this case therefore the use of a filter matched in phase and amplitude provides compressed signals with reduced sidelobes (this procedure is known as intrinsic weighting).

4.5.3.3 Ambiguity in the case of chirp signals

Chirp radars transmit a rectangular signal with linear frequency modulation and, historically, were the first pulse compression radars. Such a signal can be written in the form

$$S(t) = \sin\left(\frac{\pi t^2}{K}\right) \quad \text{for} \quad -\frac{T}{2} < t < \frac{T}{2}$$

$$S(t) = 0 \quad \text{for} \quad t < -\frac{T}{2} \quad \text{and} \quad t > \frac{T}{2}$$

The reception operation in a matched filter is equivalent to the correlation operation

$$\int_{-\infty}^{+\infty} y(u)\, S(u - t)\, \mathrm{d}u$$

where $y(t)$ is the received signal and $t = 0$ is the time of arrival of the signal. If the effect of the noise is neglected so that only the useful signal is considered, the compressed signal in the absence of the Doppler effect can be written

$$g(t) = \int_{-\infty}^{+\infty} S(u)\,S(u - t)\,\mathrm{d}u$$

$$g(t) = \int_{-T/2+t}^{+T/2} \sin\left(\frac{\pi u^2}{K}\right) \sin\left[\frac{\pi(u - t)^2}{K}\right] \mathrm{d}u \quad \text{if} \quad t > 0$$

In the presence of the Doppler effect, which displaces the received signal by a frequency F_D (assumed to be negative), the compressed signal can be written

$$g_D(t) = \int_{-T/2+t}^{+T/2} \sin\left[2\pi\left(F_D u + \frac{u^2}{2K}\right)\right] \sin\left[\pi \frac{(u - t)^2}{K}\right] \mathrm{d}u$$

These expressions can be transformed to give

$$g(t) = \frac{1}{2}\int_{-T/2+t}^{T/2} \cos\left[\frac{2\pi}{K}\left(ut - \frac{t^2}{2}\right)\right] \mathrm{d}u$$

$$g_D(t) = \frac{1}{2}\int_{-T/2+t}^{T/2} \cos\left[\frac{2\pi}{K}\left(u(t + KF_D) - \frac{t^2}{2}\right)\right] \mathrm{d}u$$

As expected, $g(t)$ has a maximum at $t = 0$ which has a value of $T/2$. If $KF_D{}^2/2 \ll 1$, $g_D(t)$ does not have a maximum at $t = 0$ but at $t = -KF_D$ which has a value of approximately $T/2$. Since $K\Delta f = T$, the inequality $KF_D{}^2 \ll 1$ can also be written as

$$F_D \ll 1.4 \frac{\Delta f}{\varrho^{1/2}}$$

where $\varrho = T\Delta f$ is the compression ratio.

Numerical examples

$-\Delta f = 10^6\,\mathrm{Hz} \quad T = 10^{-4}\,\mathrm{s} \quad \varrho = 100 \quad K = 10^{-10}\,\mathrm{s}^2 \quad 1.4\,\dfrac{\Delta f}{\varrho^{1/2}} = 140\,\mathrm{kHz}$

For a Doppler effect of 10 kHz, the compressed signal is shifted by 1 µs.

$-\Delta f = 10^5\,\mathrm{Hz} \quad T = 10^{-2}\,\mathrm{s} \quad \varrho = 1000 \quad K = 10^{-7}\,\mathrm{s}^2 \quad 1.4\,\dfrac{\Delta f}{\varrho^{1/2}} = 5\,\mathrm{kHz}$

For a Doppler effect of 1 kHz, the compressed signal is shifted by 100 µs.

$-\Delta f = 3000\,\mathrm{Hz} \quad T = 10^{-2}\,\mathrm{s} \quad \varrho = 30 \quad K = 3 \times 10^{-6}\,\mathrm{s}^2 \quad 1.4\,\dfrac{\Delta f}{\varrho^{1/2}} = 1\,\mathrm{kHz}$

For a Doppler effect of 50 Hz, the compressed signal is shifted by 150 µs.

Remark 4.10

Since a signal at $t = 0$ and $F_D = 0$ finally gives the same result as a signal of the same level modified by the Doppler effect F_D at $t = -KF_D$†, the ambiguity between these two signals is close to unity because they are impossible to distinguish. This result is due to the linear distribution of the frequency modulation used (see Chapter 3, Sections 3.10.4 and 3.10.6). If another modulation distribution of the long signal is used, a different type of ambiguity is found (see Chapter 3). If the direction of the linear frequency modulation is good, the radial range of a target is measured with an error proportional to the radial velocity of the target, i.e. the range which the target will have a short time later is measured and this is not necessarily a fault.

Remark 4.11

If the received signal is limited before it enters the reception filter, i.e. before compression, constant false alarm (CFAR) reception is obtained (see Section 4.6.7).

Remark 4.12

If a long signal modulated by increasing frequencies passes through a filter designed to compress similar signals but modulated by decreasing frequencies, it is possible as a first approximation to say that its length is doubled and that its level is reduced to 3 dB. Similarly, the level of a transmitted signal of duration $1/\Delta f$, which is not frequency modulated, will be reduced in the compression ratio $T\Delta f$. This property is useful with respect to antijamming.

4.5.4 Calculation of faults due to compression imperfection. Paired-echo method

Sidelobes caused by the difference between what is actually produced and what should have been produced in theory appear in addition to those due to the nature of the signal which can be reduced by weighting. It is necessary to know in advance the accuracy with which the theoretical performance should be reproduced on the equipment, and the paired-echo method has been developed for this purpose [10, 11]. In this section we describe the basic concepts of this method without reproducing the relevant calculations.

† This result is physically intuitive. Since this filter matched to the reception is a delay line whose transmission time increases (or decreases) with frequency, it is natural that a shift of the entire spectrum of received frequencies by F_D introduces a shift in the time of the compressed signal at the output of the delay line.

The fault introduced by the difference between the distribution obtained for $|\Phi(f)|$ and the distribution which should have been obtained can always be written as a sum of terms of the form $A \cos(2\pi T_t f + \varphi)$ (the central frequency f_0 is assumed to be zero). Such an error term gives two parasite signals situated on each side of the center of the useful signal at a distance T_t from this center, hence the name "paired echo".

4.5.5 Practical implementation of dispersive filters

This book is not intended to provide technological information (which the reader may find in the more specialized literature) but to describe theoretical concepts which may assist in understanding existing procedures or in establishing new procedures. The discussion below is undertaken in this spirit.

4.5.5.1 Use of filters with discrete components

A dispersive filter can be obtained by using a cascade of allpass filters as shown in Fig. 4.23 (or, rather, equivalent T-bridge filters). The first pulse compression radars used this solution. However, it can be shown that, for a given quality of pulse compression, the number of components with lumped constants necessary in such a filter is proportional to $\varrho^{5/3}$, where ϱ is the compression ratio, which shows that such a filter rapidly becomes too expensive and bulky as soon as ϱ is no longer small.

4.5.5.2 Use of shift registers (charge transfer devices)

When the parallel method (or the Turin method) is used, a delay line providing a delay of the order of magnitude of the duration T of the long pulse in the frequency band Δf is required. This can be done by using a shift register (or several shift registers in parallel) or a charge transfer device (of the charge-coupled device (CCD) type for example) in which the transfer from one cell to the next is made at the frequency F_R. (The first filters of this type used magnetostriction lines with very narrow pass bands ($\Delta f < 1$ MHz) as delay lines.)

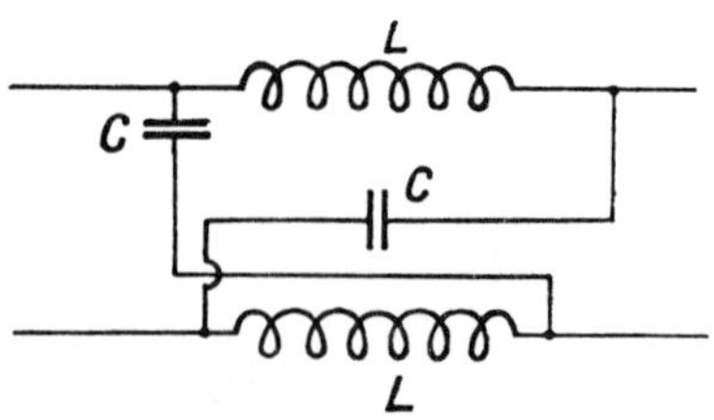

Fig. 4.23

A shift register may also provide limiting of the input signals (see Section 4.6) and the delay it produces is given by $M/F_R \approx T$, where M is the number of cells. The signal must be sampled at a sufficiently high rate, i.e. F_R must be greater than twice the highest frequency to be handled, which is in turn at least equal to Δf (or $\Delta f/2$ if an I and Q procedure is used). Therefore we have at least

$$F_R = \frac{M}{T} > 2\Delta f$$

and hence

$$M > 2T\Delta f \quad \text{or} \quad M > 2\varrho$$

It can be seen that the limitation produced in this way is not excessive.

4.5.5.3 Acoustic delay lines

The characteristics of ultrasonic acoustic propagation can be used under certain conditions (thin metal filter, Love waves) to obtain dispersive lines with the required quality for curves giving the propagation time as a function of frequency. The propagation of Rayleigh waves at the surface of a piezo-electric substrate can also be used. In this case the input electrode is not dispersive whereas the output electrode consists of an interdigital line in which the spacing between successive teeth varies along the line.

4.5.5.4 Use of Fourier transformers

Another procedure consists in taking the Fourier transform $Y(f)$ of the received signal $y(t)$, multiplying it by $\Phi(-f)$ (or by $\Phi^*(f)$) and finally taking the Fourier transform of the product $Y(f)\,\Phi(-f)$ in order to obtain $C(-t)$. The first method to be used was as follows (Fig. 4.24): $y(t)$ is transformed into an optical transparency $y(x)$† where, under certain conditions, the Fourier transform $Y(v)$ (in a plane π_1) is provided by an optical system (e.g. a simple lens) illuminated by coherent light; $Y(v)$ is multiplied by a transparency representing $\Phi(-v)$ (i.e. a transparency representing $|\Phi(v)|$ and a lens whose variation in thickness represents the variation in the phase of $\Phi(-v)$) or by a hologram which provides the same result; $Y(v)\,\Phi(-v)$ is then taken through a second optical system of the same type as the first in order to obtain a final image representing $C(-x)$. Digital systems are now used to calculate the Fourier transforms in real time (in the form of the discrete Fourier transformation (DFT) and by using fast Fourier transform algorithms).

† For example, by transforming $y(t)$ to ultrasonic waves which pass through a cell with parallel faces filled with water whose density is thus modulated by $y(x)$ giving

$$\exp[-kj\,y(x)] \approx 1 - kj\,y(x)$$

The unit term can then be eliminated using classical Schlieren methods.

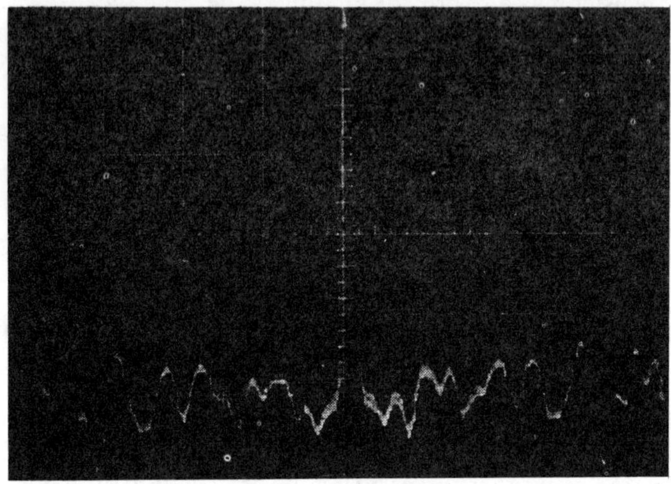

Fig. 4.24

4.6 Constant false alarm reception (CFAR)

4.6.1 Principles of constant false alarm receivers

The noise $C_p(t)$ accompanying the useful signal at the output of either a correlation radar or a matched filter radar is subject to level variations for a number of reasons, which depend on the level of the (intentional or unintentional) jamming. Furthermore, the extractor (either a classical oscilloscope or an arithmetic extractor) which follows the radar has a poor toleration of variations in the noise level which have the effect of varying the probability of false alarm. If the noise level varies sufficiently slowly, it is possible to modify the detection threshold slowly in order to keep the false alarm probability constant, but this becomes very difficult when the variations in noise level are rapid. In this case, receivers known as constant false alarm (CFAR) receivers are used.

The CFAR property is obtained by limiting the received signal at a constant very low level before passing it into the matched filter (or before correlation). The limiting level is chosen so that the noise is limited even in the absence of a useful signal and low filter noise (thermal noise from the receiver itself). The limiting is carried out on the noise–signal combination occupying a band which is larger than or equal to the width Δf of the spectrum of the transmitted signal, i.e. a band $M\,\Delta f$ ($M > 1$) generally centered on a frequency f_0 higher than $M\,\Delta f$. The result is that in the absence

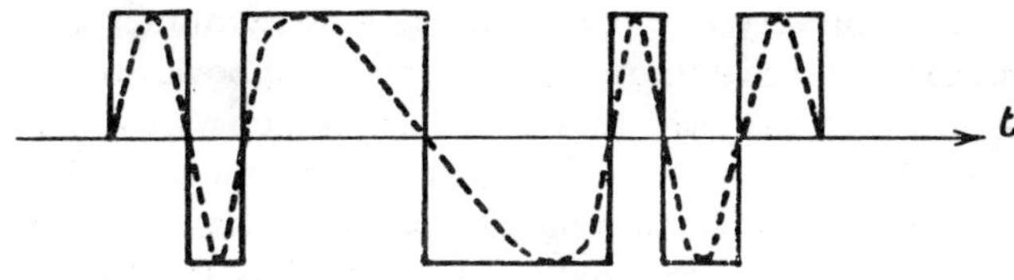

Fig. 4.25

of a useful signal we obtain a signal with zero mean which may be positive or negative but has a constant level (Fig. 4.25) and therefore cannot be gaussian.

This signal is filtered so as to eliminate components at frequencies of the order of $2f_0$, $3f_0$, . . . and in this way a signal which is always non-gaussian is obtained as shown by the broken line in Fig. 4.25. This is defined in a necessary and sufficient manner by its values at consecutive times separated by $1/M\,\Delta f$ (and by the knowledge of f_0). The operation of passing through the matched or correlation reception filter has the effect of integrating this non-gaussian noise over a time T so that

$$\frac{T}{1/M\,\Delta f} = MT\Delta f$$

is large (at least 10). Therefore the output noise of the matched filter (or correlation filter) is the sum of $MT\Delta f$ independent variables and hence is gaussian with a constant variance (power). In a simplistic manner, it can be said that the noise has a power equal to the constant power after limiting (taken as unity) divided by the ratio of the bandwidths (before limiting and after reception), i.e. a constant power equal to $1/M$.

When the useful signal is alone, its power is increased by the ratio $T\Delta f$ (i.e. by the compression ratio in the case of a matched filter radar); it therefore has a power equal to $T\Delta f$ which means that the square of its maximum is equal to $2T\Delta f$ (see Chapter 3, Remark 3.4). Thus, in crude terms, it appears that in the absence of the useful signal the output noise of the receiver is gaussian with a constant power equal to $1/M$ and when the noise is negligible compared with the useful signal (in the absence of noise) the latter has a constant power $T\Delta f$ at the output of the receiver, and thus in this case the ratio of the square of the maximum of the useful signal when it is high to the power of the noise in the absence of a useful signal is also constant and equal to $2MT\Delta f$. These general ideas are examined more closely in subsequent sections.

4.6.2 Output levels of the useful signal and noise

4.6.2.1 General considerations

Detailed results will only be given for the case when the transmission signal is a signal with constant level (constant power) throughout its duration

T. This is in fact not very different from what is encountered in practice. However, the case of a signal transmitted with varying power will be discussed briefly. It is therefore assumed that the transmission signal has the form $\sin(2\pi ft)$ in which f may vary as a function of time but such that the width of the spectrum of the transmission signal is narrow compared with its central frequency f_0. Therefore the received useful signal has the form $S\sin(2\pi ft)$ and energy $S^2T/2$. The noise signal has the form $S_N(t)\sin[2\pi ft + \varphi(t)]$, where $S_N(t)$ is a positive variable with a Rayleigh amplitude distribution and $\varphi(t)$ is equiprobably distributed between zero and 2π (this is equivalent to saying that the noise is gaussian), so that

$$N = \overline{S_N{}^2(t)\sin^2[2\pi ft + \varphi(t)]}$$

where N is the noise power. The probability density of S_N can be written (see Chapter 1, Section 1.8)

$$p(S_N) = \frac{S_N}{N}\exp\left(-\frac{S_N{}^2}{2N}\right) \quad \text{for} \quad S_N > 0$$

$$p(S_N) = 0 \quad \text{for} \quad S_N < 0$$

The ratio

$$\mathrm{SN_I} = \frac{\text{signal power}}{\text{noise power}}$$

at the input (before limiting) is therefore

$$\mathrm{SN_I} = \frac{S^2}{2N} = 2\frac{S^2T}{2N_0}\frac{1}{2MT\Delta f} = \frac{2E}{2N_0MT\Delta f} = \frac{R}{2MT\Delta f}$$

4.6.2.2 In the absence of limiting (non-CFAR reception)

After correlation over a time T we find

$$\text{useful signal} = S\int_T \sin(2\pi ft)\sin(2\pi ft)\,\mathrm{d}t = ST/2$$

$$\text{noise} = \int_T \sin(2\pi ft)\,S_N(t)\sin[2\pi ft + \varphi(t)]\,\mathrm{d}t$$

$$= \frac{1}{2}\int_T \cos\varphi(t)\,S_N(t)\,\mathrm{d}t$$

a gaussian random variable with zero mean value and variance

$$MT\Delta f\frac{1}{M^2\Delta f^2}\overline{[\tfrac{1}{2}\cos\varphi(t_i)]^2}\,\overline{S_N{}^2(t_i)} = \frac{NT}{4M\Delta f}$$

which gives at the output of the correlator a power or energy signal-to-noise ratio (information theory definition (see Section 3.4)) of

$$\frac{S^2T^2}{4}\frac{4M\Delta f}{TN} = \frac{S^2}{2N} \times 2MT\Delta f = \frac{S^2T}{2}\frac{2}{N_0} = R = \frac{2E}{N_0}$$

where $N_0 = N/M\Delta f$ is the spectral density of the noise and $S^2T/2$ is the energy of the received signal.

Remark 4.13

It should be noted that, in the absence of limiting, the same result is obtained regardless of the width of the band before it is passed into the matched filter (or correlation filter). In practice, the higher is the value of M (large pass band before matched filtering), the larger must be the dynamic range of the matched filter (or the correlation filter).

4.6.2.3 In the presence of limiting (CFAR reception)

General considerations

The combined level (signal plus noise) before limiting can be written

$$S\sin(2\pi ft) + S_N(t)\sin[2\pi ft + \varphi(t)]$$

or

$$[S^2 + S_N^{\,2}(t) + 2S_N(t)S\cos\varphi(t)]^{1/2}\sin[2\pi ft + \Phi(t)]$$

where

$$\cos\Phi(t) = \frac{S + S_N(t)\cos\varphi(t)}{[S^2 + S_N^{\,2}(t) + 2SS_N(t)\cos\varphi(t)]^{1/2}}$$

$$\sin\Phi(t) = \frac{S_N(t)\sin\varphi(t)}{[S^2 + S_N^{\,2}(t) + 2SS_N(t)\cos\varphi(t)]^{1/2}}$$

After limiting and eliminating the harmonics (it is assumed that f_0 is higher than $M\Delta f$), we are left with $\sin[2\pi ft + \Phi(t)]$. The correlation of this term with the reference $\sin(2\pi ft)$ gives the following results. In the absence of noise (or when the noise is negligible compared with the useful signal)

$$\int_T \sin(2\pi ft)\sin(2\pi ft)\,dt = T/2$$

In the absence of the useful signal ($\Phi = \varphi$)

$$\int_T \sin(2\pi ft)\sin[2\pi ft + \varphi(t)]\,dt = \frac{1}{2}\int_T \cos\varphi(t)\,dt = \frac{1}{2}\sum_{MT\Delta f}\cos\varphi(t_i)\,\frac{1}{M\Delta f}$$

i.e. a gaussian random variable with zero mean value and variance (power)

$$\tfrac{1}{4}MT\Delta f\,\overline{\cos^2\varphi(t_i)}\,\frac{1}{M^2\Delta f^2} = \frac{T}{8M\Delta f}$$

Thus the ratio of the square of the maximum of the useful signal when it is high to the power of the noise in the absence of the useful signal is given by

$$\frac{T^2}{4} \times \frac{8M\Delta f}{T} = 2MT\Delta f$$

In the simultaneous presence of the useful signal and noise, we obtain at the output of the correlator (or matched filter)

$$\int_T \sin(2\pi f t)\sin[2\pi f t + \Phi(t)]\,\mathrm{d}t = \frac{1}{2}\int_T \cos\Phi(t)\,\mathrm{d}t$$

which is a gaussian random variable whose mean value is given by

$$\frac{1}{2}\overline{\int_T \cos\Phi(t)\,\mathrm{d}t} = \frac{T}{2}\overline{\cos\Phi}$$

This random variable, which is the output signal of the radar, can therefore be considered as the sum of the useful signal $T/2\,\overline{\cos\Phi}$ and a gaussian term due to the fact that $\cos\Phi(t)$ is not constant, i.e. the sum of the output noise with zero mean and the power (variance) of the output noise.

SN_1 very small

If the signal-to-noise ratio SN_1 at the input is very small, i.e. $S^2/2N \ll 1$, we can write (with the loss of some rigor)

$$\cos\Phi \approx \frac{\cos\varphi + S/S_N}{[1 + (2S/S_N)\cos\varphi]^{1/2}} \approx \left(\cos\varphi + \frac{S}{S_N}\right)\left(1 - \frac{S}{S_N}\cos\varphi\right)$$

$$\approx \cos\varphi + \frac{S}{S_N}\sin^2\varphi$$

$$\overline{\cos\Phi} = \frac{S}{2}\overline{\left(\frac{1}{S_N}\right)} = \frac{S}{2}\int_0^\infty \frac{1}{N}\exp\left(-\frac{S_N{}^2}{2N}\right)\mathrm{d}S_N$$

$$= \frac{S(2\pi)^{1/2}}{2N^{1/2}}\frac{1}{(2\pi N)^{1/2}}\int_0^\infty \exp\left(-\frac{S_N{}^2}{2N}\right)\mathrm{d}S_N$$

$$= \frac{S}{4}\left(\frac{2\pi}{N}\right)^{1/2}$$

The useful signal can therefore be written

$$\frac{ST}{8}\left(\frac{2\pi}{N}\right)^{1/2} = \frac{T}{2}\left(\frac{\pi}{4}SN_1\right)^{1/2}$$

However, the noise power at the output is very close to the value that it has when SN_1 is zero, i.e. $T/8M\Delta f$. Thus the signal-to-noise ratio SN_2 at the output† is equal to (at the position of the signal)

$$\mathrm{SN}_2 = \frac{T^2}{4}\frac{\pi}{4}\mathrm{SN}_1\frac{8M\Delta f}{T} = \frac{\pi}{4}\mathrm{SN}_1 MT\Delta f \times 2$$

$$= \frac{\pi}{4}\frac{S^2}{2N}2MT\Delta f = \frac{\pi}{4}R$$

This is a factor of $\pi/4$ times that obtained in the case of the non-CFAR receiver (loss of 1 dB on the signal-to-noise ratio).

SN_I very large

In this case, $\cos\Phi(t) \approx 1$, the useful signal is close to $T/2$ (this result has already been found) and the signal after limiting can be written

$$\frac{S\sin(2\pi ft)}{[S^2 + 2S_N S\cos\varphi + S_N{}^2]^{1/2}} + \frac{S_N(t)\sin[2\pi ft + \varphi(t)]}{[S^2 + 2S_N S\cos\varphi + S_N{}^2]^{1/2}}$$

$$\approx \sin(2\pi ft)\left(1 - \frac{S_N}{S}\cos\varphi\right) + \frac{S_N}{S}[\sin(2\pi ft)\cos\varphi + \cos(2\pi ft)\sin\varphi]$$

where terms in $S_N{}^2/S^2$ are neglected. Alternatively it can be written as

$$\sin(2\pi ft) + \frac{S_N}{S}\cos(2\pi ft)\sin\varphi$$

where $\sin(2\pi ft)$ represents the useful signal and $S_N/S\cos(2\pi ft)\sin\varphi$ represents the noise after limiting. This shows (still without rigor) that the signal-to-noise ratio after limiting has become

$$\frac{1/2}{(1/2S^2)\overline{S_N{}^2\sin^2\varphi}} = \frac{S^2}{N} = 2\mathrm{SN}_1$$

In this case, limiting simply has the effect of doubling the signal-to-noise ratio, a result which is obviously unaltered by correlation (or reception in a matched filter). We therefore have a gain of 3 dB on the signal-to-noise ratio as a result of CFAR reception for high signal-to-noise ratios at the input (and a loss of 1 dB for low signal-to-noise ratios at the input).

This may seem surprising. At first sight, it would seem that the CFAR receiver would improve reception which contradicts the fact that, since CFAR reception is not optimum, something has to be lost or at least nothing can be gained. It will be seen in Section 4.6.3 that this contradiction is only too apparent.

† Defined as the ratio of the square of the maximum of the useful signal to the noise power.

General case: any SN_l

The useful signal is then given by $\frac{1}{2}\overline{\cos \Phi}$ and the noise appears to be given by

$$\int_T (\{\sin(2\pi ft)[\sin 2\pi ft + \Phi(t)]\} - \tfrac{1}{2}\overline{\cos \Phi})\,\mathrm{d}t$$

or by

$$\frac{1}{M\Delta f}\sum [f(t_i) - \overline{f(t)}]$$

where the t_i are sampling times at intervals of $1/M\Delta f$ and

$$f(t) = \sin(2\pi ft)\sin[2\pi ft + \Phi(t)] = \tfrac{1}{2}\cos \Phi(t)$$

The noise power is the variance of this term, i.e.

$$\frac{1}{M^2\Delta f^2}\sum \overline{[f(t_i) - \overline{f(t)}]^2} = \frac{1}{M^2\Delta f^2}\sum \{\overline{f(t_i)^2} - [\overline{f(t)}]^2\}$$

$$= \frac{T}{4M\Delta f}[\overline{\cos^2 \Phi} - (\overline{\cos \Phi})^2]$$

The signal-to-noise ratio at the signal position is therefore given by

$$\mathrm{SN}_2 = \frac{T^2}{4}\frac{4M\Delta f}{T}\frac{(\overline{\cos \Phi})^2}{\overline{\cos^2 \Phi} - (\overline{\cos \Phi})^2}$$

$$= MT\Delta f\frac{(\overline{\cos \Phi})^2}{\overline{\cos^2 \Phi} - (\overline{\cos \Phi})^2}$$

If SN_1 is very small, we obtain

$$(\overline{\cos \Phi})^2 \approx \frac{S^2}{8}\frac{\pi}{N} \quad \text{and} \quad \overline{\cos^2 \Phi} = \tfrac{1}{2}$$

$$\mathrm{SN}_2 \approx \frac{\pi}{4}R\left(1 + \frac{\pi}{4}\frac{R}{MT\Delta f}\right)$$

The complete results are given in Fig. 4.26 where the variations in SN_2 and in the useful signal S_u (divided by $0.5T$), i.e. $\cos \Phi$, are plotted as functions of $R/2MT\Delta f$. These results are shown in a more meaningful manner in Fig. 4.27 where the output signal of a CFAR receiver is plotted as a function of the time t for $MT\Delta f = 60$ ($T\Delta f = 10$ and $M = 6$) when there is a useful signal at t_1 with $R = 10$ dB (Fig. 4.27(a)) or $R = 30$ dB (Fig. 4.27(b)). The noise power is reduced on each side of the useful signal over a total width of $2T$: for $R = 30$ dB ($R/2MT\Delta f = 9$ dB) it goes from -33 dB at the signal position ($-R - 3$ dB) to -21 dB outside the interval (i.e. $0.5/MT\Delta f$), whereas for $R = 10$ dB ($R/2MT\Delta f = -11$ dB) it goes from

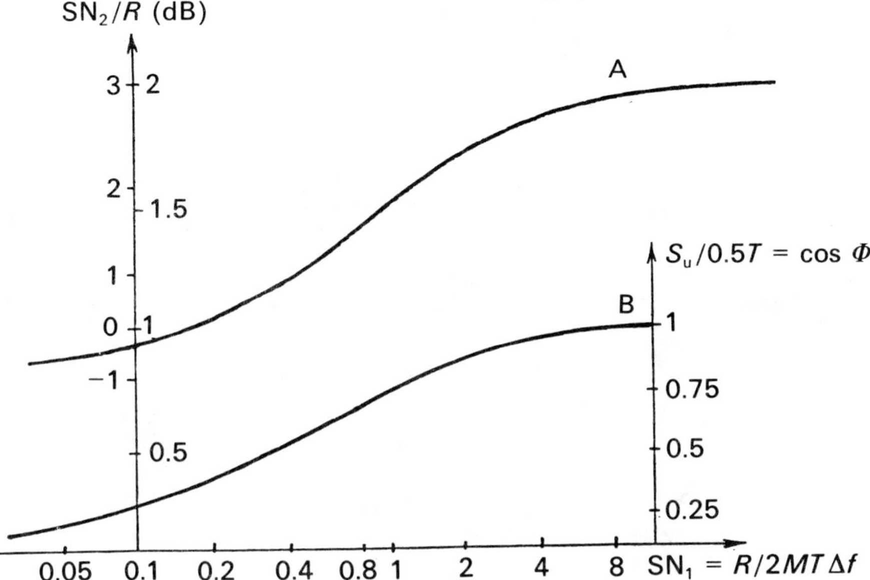

Fig. 4.26 Variations in SN_2/R (curve A) and $S_u/0.5T = \cos \Phi$ (curve B) as functions of $R/2MT\Delta f$.

-21.5 dB at the position of the signal with a level of -12 dB $(-12 - 10 + 0.5)$ to a constant -21 dB outside the interval $2T$. If $T\Delta f$ were equal to unity (classical radar), the figures would not show the noise attenuation at the signal position.

4.6.3 Detection probability

4.6.3.1 General considerations

It has been shown in Section 4.6.2 that when the transmission signal has a constant power over its duration, the signal-to-noise ratio at the output of a CFAR receiver is equal to R reduced by 1 dB if $R/2MT\Delta f$ is small and it is equal to $2R$ if $R/2MT\Delta f$ is large. However, what is important in practice is the detection probability (associated with a given false alarm probability).

In the case of a CFAR receiver, the false alarm probability is fixed by defining the threshold at a given level with respect to the standard deviation of the noise (its effective value) in the absence of a useful signal. In this way, a false alarm probability of approximately 10^{-3} is obtained with a threshold at a level of three times the standard deviation of the noise in the absence of a useful signal, i.e. at a level of approximately

$$\frac{3}{2}\left(\frac{T}{2M\Delta f}\right)^{1/2}$$

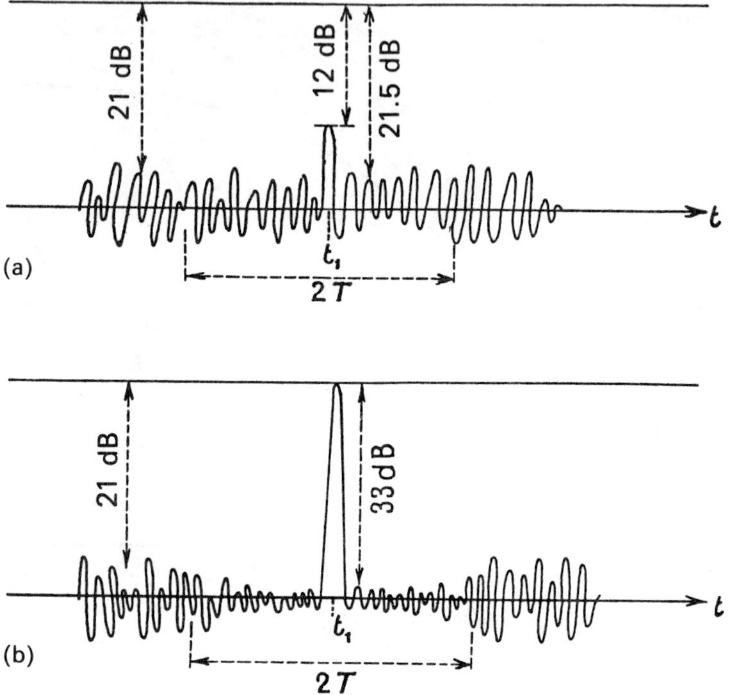

Fig. 4.27 Output signal of a CFAR receiver as a function of the time t for a useful signal at t_1 with (a) $R = 10\,\text{dB}$ and (b) $R = 30\,\text{dB}$.

The detection probability is therefore the probability that the sum of a signal of level $T/2\ \overline{\cos\,\Phi}$ and a gaussian noise with zero mean and variance

$$\frac{T}{4M\,\Delta f}\left[\overline{\cos^2\,\Phi} - (\overline{\cos\,\Phi})^2\right]$$

is greater than

$$\frac{3}{2}\left(\frac{T}{2M\,\Delta f}\right)^{1/2}$$

4.6.3.2 Detection probability of 0.5

A detection probability of 0.5 is obtained if

$$\frac{T}{2}\,\overline{\cos\,\Phi} = \frac{3}{2}\left(\frac{T}{2M\,\Delta f}\right)^{1/2}$$

$$\frac{S_u}{0.5T} = \overline{\cos\,\Phi} = \left(\frac{9}{2MT\,\Delta f}\right)^{1/2}$$

Table 4.2 Loss due to CFAR for a detection probability of 0.5 and a false alarm probability of 10^{-3}

$2MT\Delta f$	100	50	20	15	10	9
Loss (dB)	1	1.3	2.2	3	7	∞

If $MT\Delta f$ is large,

$$\overline{\cos \Phi} \approx \left(\frac{\pi}{4}\,\mathrm{SN}_1\right)^{1/2} \approx \left(\frac{\pi}{8}\,\frac{R}{MT\Delta f}\right)^{1/2}$$

and we obtain

$$\left(\frac{\pi}{4}\,R\right)^{1/2} = \sqrt{9}$$

$$R = 9 \times \frac{4}{\pi}$$

For the non-CFAR receiver, we found $R = 9$. More generally, for any false alarm probability we must produce a threshold at a level equal to K times the standard deviation of the noise in the absence of the signal, and we find that a detection probability of 0.5 is obtained when $MT\Delta f$ is sufficiently large for A to be equal to $4K^2/\pi$ with the CFAR instead of K^2 with the matched receiver. This corresponds to a loss of 1 dB for $MT\Delta f$ sufficiently large and a detection probability of 0.5.

If $MT\Delta f$ is not large, the loss due to CFAR for a detection probability of 0.5 increases. Table 4.2 gives the loss introduced by the CFAR receiver as a function of $2MT\Delta f$ for a detection probability of 0.5 and a false alarm probability of 10^{-3}.

Remark 4.14

It is asssumed that detection is based on a single measurement of duration T. If n identical measurements are used (particularly with coherent integration), the results are improved and the losses are smaller (in the case of coherent integration the system behaves as if $MT\Delta f$ were multiplied by n).

4.6.3.3 General case

The detection probability can be calculated as a function of R for a given value of $MT\Delta f$ by using the relationship given at the end of Section 4.6.3.1. This method was used to calculate the curve shown in Fig. 4.28 for $MT\Delta f = 50$ and $P_f = 10^{-3}$. It can be seen that, for reasonable values of the detection probability (less than 0.999), the CFAR receiver introduces a loss of the order of 1.3 dB although the signal-to-noise ratio (at the signal position) is, for a detection probability of 0.999, 1 dB higher. Although the CFAR

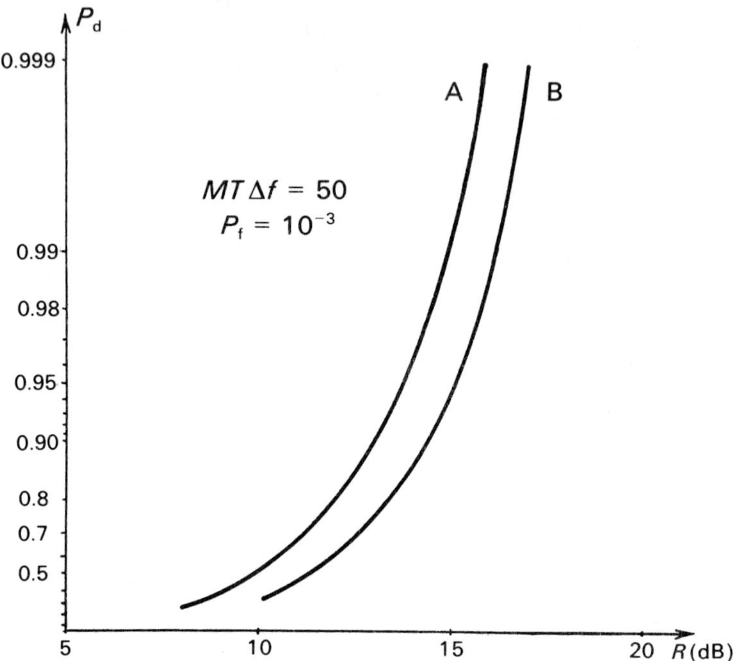

Fig. 4.28 Detection probability as a function of R for $MT\Delta f = 50$ and $P_f = 10^{-3}$: curve A, non-CFAR reception; curve B, CFAR reception.

receiver sometimes improves the signal-to-noise ratio, this is because the signal-to-noise ratio at the signal position does not just have a physical meaning. The important factor is the detection probability which is only related simply to the signal-to-noise ratio if the noise has the same characteristics in both the presence and absence of a useful signal. This is because the detection probability depends on both the signal-to-noise ratio at the signal position and the ratio of the signal to the noise outside the signal.

Remark 4.15

The assumption that $MT\Delta f$ is large enough for the noise to be gaussian is only valid if detection probabilities very close to unity or low false alarm probabilities are ignored. If the central frequency f_0 of the signal at the limiting position is not high with respect to $M\Delta f$, the losses obtained may be increased by as much as 1 dB.

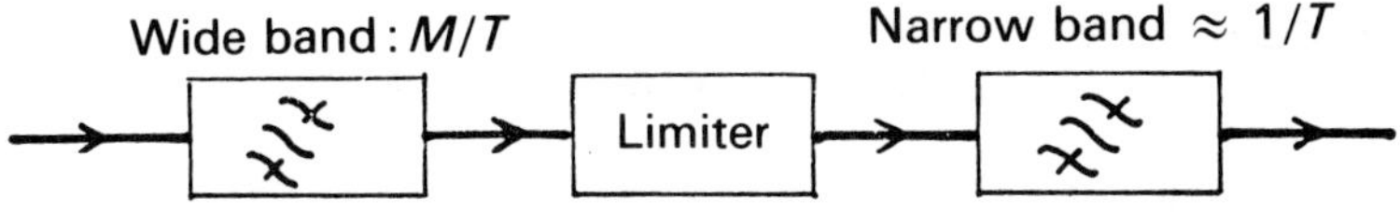

Fig. 4.29 Dicke–Fix receiver (T, pulse length).

4.6.4 Examples of applications

4.6.4.1 Classical radar

In the case of classical radar $T\Delta f \approx 1$. Then, $MT\Delta f = M$, where M is the ratio of the pass band before limiting to the matched band. A receiver produced in this way is usually called a Dicke–Fix receiver (Fig. 4.29).

4.6.4.2 Pulse compression radar

In the case of pulse compression radar with normal bandpass limiting before compression (Fig. 4.30) $M = 1$ and $T\Delta f$ is large. However, this system has a disadvantage which does not exist in the Dicke–Fix receiver. Let us assume that there are two identical targets which are very close but can be resolved (radial distance of the order of $1/\Delta f$), each giving a value R. It can be assumed without performing rigorous calculations that the power of each of the signals at the output is equal to half that which would be obtained from a single target which was twice as powerful. However, the noise power at the output at the position of the signals is almost unchanged if R is small and is halved if R is very large. This can be demonstrated using the following crude arguments. Because of the limiting, the sum of the noise power P_N and the signal power P_S corresponding to a target is equal to unity:

$$P_N + P_S = 1$$

Since

$$P_S = P_N \mathrm{SN}_1$$

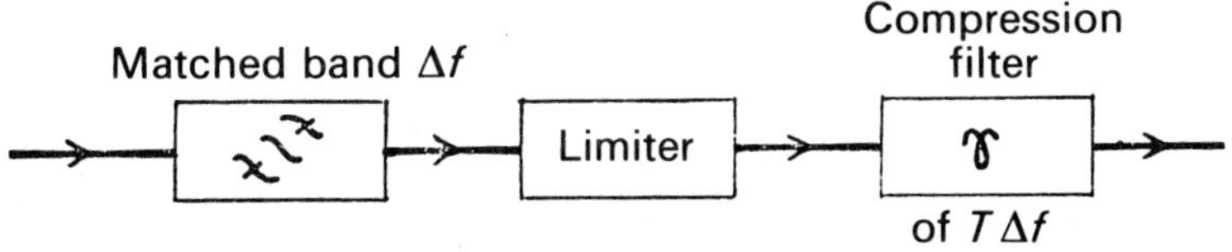

Fig. 4.30 Pulse compression radar with normal bandpass limiting before compression (frequency-modulated signals with duration T and pulse width Δf).

we can deduce that

$$P_N(1 + SN_1) = 1$$

$$P_N = \frac{1}{1 + SN_1}$$

and hence when there are two targets,

$$P_N = \frac{1}{1 + 2SN_1}$$

Therefore if $SN_1 \ll 1$, $P_N \approx 1$ in both cases. However, if $SN_1 \gg 1$, P_N in the second case is half its value for the case of a single target. Similarly we can write

$$P_u\left(1 + \frac{1}{SN_1}\right) = 1$$

$$P_u = \frac{1}{1 + 1/SN_1}$$

for one target and

$$P_u = \frac{1}{1 + 1/2SN_1}$$

for two targets (where P_u is the useful power). Then if SN_1 is small, $P_u = SN_1$ for one target and $P_u = 2SN_1$ for two targets, i.e. SN_1 for each. However, if SN_1 is large, $P_u = 1$ for one target and $P_u = 1$ for two targets, i.e. 0.5 for each. These results are shown diagrammatically in Fig. 4.31.

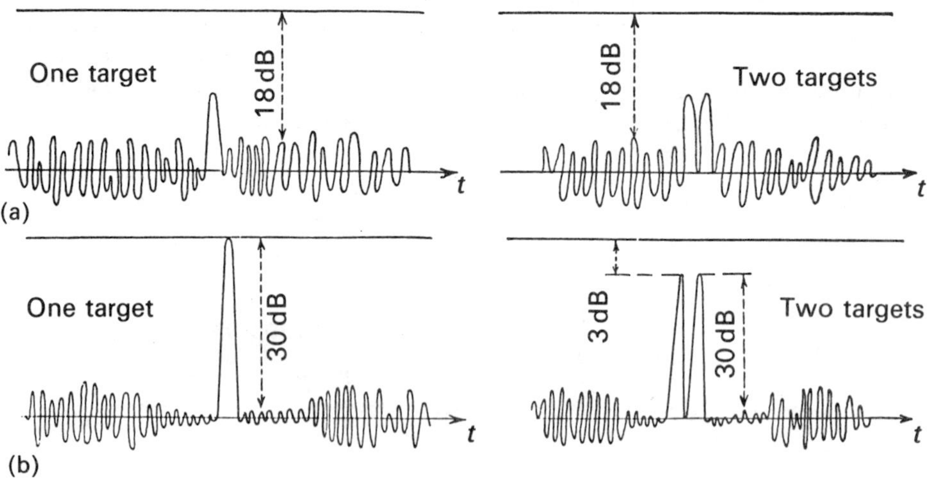

Fig. 4.31 Output for two identical targets which are very close but resolvable: (a) small SN_1; (b) large SN_1.

Table 4.3 Losses due to CFAR for two targets close together

$2MT\Delta f$	100	50	30	20	18
Loss (dB)	1.3	1.7	3	7	∞

The losses due to CFAR for a detection probability of 0.5 and a false alarm probability of the order of 10^{-3} for two targets close together are given in Table 4.3. When the targets move away from each other, the loss due to CFAR is reduced.

We now consider two targets which are very close together but have very different levels: one is very powerful and the other is a factor of B^2 less powerful ($B \gg 1$). The very powerful target will give a signal with maximum amplitude $T/2$, and the less powerful target will give a signal with amplitude limited to $T/2B$ although the noise power will be lower at the position of the targets than it would have been in the absence of the powerful target. The behavior of the system can be demonstrated using the same crude arguments as above:

$$P_{u1} = B^2 \mathrm{SN}_1 P_N \qquad P_{u2} = \mathrm{SN}_1 P_N$$

$$P_N[1 + (B^2 + 1)\mathrm{SN}_1] = 1 \qquad P_N = \frac{1}{1 + (B^2 + 1)\mathrm{SN}_1}$$

$$P_{u2}\left(1 + \frac{1}{\mathrm{SN}_1} + B^2\right) = 1 \qquad P_{u2} = \frac{1}{1 + 1/\mathrm{SN}_1 + B^2}$$

where SN_1 is the signal-to-noise ratio for the small target before limiting, P_N is the noise power after limiting, P_{u1} is the signal power of the large target after limiting and P_{u2} is the signal power of the small target after limiting. If the large target is sufficiently large, $B^2\mathrm{SN}_1 \gg 1$ and therefore

$$P_N = \frac{1}{B^2 \mathrm{SN}_1} \qquad P_{u2} = \frac{1}{B^2}$$

In the absence of the large target, we would have had

$$P_N = \frac{1}{1 + \mathrm{SN}_1} \qquad P_{u2} = \frac{\mathrm{SN}_1}{\mathrm{SN}_1 + 1}$$

Therefore a detection probability of 0.5 for the small target with $P_f \approx 10^{-3}$ is only possible if

$$\frac{T}{2B} > \frac{3}{2}\left(\frac{T}{\Delta f}\right)^{1/2}$$

$$\frac{1}{B^2} > \frac{9}{T\Delta f} \qquad B^2 < \frac{T\Delta f}{9}$$

(These relations are valid for $B^2 \gg 1$.) Table 4.4 gives the maximum value of B^2.

Table 4.4 Maximum value of B^2

$2MT\Delta f$	100	50	30
B^2 (dB)	10.5	7.5	6

Table 4.5

$2MT\Delta f$	100	50	30	20
R (dB)	14.5	15.0	15.5	20.0

If the distance between the two targets is increased, the results are improved. Thus, if they are separated by a distance $T/2$, it will always be possible to obtain a detection probability of 0.5 for the small target, even if the large target is very large, provided that the value of R for this small target is no less than the value (in decibels) given in Table 4.5 (for a false alarm probability of 10^{-3}) which corresponds to an additional loss of 3 dB with respect to the value required to obtain the same result for two identical targets very close together.

4.6.4.3 General case

When M and $T\Delta f$ are both greater than unity, the above discussion applies to both pulse compression radar and Dicke–Fix radar. Obviously, all the above discussion applies to correlation radar.

Remark 4.16

The above discussion applies exclusively to radar systems transmitting a signal with constant power throughout its duration. If the transmitted signal is amplitude modulated, the CFAR limiting causes the information contained in the amplitude to be lost and the losses in detection probability are therefore increased. Thus if we take as an example a transmitted signal with random amplitude following a gaussian distribution (similar to thermal noise), we find a loss of the order of 5 dB instead of 1 dB.

4.7 Elimination of clutter

4.7.1 Introduction

A radar system is usually designed for detecting moving targets with small dimensions (small radar cross-section) moving along the ground or at an altitude. The number of these targets is small and their echoes are often immersed in a very powerful background of parasitic echoes called clutter (which normally designates an accumulation of disordered irregular objects).

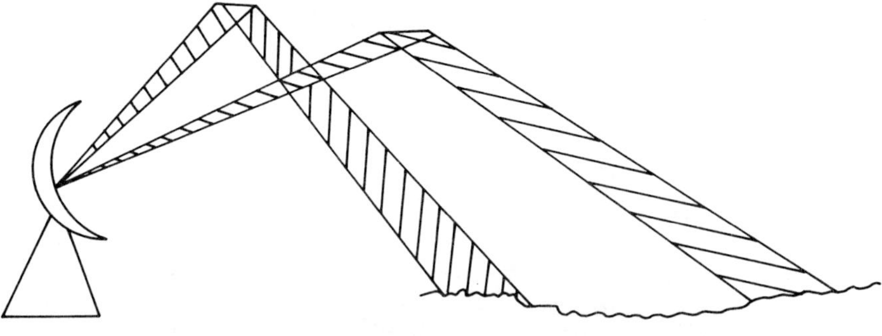

Fig. 4.32

Clutter can consist of echoes from atmospheric scatterers or echoes from objects situated on the surface of our planet.

Atmospheric scatterers are either water droplets in the form of cloud or rain or what have come to be called "angels". Angels give the impression of targets with quite considerable depth sometimes moving at quite high velocities and are probably due to abnormal reflections (caused by rapid local variations in the refractive index of the air) which return the radar beams to the ground and provide images of ground spots which appear to move at a velocity twice that of the atmosphere (wind velocity) (Fig. 4.32). A large number of angels are observed when weather conditions favor abnormal propagation, and they are larger in number and more likely to occur for short-wavelength radar.

Echoes from rain and clouds are also more powerful at short wavelengths. If all other things are equal, their radar cross-section varies as the reciprocal of the fourth power of the wavelength (the following orders of magnitude can be adopted for radar cross-sections: $5 \times 10^{-10}\,\mathrm{m}^2$ per cubic meter of rain at $\lambda = 0.1$ for rain at $1\,\mathrm{mm\,h^{-1}}$ and $2 \times 10^{-8}\,\mathrm{m}^2$ per cubic meter of rain at $\lambda = 0.1\,\mathrm{m}$ for rain at $10\,\mathrm{mm\,h^{-1}}$).

Artificial parasitic targets called chaff or window which act rather like rain can also be classified as atmospheric scatterers. Surface objects that cause clutter are also highly varied (bushes, ploughed earth, rocks, buildings, surface of the sea).

4.7.2 Restriction of the level of parasitic signals entering the radar system

It is always desirable to restrict the power of the clutter entering the radar system in order to avoid having to mount an expensive cleaning operation inside it. This commonsense approach implies a certain number of actions.

It is strongly recommended that circular polarization is used in radar transmission and reception to minimize echoes from rain or clouds. A water droplet is nearly spherical and therefore almost reverses the radar waves. This means that the wave reflected from a water droplet illuminated with circular polarization is also circularly polarized but in the opposite direction. Thus the use of the same high-quality circular polarization at transmission and reception considerably reduces clutter due to rain or cloud (a gain of 20 dB is normally obtained at $\lambda = 0.1$ m). The myth that the radar cross-section of airplanes is 3 dB smaller in circular polarization than in linear polarization is a product of a fertile imagination; the loss of range as a result of circular polarization (except against almost spherical targets) is on average zero.

A good antenna with very small sidelobes against all clutter echoes should always be used because the level of the clutter returned by these sidelobes is substantially reduced.

The elevation angle at which a radar system sees targets flying at levels which are not too low is generally larger than the elevation at which the Earth's surface is seen. Therefore it is possible to produce an antenna whose gain in the direction of the ground is significantly less (by M dB) than the gain in useful directions. This type of antenna is said to have good low angle cut-off, and it is perfectly clear that ground or sea clutter will be reduced by $2M$ dB with respect to useful targets. Thus two beams are often used for scanning surveillance radar systems: one, called low coverage, illuminates ground targets with a gain close to the maximum and is used for detection at low elevation angles and long range; the second beam, called high coverage, has a considerable low angle cut-off and is used to detect targets with high elevation angles (nearby targets or targets far away flying at high

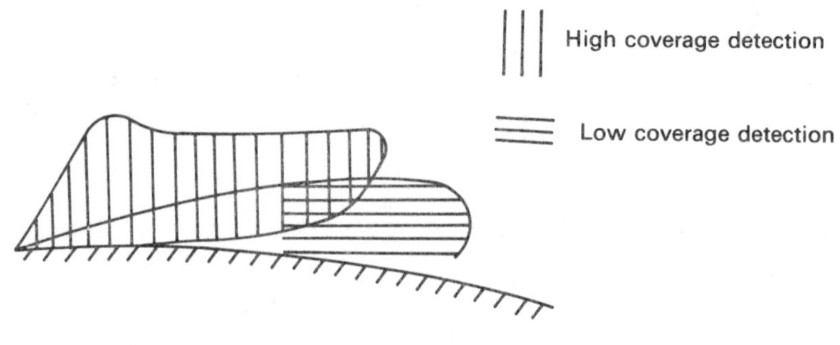

Fig. 4.33

altitude) (Fig. 4.33). This two-beam technique is sometimes called area target indication (ATI).

4.7.2.1 Reduction of the resolution cell

Parasitic echoes which interfere with the detection of a given useful target are those which cannot be resolved in terms of range and angle. In other words, if the range resolution of the radar is denoted by τ (with all the lack of accuracy with which τ may be defined) and the azimuth width (at 3 dB) of the beam of a scanning radar is denoted by $\theta_{3\,\mathrm{dB}}$, we can, to a first approximation, define the resolution cell C of the radar by

$$C \triangleq \theta_{3\,\mathrm{dB}} D \tau$$

where $\theta_{3\,\mathrm{dB}}$ is in radians and D is the range of the useful target. In addition, the disturbing power of the clutter is defined as that of the parasitic echoes located in C. It is obvious that if the clutter is diffuse, i.e. if there are a very large number of clutter echoes in C, the power of the clutter will be proportional to C:

$$\sigma_{\mathrm{cl}} = \sigma_0 C \tag{4.3}$$

where σ_{cl} is the radar cross-section of the clutter and σ_0 is a coefficient (square meters of radar cross-section per square meter of area) whose value depends on a large number of factors (type of clutter, average angle of incidence of the radar beams on the clutter surface). Clutter from countryside with few woods and valleys has an order-of-magnitude range of -20 to -40 dB. If the disturbing clutter is atmospheric, it is necessary to define the resolution volume λ. In the case of a conical beam with aperture α at 3 dB, we have

$$\lambda \triangleq \alpha^2 D^2 \tau$$

and of course

$$\sigma_{\mathrm{cl}} = \sigma_0' \lambda \tag{4.4}$$

(Order-of-magnitude values of σ_0' for rain were given in Section 4.7.1.)

Equations (4.3) and (4.4) clearly show all the advantages to be gained by using a resolution cell with reduced dimensions for combating clutter. For example, if we compare radar A with $\tau = 20\,\mu\mathrm{s}$ operating in L band ($\lambda = 0.25$ m) with radar B with $\tau = 0.5\,\mu\mathrm{s}$ operating in S band ($\lambda = 0.1$ m) (the two radars have the same antenna reflector), the resolution cell of radar A will be 20 dB larger than that of radar B. This means that, in the case of diffuse clutter, the power of the clutter will be 20 dB lower for radar B than for radar A, all other things being equal (distance, power etc.).

In practice, however, clutter can rarely be considered as diffuse and the practical benefit obtained by a reduction in the resolution cell is much greater than the benefit resulting from the use of eqns (4.3) or (4.4). When the clutter is not sufficiently diffuse (which means that on average there are not enough clutter-independent echoes in C), the average value of σ_{cl} is still proportional to C:

$$\bar{\sigma}_{cl} = \sigma_0 C$$

However, this average value has no immediate physical meaning as the following example shows.

Let us assume that we have a line of towers, each with a radar cross-section of $10^7\,m^2$ in the azimuth beam of the radar, and that the radial distance between two adjacent towers is 600 m. If the range resolution is 300 m, whatever the position of the useful target, it will always be mixed with a fixed target of radar cross-section $10^7\,m^2$: $\sigma_{cl} = 10^7\,m^2$. It will only be possible to detect a useful target with a radar cross-section of $1\,m^2$ if the rejection rate of fixed targets is higher than 70 dB. If the resolution cell is reduced to $\tau = 30$ m (0.2 μs), there is a 90% chance that the useful target is not mixed with clutter ($\sigma_{cl} = 0$) and a 10% chance that it is ($\sigma_{cl} = 10^7\,m^2$) so that the average value of the clutter is given by $(0.9 \times 0 + 0.1 \times 10^7) = 10^6\,m^2$ (reduced by a factor of 10 because τ is reduced by a factor of 10). However, it is clear that there is no need to use a high performance system to eliminate the parasitic echo if we can be satisfied with a detection probability of less than 0.9. Thus we can say that the reduction of σ from 2 μs to 0.2 μs has caused 70 dB to be gained on the clutter in this particular case.

4.7.3 The main systems using Doppler filtering

4.7.3.1 General principles

Spectrum of the signal received with a fixed antenna

Although the width Δf of the spectrum transmitted by radar is small compared with the transmission frequency, the frequency of the signal received by a fixed antenna from a target with a radial velocity V_r is shifted with respect to the transmitted signal by a quantity called the Doppler frequency:

$$f_D = \frac{2V_r}{\lambda}$$

Thus if the transmission signal is a periodic signal with repetition frequency f_R, the transmitted signal has a spectrum consisting of a large number of lines (approximately $\Delta f / f_R$, i.e. $1/\tau f_R$, which is equal to the reciprocal of the duty cycle of the radar if the transmitted pulses are not frequency modulated) separated from one another by f_R (Fig. 4.34, full lines).

A signal from a fixed target received by an antenna which is assumed to be fixed is obviously of the same type, whereas a signal from a moving target with velocity V_r will have a spectrum which has been frequency shifted by f_D,

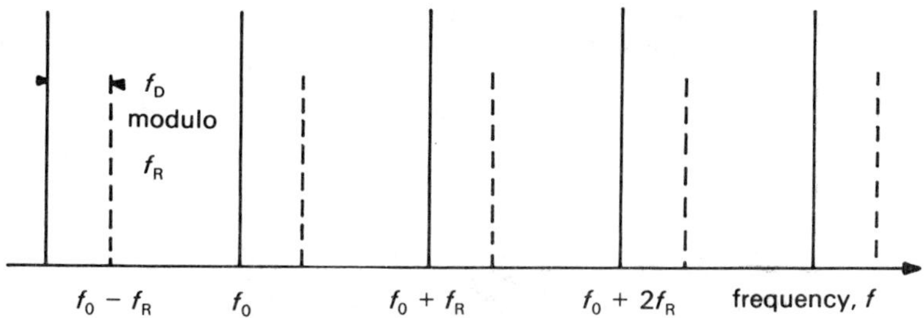

Fig. 4.34 Signal received by a fixed antenna.

i.e. in practice it will consist of the same lines but shifted by f_D modulo f_R (Fig. 4.34, broken lines). Thus, it is obvious that in order to eliminate fixed targets while preserving moving targets, it is sufficient to filter the received signal in a system (comb filter) which eliminates frequencies equal to f_0 modulo f_R and preserves all other frequencies. It is then clear that if the radial velocity of the moving target is such that $f_D = 0$ modulo f_R, it will also be eliminated (this velocity is known as the blind speed).

Spectrum of the signal received with a rotating antenna

If the antenna is not fixed, in particular if we are dealing with a scanning radar, the signal received from both a fixed target and a moving target is that which would have been received by the fixed antenna but with its amplitude modulated. Thus the signal received from either a fixed or moving target is no longer composed of lines but of "broadened lines". In the case of a scanning radar, where the gain of the antenna is a function $g(\theta)$ of the azimuth θ and the gain in the corresponding power is given by $G(\theta) = g^2(\theta)$, the amplitude of the signal received from a fixed target varies as $G(vt)$, where v is the velocity of rotation of the antenna (in order to avoid errors θ is expressed in radians and v in radians per second). The illumination at the aperture of the antenna system, which is generally a reflector, is denoted by $A(v)$, i.e. it is expressed as a function of its spatial frequency $v = x/\lambda$ where λ is the radar wavelength (cf. Chapter 6, Section 6.2) (Fig. 4.35). It is known that $g(\theta)$ is the Fourier transform of $A(v)$ and therefore that $G(\theta)$ is the Fourier transform of the autocorrelation function $\varrho_A(v)$ of $A(v)$. The Fourier transform of $G(vt)$ is therefore $\varrho_A(f/v)$ multiplied by a factor.

This makes it easy to calculate the shape of the broadened line. If the antenna is uniformly illuminated, ($A(v) = 1$ for $|v| < L/\lambda$, where $2L$ is the span of the antenna, and $A(v) = 0$ for $|v| > L/\lambda$), the function $\varrho_A(v)$ is triangular and so are the "broadened lines" returned by a fixed or moving target (Fig. 4.36). The width of the base of the triangle is $2f_c$, where f_c is the

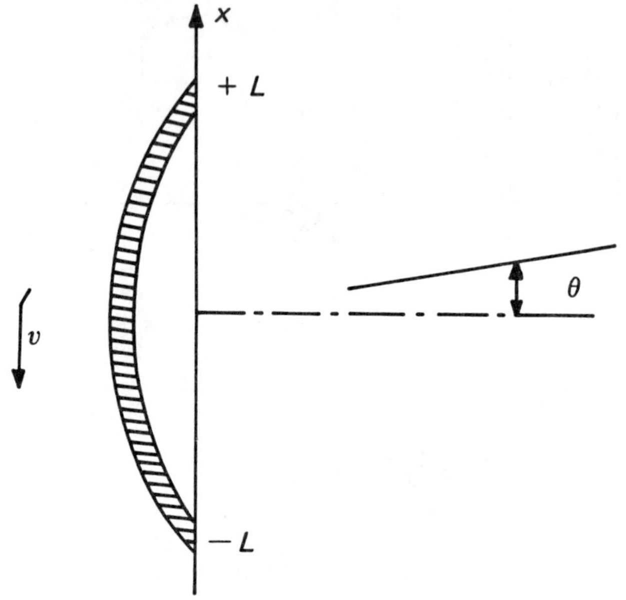

Fig. 4.35 Diagrammatic representation of a moving antenna.

cut-off frequency of the ground clutter and is given by

$$f_c = \frac{2L}{\lambda} v$$

When the antenna has a span of $2L$ and is illuminated nonuniformly a signal from a fixed target still has a zero Fourier transform outside the intervals

$$[f_0 - f_c, f_0 + f_c] \text{ modulo } f_R$$

Similarly, a signal from a moving target has a zero Fourier transform outside the intervals

$$[f_0 - f_c + f_D, f_0 + f_c + f_D] \text{ modulo } f_R$$

An intuitive argument leads to the same result: in fact, a fixed target is only fixed with respect to the axis of rotation of the antenna since the latter is rotating. With respect to the ends of the antenna (which give the highest relative velocities), a distant fixed target has a relative radial velocity vL (end-of-antenna velocity) corresponding to a maximum Doppler frequency of $2vL/\lambda = f_c$.

Examination of Fig. 4.36 shows that in order to eliminate signals from fixed targets it is necessary to filter the received signal in a system which

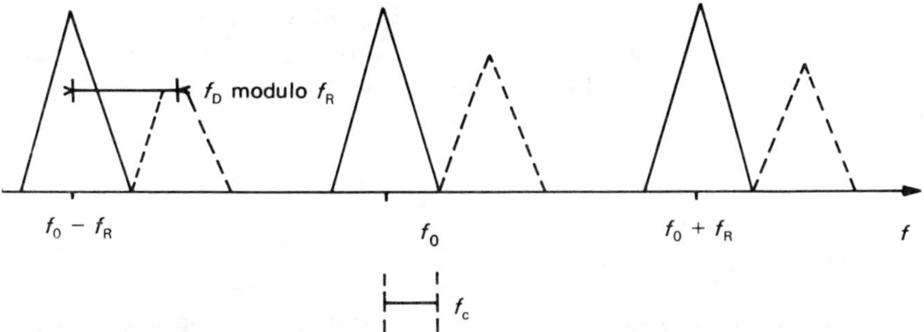

Fig. 4.36 Signal received by a uniformly illuminated rotating antenna: ——, Fourier transform of the signal from a fixed target; – – –, Fourier transform of the signal from a moving target.

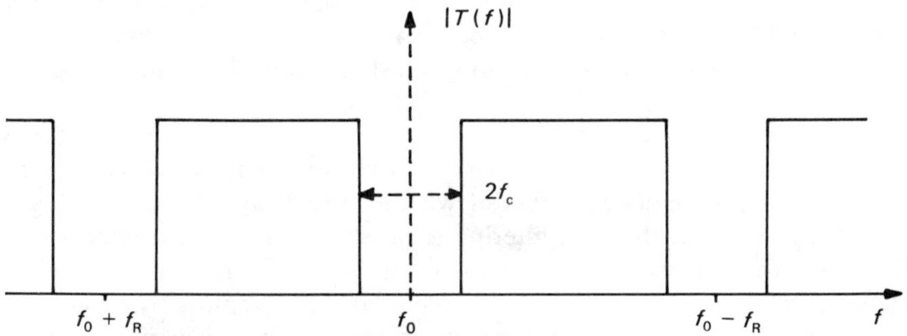

Fig. 4.37 Response $T(f)$ of a system which eliminates frequencies lying between $f_0 - f_c$ and $f_0 + f_c$.

discards frequencies lying between $f_0 - f_c$ and $f_0 + f_c$ (modulo f_R). Figure 4.37 shows the response $T(f)$ of such a system. It follows very clearly that the signal from a moving target whose Doppler frequency f_0 (modulo f_R) is less than $2f_c$ will be partially eliminated, and thus will increase and aggravate the blind speed phenomenon.

Elimination of blind speeds

This dreadful expression (which, however, has become commonplace) designates the attempts which are made to avoid eliminating targets whose speeds are close to blind speeds. The general idea is always as follows: different repetition frequencies f_{R1}, f_{R2}, f_{R3} are used so that the targets eliminated with f_{R1} are not eliminated when f_{R2} and/or f_{R3} is used. This can be done either by transmitting trains of pulses first at the rate f_{R1}, then at the rate f_{R2} and finally at the rate f_{R3} or by sweeping the repetition frequency so that two consecutive intervals between transmitted pulses are always unequal. The first procedure is often preferred when it is feasible since it also enables signals obtained from fixed targets during the second scan to be eliminated (the radar range of these targets is greater than the repetition period).

Intermediate frequency and homodyne filtering

The elimination of signals from fixed targets by filtering is almost never possible for the received signal at the input of the radar but it can be achieved for the signal after frequency conversion, i.e. for the received signal at intermediate frequency. Technological problems are encountered with the filtering process if the intermediate frequency is high. In fact, in order to have comb filtering, i.e. to provide a response which is periodic in frequency (with period f_R), delay lines must be used. This is because the transfer function $F(p)$ of the filter, where p is the Laplace operator, must reproduce itself identically when p $(=2\pi \mathrm{j} f)$ is replaced by $p + 2\pi \mathrm{j} f_R$, i.e. it must have the form $G[\exp(-p/f_R)]$ and the filter must be a combination of delay lines giving a multiple delay of $1/f_R$. In what follows we shall adopt the common practice of replacing $\exp(+p/f_R)$ by z to simplify the notation.

We denote the intermediate frequency by f_I. The delay lines used must provide a delay of $1/f_R$ with an error much less than $1/f_I$, which is therefore easier when f_I is small regardless of whether the delay lines are analog or digital. For most of the time filtering is carried out for a zero intermediate frequency (homodyne filtering), i.e. the received signal corresponding to transmission centered on f_0 is either demodulated at reception by a frequency f_0 synchronous with the central transmission frequency or converted to an intermediate frequency and then demodulated for this intermediate frequency. We denote the carrier of the transmission signal by $\sin(2\pi f_0 t)$; the signal corresponding to it at reception is

$$\sin[2\pi(f_0 + f_D)t + \varphi_1]$$

where f_D is the Doppler frequency corresponding to the radial velocity of the target. Demodulation consists in multiplying this received signal by $\sin(2\pi f_0 t)$ to give

$$\tfrac{1}{2}[\cos(2\pi f_D t + \varphi_1)]$$

if we neglect the term in $2f_0 + f_D$ which is certainly eliminated. It can clearly be seen that after demodulation the signal is zero or very weak whenever

$$2\pi f_D t + \varphi_1 \approx \pi/2$$

This is the blind phase phenomenon which simply corresponds to the fact that, after homodyne demodulation, the useful received signal from a moving target appears as a sequence of pulses whose amplitude varies sinusoidally at the Doppler frequency and that a sinusoidal wave has to pass through zero

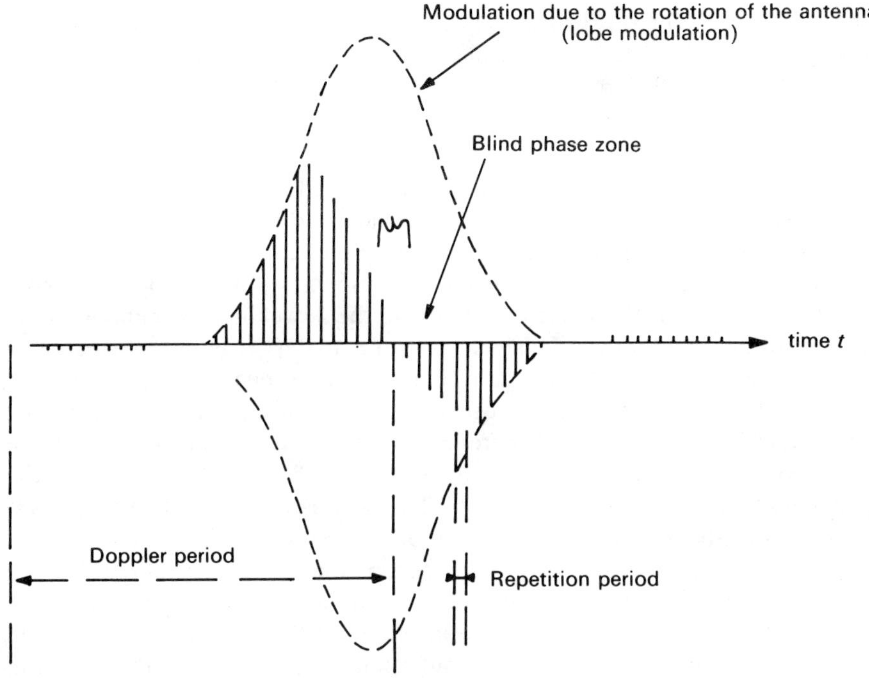

Fig. 4.38 Blind phase phenomenon.

every half-period. Figure 4.38 shows the shape of this sequence of pulses for a low Doppler frequency with respect to the repetition frequency of the radar, taking into account the modulation due to the rotation of the antenna on a scanning radar (13 pulses in the beam at 3 dB). The disadvantage of this phenomenon is obvious: it clearly contributes to a reduction in the sensitivity of the radar to certain moving targets.

In order to overcome this handicap, it is necessary in some cases to double the chain by using in-phase demodulation (multiplication by $\sin(2\pi f_0 t)$ and quadrature demodulation (multiplication by $\cos(2\pi f_0 t)$) in parallel. These two parallel channels are not simultaneously disturbed by the blind phases in a significant manner. When it is not possible to obtain two parallel channels which are identical (analog and nondigital processing), it is necessary to use a single-sideband mixer which also has two outputs (the targets are sorted in terms of their radial velocity).

Coherence of the radar

In the final analysis, Doppler filtering consists in measuring the variation in the phase of the received signal, pulse by pulse. It is obvious that if the frequencies used in the processing have no phase relationship with each other, Doppler filtering is impossible. In practice, two solutions are adopted to ensure that all the frequencies used have a good phase relationship and that the radar is coherent.

(1) All the frequencies required are generated at low level by synthesis from a single oscillator so that they are synchronous, and the transmission frequency is then amplified in a power amplifier chain. For example, it is possible to use a very stable microwave oscillator, whose output signal supplies both the local oscillator system and the power amplifier chain, in the case of a homodyne radar. It is also possible to start from a low frequency obtained using a quartz oscillator, which is multiplied and amplified in semiconductor circuits until a frequency $f_0 + f_I$ equal to the local oscillator frequency f_{LO} is obtained. This frequency, which is synchronous with the intermediate frequency f_I provided by another stable oscillator, supplies the transmission frequency f_0 to a mixer. The signal is then amplified in a chain of amplifier tubes (see Fig. 4.39). In a more economic version, which is often employed, a self-excited oscillator (triode or magnetron oscillator) is used instead of the power amplifier chain. The transmission frequency f_0 is injected into this oscillator at a relatively low level, and if this procedure is managed well the transmission phase is brought under the control of the injection phase. This is called a coherent auto-oscillator (CAO).

(2) When the transmitter is a noncoherent self-excited oscillator and when (for financial reasons) the use of an amplifier transmission chain is not an option, the phase transmitted by the transmitter normally varies randomly from pulse to pulse. This fault can be overcome by using the system shown in Fig. 4.40. As each pulse is transmitted a sample of the transmission frequency f_0 generated by the magnetron is mixed with the local oscillator frequency generated by a very stable microwave oscillator (STALO), and the pulse obtained in this way adjusts the phase of the intermediate frequency generated by a second oscillator, thus making it coherent. This oscillator,

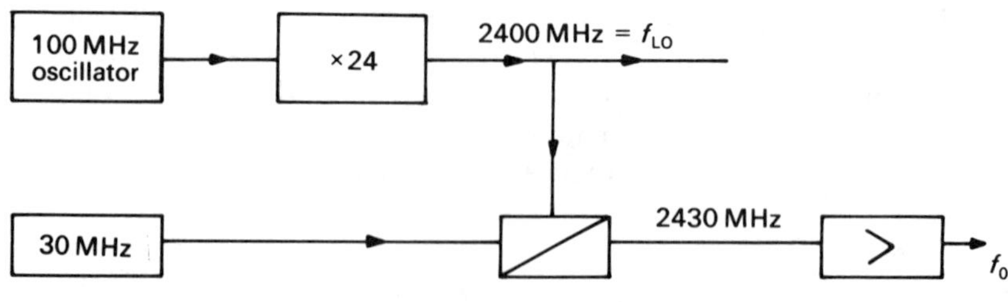

Fig. 4.39 Coherent transmitter.

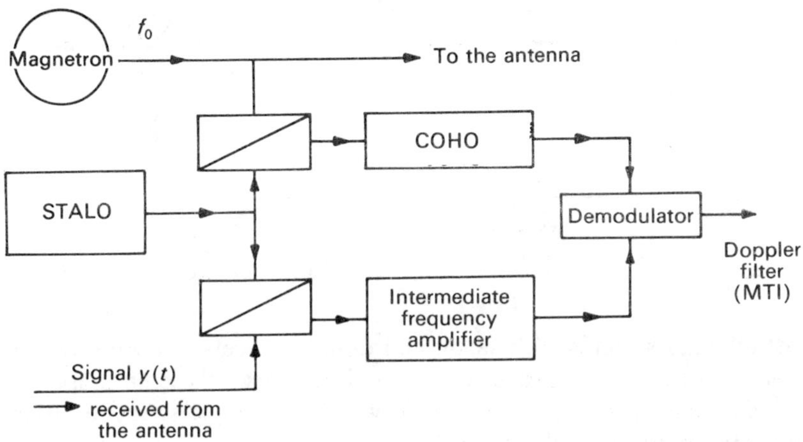

Fig. 4.40 Incoherent oscillator transmitter.

which is known as the coherent oscillator (COHO), maintains the same reference phase throughout the next repetition period. Therefore the intermediate frequency provided by the COHO is used to demodulate the intermediate frequency signal of the normal radar channel. This system has an obvious handicap: only signals received in a time interval shorter than the repetition period benefit from the coherence. This means that very powerful signals from fixed targets located further away than the distance corresponding to the repetition period (first ambiguous distance) cannot be eliminated by Doppler filtering. In other words, the fixed targets of the second (and *a fortiori* third) scan are not eliminated.

4.7.3.2 Principal systems for eliminating fixed targets for Doppler filtering

There are basically two types of moving target indicator (MTI) systems: delay line MTIs and range-gated MTIs.

Delay line MTIs

These are filters whose transfer function is such that the transmittance is similar to that shown in Fig. 4.37. Therefore delay lines are essential.

The first MTIs were analog devices which used ultrasonic delay lines (mercury lines and then quartz lines) providing an approximately constant fixed delay (i.e. constant in terms of the pulse duration). The signal to be delayed is often modulated by a carrier (e.g. at 20 MHz), and this modulated signal, which is delayed in the form of an ultrasonic wave, is then reconverted to an electric signal and demodulated in order to eliminate the carrier.

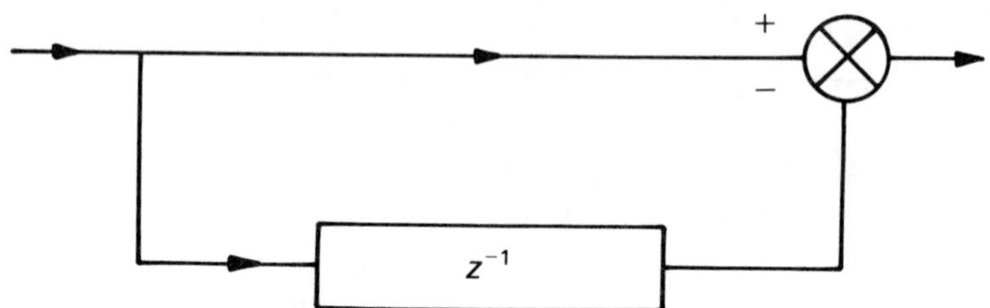

Fig. 4.41 Diagrammatic representation of a single-cancelation MTI.

Almost all MTI systems with single and double cancelation use such delay lines. An obvious disadvantage is that the delay which these lines provide is fixed and equal to the repetition period which cannot be swept in order to eliminate the problem of blind speeds. This drawback can be overcome either by obtaining several delays using delay lines in series, some of which may be short-circuited (staggered pulse repetition frequency MTIs), or by using an aperiodic delay unit consisting of a storage tube providing a delay time which can be controlled.

Figure 4.41 shows a diagram of a single-cancelation MTI in which the delay T_R is denoted z^{-1}, where

$$z^{-1} = \exp(-T_R p)$$

The transfer function of such a filter can be written as

$$F(p) = 1 - z^{-1} = 1 - \exp(-T_R p)$$

which gives a response of

$$T(f) = 1 - \exp(-2\pi j T_R f)$$

where $T_R = 1/f_R$. Thus

$$|T(f)| = \left\{\left[1 - \cos\left(2\pi \frac{f}{f_R}\right)\right]^2 + \sin^2\left(2\pi \frac{f}{f_R}\right)\right\}^{1/2}$$

$$= 2\left|\sin\left(\pi \frac{f}{f_R}\right)\right|$$

The behavior of $|T(f)|$ is shown in Fig. 4.42 and by the dotted curve in Fig. 4.43. Comparison of this curve with the ideal curve (Fig. 4.37) shows that the result obtained is not very good.

The elimination of fixed targets is improved by using two single-canceler MTIs in cascade, thus obtaining a double-canceler MTI (see Fig. 4.44) whose transfer function is obviously

$$(1 - z^{-1})^2$$

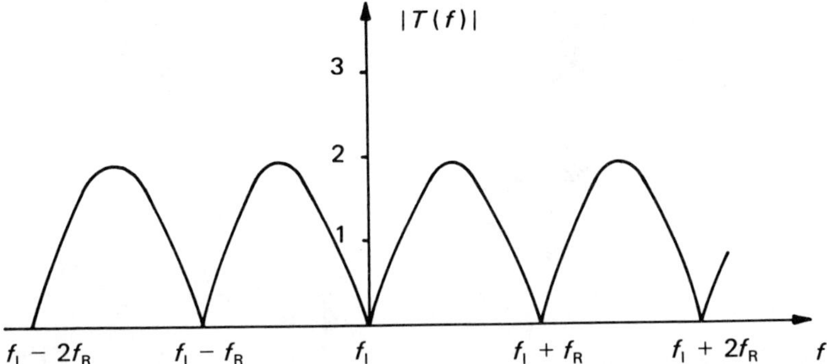

Fig. 4.42 Variation of $|T(f)|$ with f for a single-canceler MTI.

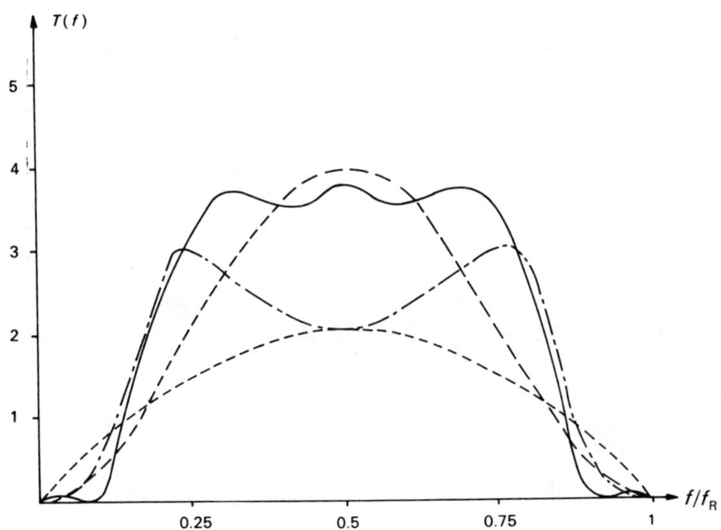

Fig. 4.43 Variation of $|T(f)|$ with f/f_R: - - - - -, single-cancelation MTI; — — — double-cancelation MTI; – · –, double-cancelation MTI with feedback where

$$T(z^{-1}) = \frac{(1 - z^{-1})^{1/2}}{1 - 0.5z^{-1} + 0.5z^{-2}}$$

——, linear MTI characterized by

$$T(z^{-1}) = \frac{(1 - z^{-1})(1 - 1.732z^{-1} + z^{-2})}{1 - 0.5z^{-1} + 0.5z^{-2}}$$

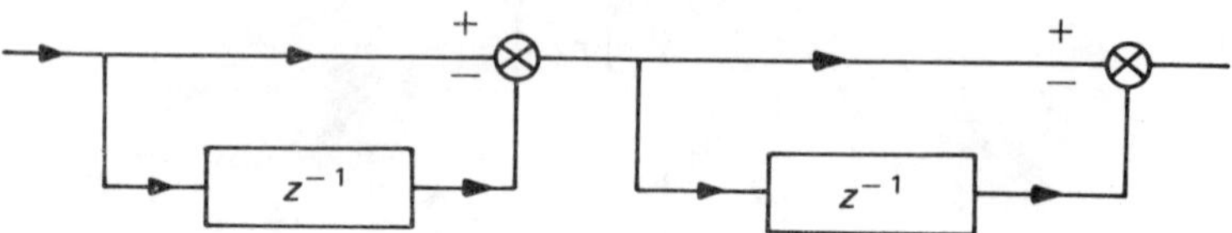

Fig. 4.44 Diagrammatic representation of a double-canceler MTI.

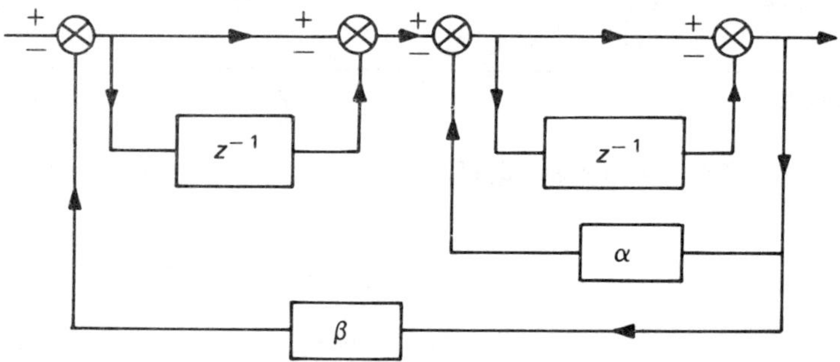

Fig. 4.45 Diagrammatic representation of a double-canceler MTI with feedback.

and whose response is

$$T(f) = 4 \sin^2\left(\pi \frac{f}{f_R}\right)$$

The behavior of $|T(f)|$ is shown by the broken line in Fig. 4.43. The gain of the filter is reduced around the frequency f_I (modulo f_R) which clearly improves the elimination of fixed targets. Unfortunately, it is also low at frequencies far removed from f_I which produces an appreciable attenuation in the number of moving targets.

This drawback can be overcome by introducing feedback as shown in Fig. 4.45 in which α and β are multiplying coefficients. If α and β are suitably chosen, the curve of $|T(f)|$ against f is unchanged for f close to f_D (very low) and is improved for f far from f_D. The transfer function of this filter is

$$\frac{(1 - z^{-1})^2}{(1 + \alpha + \beta) - z^{-1}(\alpha + 2\beta) + \beta z^{-2}}$$

We usually say that it has two zeros $z^{-1} = 1$ (in fact it has a double zero) and two poles

$$z^{-1} = 2 + \frac{\alpha}{3} \pm \left(\frac{\alpha^2}{\beta^2} - \frac{4}{\beta}\right)^{1/2}$$

The zeros are true zeros since they make $|T(f)|$ zero. The existence of the poles results in the appearance of two maxima in the curve of $|T(f)|$ and therefore it can be adjusted more rapidly if they are suitably chosen. The behavior of $|T(f)|$ for $\alpha = -0.5$ and $\beta = +0.5$ is shown by the chain curve in Fig. 4.43.

It is tempting to assume that $|T(f)|$ could be improved even further in order to bring it nearer to the ideal shown in Fig. 4.37 by using triple- or quadruple-canceler analog MTIs (with feedback), and this has been done. However, the benefit obtained is generally illusory because the quality of the analog calculation is inadequate. Therefore, if a high quality MTI is required, i.e. if the filtering calculation is required to be very accurate, it is necessary to use a hard-wired digital computer, specially constructed for the particular requirements, to carry out the processing. In this case, it is obviously necessary to transform the analog signal to be processed to a digital signal in an analog–digital converter and to replace the delay lines by shift registers. The sweeping of the repetition frequency is then very convenient because the delay of a shift register can be changed by changing the clock frequency controlling it. The coding accuracy (number of bits) will be adequate taking into account the accuracy of the calculation, i.e. taking into account the desired feasible rejection rate. It is necessary to code to at least 8 bits if we wish to have an $8 \times 6 = 48$ dB cancelation ratio (plus the sign bit and one or two security bits).

Figure 4.46 shows a block diagram of a digital filter in the canonical representation. It corresponds to the transfer function

$$\frac{V_o}{V_i} = \frac{1 + a_1 z^{-1} + a_2 z^{-2} + \cdots + a_n z^{-n}}{1 + b_1 z^{-1} + b_2 z^{-2} + \cdots + b_n z^{-n}}$$

which has n zeros and n poles. If the coefficients a_i (which are real) are suitably chosen, the zeros will be either of the form $z^{-1} = 1$ or of the form $z^{-1} = \exp(\pm j\Psi)$. The zeros of the first family will cause $|T(f)|$ to vanish

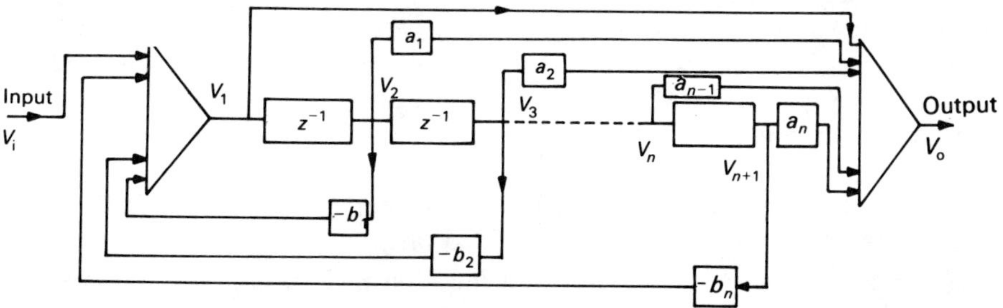

Fig. 4.46 Canonical representation of a digital filter.

(once or several times depending on their number) at $f = f_I$ (modulo f_R), where f_I is the intermediate frequency. The zeros of the second family (which come in pairs) cause $T(f)$ to vanish at

$$f = f_I \pm \frac{\Psi}{2\pi} f_R \text{ (modulo } f_R)$$

The poles make it possible to modify the form of the response $|T(f)|$.

It is easy to show that the more memories that are used, the closer $|T(f)|$ approaches the ideal form shown in Fig. 4.37. For example, the transmittance of a digital MTI with three memories (three zeros and two poles) is shown by the full curve in Fig. 4.43. However, if there are more than two shift registers, implementation in the canonical form is generally not recommended because the shape of the curve $T(f)$ obtained is very sensitive to slight errors in the value of the coefficients. For example, the transfer function

$$\frac{1 - \sqrt{3}z^{-1} + z^{-2}}{1 - 0.5z^{-1} + 0.5z^{-2}}$$

can certainly be obtained using the canonical diagram of Fig. 4.46 with the following values for the coefficients: $a_1 = \sqrt{3}$, $a_2 = 1$, $b_1 = 0.5$ and $b_2 = 0.5$. However, it is preferable to use the intermediate variables v_1 and v_2 in the algorithms

$$v_1(1 - z^{-1}) + v_2(1.5 - 0.5z^{-1}) = V_i$$

$$v_2(1 - z^{-1}) = v_1 z^{-1}$$

$$V_o = v_1(1 - z^{-1}) + v_2(2 - \sqrt{3})$$

because it is permissible to define the coefficient $2 - \sqrt{3}$ with a relative acurracy six times poorer than that required for the coefficient $\sqrt{3}$ of the canonical implementation. (These algorithms are due to J.E. Picquendar and M.G. de Coligny.)

Range-gated MTIs

In range-gated MTIs a filter whose response is as shown in Fig. 4.47, i.e. it is zero for $f > f_I + f_R$, is used instead of a filter whose response is periodic in f with period f_R. The MTI filter only allows a single "line" of the spectrum received from a useful signal tc pass through it. This has two consequences:

(a) the filter is easy to implement (and is therefore economical), even for obtaining very good cancelation for fixed targets;

(b) range measurement is lost during filtering because only a single line is received, and it is necessary to receive a large number of lines in order to have a valid range measurement.

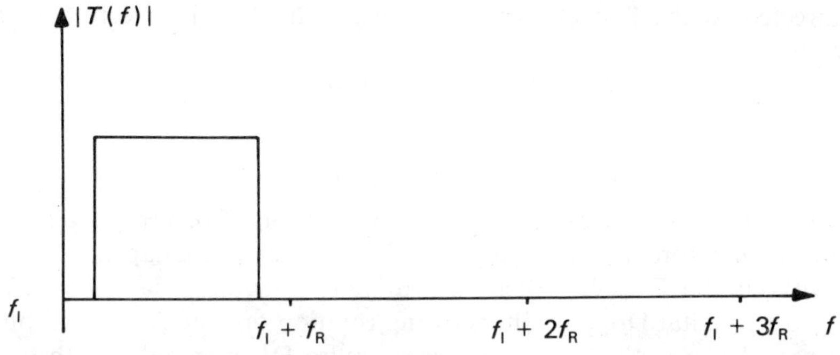

Fig. 4.47

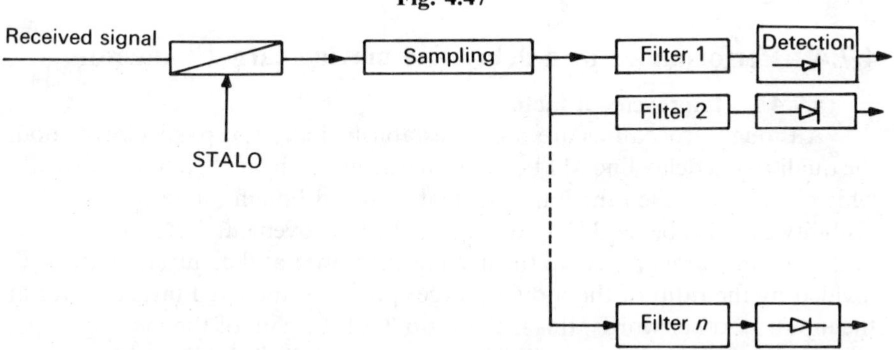

Fig. 4.48 Diagrammatic representation of a range-gated MTI.

In order to avoid this loss of range measurement, which is unacceptable in most cases, it is necessary to measure the range before the Doppler filtering is performed (Fig. 4.48). This is done by separating the received signals (if necessary after pulse compression) by range gates of the same width as the resolution cell τ. Thus if the starting times of the signals at zero range are denoted $t_0, t_1, t_2, \ldots$ (if the repetition frequency is constant and equal to f_R),

$$t_1 = t_0 + \frac{1}{f_R}$$

$$t_2 = t_1 + \frac{1}{f_R} \text{ etc.}$$

The received signals between

$$t_0 \text{ and } t_0 + \mathscr{T}$$

$$t_1 \text{ and } t_1 + \mathscr{T}$$

$$t_2 \text{ and } t_2 + \mathscr{T}$$

are directed to the first clutter filter (range gate 1). The received signals between

$$t_0 + (n - 1)\mathscr{T} \text{ and } t_0 + n\mathscr{T}$$

$$t_1 + (n - 1)\mathscr{T} \text{ and } t_1 + n\mathscr{T}$$

$$t_2 + (n - 1)\mathscr{T} \text{ and } t_2 + n\mathscr{T}$$

are directed to the *n*th clutter filter (range gate *n*). The range-gated MTI receiver is therefore a pulse Doppler radar receiver (see Chapter 4, Section 4.2) without the Doppler filter. Similarly, if a digital delay line MTI is followed by digital Doppler filters (using rotation operators or fast Fourier transform devices), the radar becomes a pulse Doppler radar with digital processing.

4.7.4 Performance of a delay line moving target indicator

4.7.4.1 Improvement factor

Although procedures are poorly established and not particularly good, the quality of a delay line MTI can be characterized by its "gain in contrast", which is usually called the improvement factor although the term subclutter visibility can also be used [12]. We define the improvement factor as the ratio of the moving target power to the fixed target power at the output of the MTI divided by the ratio of the moving target power to the fixed target power at the input. In other words, this is the ratio T_i of the gain of the moving target to the gain of the fixed target, which is expressed in decibels. The improvement factor depends on a large number of parameters, of which the most important are the velocity of the moving target, the type of MTI being used (single canceler, double canceler with or without feedback, digital linear MTI), the type of clutter, the width of the spectrum of the fixed targets (in other words the number of pulses in the beam at 3 dB for a scanning surveillance radar) and the stability of the transmitter–receiver.

4.7.4.2 Velocity of the moving target

It has been seen that the response of an MTI varies with the Doppler frequency, and that the best response with a single-canceler or double-canceler MTI without feedback was obtained for Doppler frequencies equal to $f_R/2$ (modulo f_R) which are known as optimal frequencies (optimal velocities). Designers often give an approximate value of the improvement factor for the optimal velocity.

4.7.4.3 Type of moving target indicator and number of pulses per transit over the target (limitations due to scanning)

It is clear that the wider is the spectrum of the fixed targets with respect to the repetition frequency f_R, the poorer will be the visibility level. In the case of a scanning surveillance radar the visibility level increases with the number of pulses obtained when the beam (defined at 3 dB) passes over the target. The number of pulses is given by

$$n = \frac{\alpha}{\Delta\theta} = \frac{\alpha f_R}{v}$$

where $\Delta\theta$ is the angle through which the antenna turns during a repetition period, α is the width of the beam (in azimuth) of a scanning radar at 3 dB and v is the velocity of rotation of the antenna. For example, if $\alpha = 1.5°$, $f_R = 1500$ Hz and $v = 90°\,\mathrm{s}^{-1}$ (15 rev min^{-1}), $n = 25$.

Diffuse clutter and single-canceler MTI

Let us first assume that the clutter is diffuse, i.e. in a given range interval corresponding to the resolution cell of the radar we receive a very high number of fixed target signals with all orientations. An isolated fixed target would generate a pulse train with Fourier transform $\varphi(f)$. A very large number of fixed targets located in all directions generate a signal with a noise structure whose spectral density is of the form $k|\varphi(f)|^2$, where k depends on the power of the clutter.

The power of the clutter before the MTI is given by

$$k \int |\varphi(f)|^2 \, \mathrm{d}f$$

and the power of the clutter after the MTI is given by

$$k \int |\varphi(f)|^2 |T(f)|^2 \, \mathrm{d}f$$

where $|T(f)|$ is the modulus of the response of the filter constituting the linear MTI. The gain for fixed targets, for given periodicity of $\varphi(f)$ and $T(f)$, can be written as

$$G_{FT} = \frac{\int_0^{f_R/2} |\varphi(f)|^2 |T(f)|^2 \, \mathrm{d}f}{\int_0^{f_R/2} |\varphi(f)|^2 \, \mathrm{d}f} \tag{4.5}$$

In the case of a single-canceler MTI

$$|T(f)| = 2\left|\sin\left(\pi \frac{f}{f_R}\right)\right|$$

which gives

$$G_{FT} = \frac{4\int_0^{f_R/2} |\varphi(f)|^2 \sin^2(\pi f/f_R) \, \mathrm{d}f}{\int_0^{f_R/2} |\varphi(f)|^2 \, \mathrm{d}f}$$

This expression for G_{FT} can be simplified as soon as n is sufficiently large (i.e. when $|\varphi(f)|$ is only significant for $f \ll f_R$) by replacing

$$\sin^2\left(\pi \frac{f}{f_R}\right) \text{ by } \frac{\pi^2}{f_R{}^2} f^2$$

thus obtaining

$$G_{FT} \approx \frac{1}{f_R{}^2} \frac{\int_0^{f_R/2} (2\pi f)^2 |\varphi(f)|^2 \, df}{\int_0^{f_R/2} |\varphi(f)|^2 \, df} \tag{4.6}$$

If we denote the antenna gain as a function $g(\theta)$ of the azimuth θ and the corresponding power gain by $G(\theta) = g^2(\theta)$, Parseval's theorem can be used to transform expression (4.6) to

$$G_{FT} = \frac{(\Delta\theta)^2 \int_{-\infty}^{+\infty} [G'(\theta)]^2 \, d\theta}{\int_{-\infty}^{+\infty} [G(\theta)]^2 \, d\theta} \tag{4.7}$$

In the case of a gaussian lobe it is assumed that

$$G(\theta) = G(0) \exp\left(-\frac{2.8\theta^2}{2}\right)$$

and application of eqn (4.7) gives

$$G_{FT} = \frac{2.8}{n^2}$$

Having calculated the gain for the clutter, it is now necessary to calculate the gain for moving targets. If we again assume that n is large enough for $T(f)$ to be almost constant in the frequency band occupied by the useful signal, it is clear that the power gain for moving targets is given by $|T(f_D)|^2$ modulo f_R, where f_D is the Doppler frequency of the moving target. In the case of single cancelation this is

$$4 \sin^2\left(4\pi \frac{f_D}{f_R}\right)$$

and the improvement factor is given by

$$T_i = \frac{4 \sin^2[\pi(f_D/f_R)]}{2.8/n^2} = \frac{4n^2 \sin^2[\pi(f_D/f_R)]}{2.8}$$

It can be seen that the improvement factor varies with f_R. Although it is unwise to use average values, particularly in this case, in the absence of anything better we calculate the average value of T_i by making the assumption, which is generally valid, that all values of f_D are equiprobable, and we find

$$\overline{T_i} \approx \frac{n^2}{1.4} \tag{4.8}$$

For example, if $n = 7$, $\overline{T_i} = 15\,\text{dB}$.

Nondiffuse clutter and single-canceler MTI

If we assume that there is never more than one clutter signal in the azimuth beam of the radar (in a range cell), or that there are very few, and one of these is much more powerful than the others, the preceding argument is no longer valid. Before the MTI, the clutter signal has the form $kG(vt)$. After the MTI, if n is large, it has the form $k\,\Delta\theta G'(vt)$. The clutter gain is then given by

$$G_{FT} = \frac{\Delta\theta^2[\max G'(\theta)]^2}{[G(0)]^2} \tag{4.9}$$

In the case of a gaussian lobe,

$$G_{FT} = \frac{2.5}{n^2}$$

which gives a value for T_i which is very close to the value given by expression (4.8).

Diffuse clutter and double-canceler MTI without feedback

In this case the calculation gives

$$G_{FT} = \frac{23}{n^4}$$

for a gaussian lobe. The improvement factor varies considerably with the radial velocity of the target. Its average value (which is only exceeded for a third of the time) is

$$\overline{T_i} = \frac{6n^4}{23} \approx \frac{n^4}{4}$$

For example, if $n = 7$, $\overline{T_i} \approx 28$ dB.

Nondiffuse clutter and double-canceler MTI without feedback

In this case

$$G_{FT} \approx \frac{\Delta\theta^4[\max G''(\theta)]^2}{[G(0)]^2} \tag{4.10}$$

and for a gaussian lobe

$$G_{FT} = \frac{32}{n^4}$$

and

$$\overline{T_i} \approx \frac{n^4}{5}$$

For example, if $n = 7$, $\overline{T_i} \approx 27$ dB.

Diffuse clutter and double-canceler MTI with feedback

The fact that the average value of T_i is only exceeded for a third of the time in a double-canceler MTI when diffuse clutter is present is a major drawback and necessitates the introduction of feedback. It is no longer possible to find a general expression for T_i or $\overline{T_i}$ because they depend on the type of feedback. Therefore we give below, as an example, the result obtained in the presence of diffuse clutter for a gaussian lobe with $n = 7$ and a double-canceler MTI with feedback whose transfer function in z^{-1} is given by (chain curve in Fig. 4.43)

$$\frac{(1 - z^{-1})^2}{1 - 0.5z^{-1} + 0.5z^{-2}}$$

The application of eqn (4.5) gives the following results:

$$G_{FT} = -18\,\text{dB}$$

$$\overline{T^2(f)} = 4.7\ (7\,\text{dB})$$

$$\overline{T_i} = 25\,\text{dB}$$

Although $\overline{T_i}$ is 2 dB less than its value without feedback, the probability that T_i exceeds its average value is now 63% instead of 33%.

Diffuse clutter and delay line MTI with three memories

The results obtained under the same assumptions as in the previous example ($n = 7$) for a delay line MTI with three memories characterized by (full curve in Fig. 4.4.3)

$$T(z^{-1}) = \frac{(1 - z^{-1})(1 - 1.732z^{-1} + z^{-2})}{1 - 0.5z^{-1} + 0.5z^{-2}}$$

are as follows:

$$G_{FT} = 25\,\text{dB}$$

$$T^2(f) = 8.4\ (9\,\text{dB})$$

$$\overline{T_i} = 38\,\text{dB}$$

The probability that T_i exceeds its average value is 56%.

4.7.4.4 Stability of the transmitter–receiver

In the preceding calculations it has been assumed that the signal transmitted by the transmitter is perfect. In fact, a transmitter is never perfectly stable so that the transmitted pulses are randomly modulated in amplitude and particularly in phase. Since these instabilities vary from one pulse to another, the signal received from a clutter consists of the signals with low Doppler frequencies already considered plus a white noise. The MTI filtering has little effect on this white noise, which then reduces the gain G_{FT}.

If we consider only fluctuations in the pulse-to-pulse phase (the amplitude fluctuations in good transmitters, although not zero, are less significant), we can measure and therefore define the effective value of the fluctuation θ_t in the transmitter phase (the effective value of the pulse-to-pulse fluctuation which is a factor of $\sqrt{2}$ larger is sometimes measured). The parasitic power received from the clutter because of this fluctuation is equal to θ_t^2 multiplied by the power which would have been received from the same clutter without the instability. This power is uniformly distributed over frequency so that, even with a fixed antenna, the gain for fixed targets is limited to

$$G_{FTL} = \theta_t^2 \overline{T(f)^2}$$

Thus the average value of the improvement factor is generally

$$\overline{T_i} = \frac{\overline{T^2(f)}}{G_{FTL} + \theta_t^2 \overline{T^2(f)}}$$

where G_{FTL} is the gain for fixed targets which was given above.

In practice, other phase instabilities may occur in the local oscillator, the coherent oscillator etc., and it is necessary to define the total phase instability of the transmitter–receiver by its effective value θ_{tr}:

$$\theta_{tr}^2 = \theta_t^2 + \theta_{LO}^2 + \theta_{COHO}^2$$

where θ_{LO} and θ_{COHO} are the effective values of the instabilities of the local oscillator and the coherent oscillator respectively.

Remark 4.17

When the antenna is fixed (or, in general, when n is so large that G_{FTL} is negligible with respect to θ_{tr}^2), we have

$$\overline{T_i} = \frac{1}{\theta_{tr}^2}$$

Values of $\overline{T_i}$ under various conditions for $n = 7$ with θ_{tr} at 31 and 10 mrad are given in Table 4.6.

4.7.4.5 Stability of the clutter

In the discussions in Sections 4.7.4.3 and 4.7.4.4 it has been assumed that the clutter consists of fixed targets. In fact, clutter may move with both an

Table 4.6 Values of $\overline{T}_i$ under various conditions

	$\overline{T}_i$ (dB)
$n = 7$, $\theta_{tr} = 31$ mrad	
Fixed antenna	30
Single cancelation	15
Double cancelation without feedback	26
Double cancelation with feedback	24
Three-memory MTI	29
$n = 7$, $\theta_{tr} = 10$ mrad	
Fixed antenna	40
Single cancelation	15
Double cancelation without feedback	27
Double cancelation with feedback	25
Three-memory MTI	37

average radial velocity (wind-borne clutter such as rain or clouds) and by turbulence about its average velocity (rain and clouds, falling leaves etc.).

The first effect produces a frequency shift in the spectrum of the clutter signal. This shift can be almost completely compensated by a shift in the local oscillator frequency. However, it always produces a deterioration in the performance of the MTI.

The second effect causes a broadening of the clutter spectrum which in the final analysis produces the same effect as a reduction in n. The resulting disturbance becomes more significant as the repetition frequency f_R decreases and as the transmission frequency of the radar increases. An increase in the transmission frequency of the radar causes an additional problem because the radar cross-section of clutter due to rain or clouds is proportional to the square of the transmission frequency, all other things being equal.

Remark 4.18

The preceding discussions of the improvement factor are only valid if the level of the parasitic signals is not so high that it lies outside the dynamics of the MTI. In order to avoid having unwanted residues of fixed signals left over after MTI filtering, limiting is frequently used at the MTI input so that the residues of powerful fixed signals remaining after filtering are negligible. If a moving target is then superimposed on such a fixed target, it will be eliminated at the same time as the fixed target unless the ratio of the power of the moving target to the power of the fixed target is greater than f_R/f_c. This can be shown by a laborious nonlinear calculation.

Chapter 5

Behavior of Real Targets. Fluctuation of Targets

5.1 General considerations

In the preceding chapters we have considered the simple case when the target behaves in a manner convenient for calculations. It has been assumed that if the radar transmits a signal $S(t)$ whose spectrum is $\Phi(f)$, it receives a reflected signal from the target (attenuated in a known given ratio) which is obtained by shifting $S(t)$ by t_{01} in time and f_D in frequency. (This applies to the general case when the width Δf of the spectrum $\Phi(f)$ is small with respect to the carrier frequency.) When we began in a very general manner to write down the radar equations in order to determine the energy to be transmitted, it was necessary to define the radar cross-section as follows: if the power density (in W m^{-2}) at the position of the target is p and if the target reradiates a particular power density (in W m^{-2}) in the direction of the radar, the radar behaves as if the target were reradiating a power $p\sigma_r$ isotropically, where σ_r is, by definition, the radar cross-section of the target. In other words, we can assume that the actual target is replaced by a point target, which is generally called the radiant point, which can consume a power $p\sigma_r$ and reradiate it isotropically. The coordinates of the radiant point are x, y and z.

The concept of the radiant point is valid provided that it is accepted that its position varies substantially with time and is only loosely related to the target: sometimes it lies to the right of the target, sometimes to the left and sometimes at the centre of gravity of the target. Similarly, the concept of the radar cross-section is valid, for the want of anything better, if it is accepted that it has very little relationship with the surface area of the target and that it varies with time. This is not surprising if we remember that targets are very large with respect to the wavelength (1000 times larger, for example) and have irregular shapes, and in general do not resemble a radiant point consuming $p\sigma_r$. However, this pessimistic picture has at least one point in its favor. If

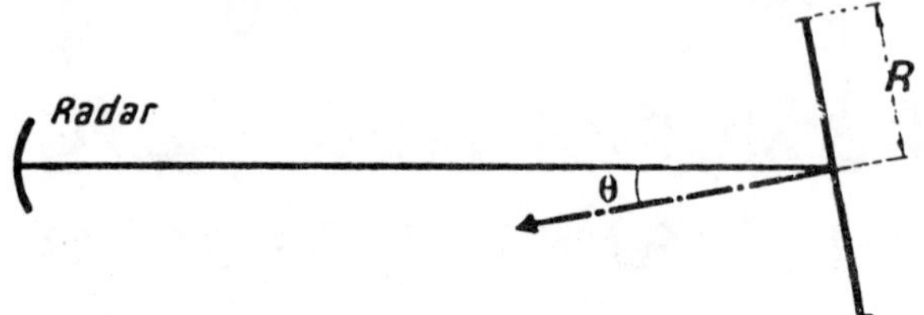

Fig. 5.1 Reradiation from a plane circular mirror.

both the radar cross-section and the position of the radiant point are random variables, they are not independent but are correlated in some way. In this chapter we examine this correlation.

5.2 Radar cross-section of a planar plate

Let us consider a target composed of a plane circular mirror which is a perfect reflector. Commonsense tells us that this mirror reflects the received waves very well when they are normally incident and very badly otherwise. In other words, the radar cross-section of the mirror is very large if the angle θ between the normal and the radar–target direction is zero or very small. However, as soon as θ increases to any extent the radar cross-section becomes very small.

If $\theta = 0$, all the points on the plate are illuminated by electric fields with the same phase, and the power captured by the mirror is

$$\pi R^2 p$$

This power is reradiated with different gains depending on the direction. In particular, it is reradiated in the direction of the radar with a gain

$$G = \frac{4\pi}{\lambda^2} \pi R^2$$

The radar cross-section of the plate for $\theta = 0$ is therefore

$$\sigma_r = \frac{4\pi S^2}{\lambda^2}$$

where S is the area of the plate. For example, if $R = 1$ m and $\lambda = 0.03$ m, $S = 3\,\text{m}^2$ and hence $\sigma_r = 100\,000\,\text{m}^2$. In this case the radiant point is obviously the center of the plate.

If $\theta = \lambda/4R$, the gain of the plate in the direction of the radar is zero, the radar cross-section is zero and the radiant point no longer exists. For

Fig. 5.2 Reradiation diagram for a planar plate.

example, if $R = 1$ m and $\lambda = 0.03$ m, the radar cross-section will be zero for $\theta = \lambda/4R = 0.008$ rad $= 0.5°$. Thus, if the plate is a moving radar target in space, which on average is oriented perpendicular to the radar–target direction but is subject to small rotational movements so that θ varies by a few degrees about an average value of $\theta = 0$ (this is very difficult to avoid), the radar cross-section will vary between zero and $100\,000\ \mathrm{m}^2$. However, if the radar used has a wavelength of 10 cm, the radar cross-section of the plate will be $10\,000\ \mathrm{m}^2$ for $\theta = 0$ and zero for $\theta = \lambda/4R = 0.025$ rad $= 1.5°$, and on average will vary three times more slowly than in the previous case.

It is convenient to represent the variation in the radar cross-section as a function of the angle θ between the normal to the plate and the direction of the radar in polar coordinates (Fig. 5.2). In this way, a diagram of reradiation from the target is obtained which is similar to the classical radiation diagrams for an antenna. (However, this analogy should not be pushed too far: a radar target is not an antenna but a mobile reflector with respect to its primary source which is the radar situated a large distance away.)

5.3 Radar cross-section of a metal sphere

We consider a target formed by a metal sphere with a radius R which is very much greater than the radar wavelength λ (Fig. 5.3). Then the part of the sphere normal to the radar–target direction reradiates mainly in the direction of the radar (angle of incidence close to zero), whereas the other

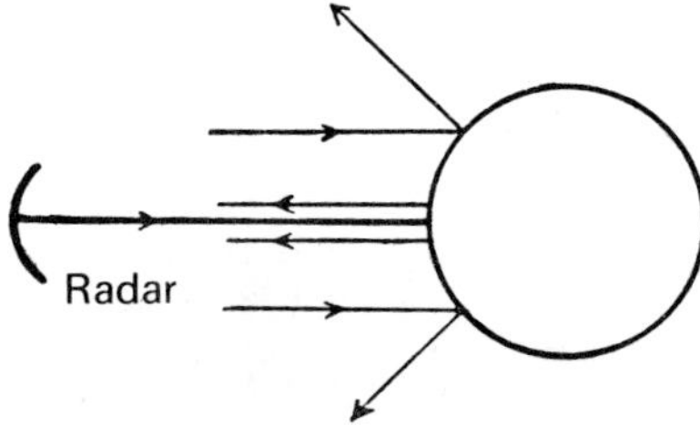

Fig. 5.3 Reradiation from a metal sphere.

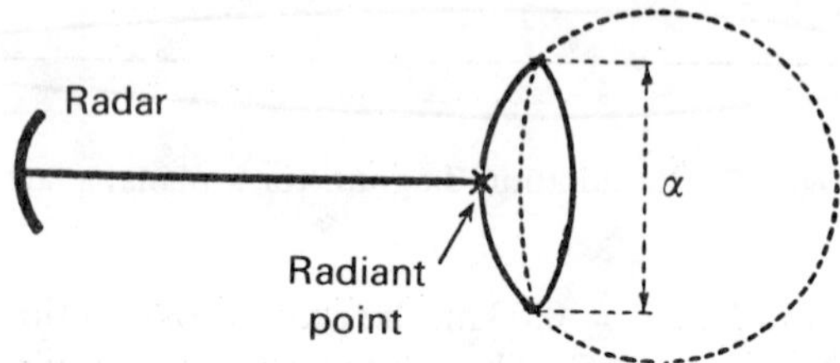

Fig. 5.4 Radar cross-section of a sphere.

parts of the sphere reradiate mainly in different directions. It therefore makes sense to assume that the part of the sphere which is closest to the radar contributes most to reflection, i.e. the radiant point is the closest point of the sphere to the radar. It can be proved that the radar cross-section of the entire sphere is no larger than the radar cross-section of a segment of the sphere centered on the radiant point if the largest dimension α of this segment is greater than $(2\lambda R)^{1/2}$ (Fig. 5.4). If this condition is satisfied, the segment of the sphere shown in Fig. 5.4 has the same radar cross-section as the entire sphere shown in Fig. 5.3, i.e. πR^2. For example, if $R = 20$ m and $\lambda = 0.1$ m, $(2\lambda R)^{1/2} = 2$ m. Then the entire sphere has a radar cross-section of 1200 m^2. A spherical segment with an aperture of 3 m also has a radar cross-section of 1200 m^2 although it has an area of the order of 10 m^2.

This property can be generalized. If we are dealing with a sufficiently large spherical segment of a convex metal surface which is normal to the radar–target direction at some point, the radiant point is the closest point to the radar and the radar cross-section of the object is $\pi R_1 R_2$, where R_1 and R_2 are the principal radii of curvature. In this case, the radar cross-section has little relationship to the area of the object.

Remark 5.1

It is obvious that, since the sphere is isotropic, its radar cross-section does not depend on its orientation. Figure 5.5 shows the reradiation diagram of a metal sphere.

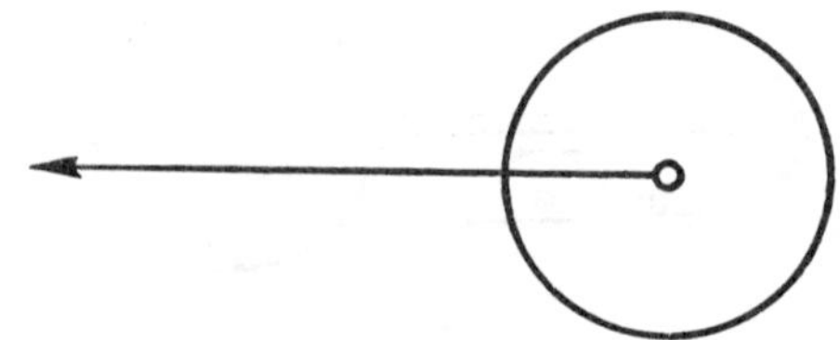

Fig. 5.5 Reradiation diagram of a metal sphere.

5.4 Field reradiated by a hypothetical isotropic target

As an aid to a proper analysis of the main differences between actual targets and theoretical targets, we now review the properties of an ideal target, i.e. a point target situated at O which receives power $p\sigma_r$ from the radar radiation and effectively reradiates it isotropically. In this case, if we consider all the points of a sphere located a distance D from the target O, the amplitude and phase of the electrical field are the same at every point of this sphere. It is said that the sphere with center O is an equiphase surface.

At a sufficiently large distance D from the target with respect to the dimensions of the radar antennas used, it can be assumed that the portion of the equiphase surface which is of interest is a plane along which the electric field has a constant amplitude. Two equiphase surfaces π_1 and π_2 separated from each other by d (d is small with respect to the wavelength λ) correspond to phases Φ_1 and Φ_2 such that

$$\Phi_2 = \Phi_1 - \frac{4\pi d}{\lambda}$$

In addition, since D is assumed to be extremely large with respect to λ and so is also large with respect to d, the amplitude of the field along π_1 is the same as the amplitude of the field along π_2 (since $1/d^2 \approx 1/(D + d)^2$) (Fig. 5.6). In other words, the normal to the equiphase surface is in the direction of the hypothetical point isotropic target.

5.5 Target consisting of two identical point targets

Let us consider a target consisting of two identical point sources situated at O_1 and O_2 such that the distance O_1O_2 is fixed and equal to r (r is extremely small with respect to the distance D from the target to the radar and very large with respect to the wavelength λ). It is also assumed that the mid-perpendicular plane of O_1O_2 makes a small angle θ with the radar–target direction (Fig. 5.7). Then the wave from O_1 is given by

$$A \exp\left(-\frac{4\pi j d_1}{\lambda}\right)$$

and the wave from O_2 is given by

$$A \exp\left(-\frac{4\pi j d_2}{\lambda}\right)$$

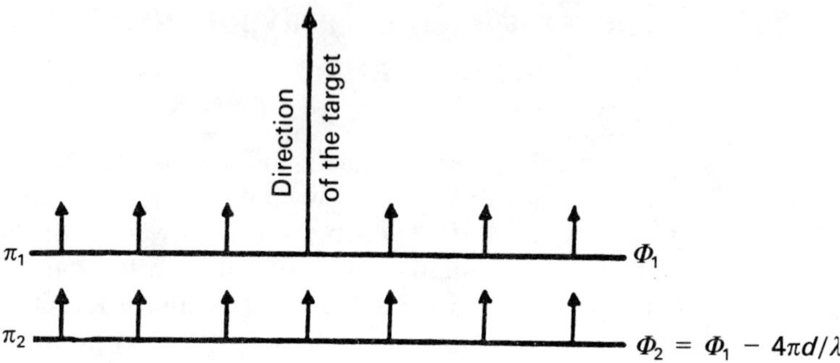

Fig. 5.6 Sections of two equiphase surfaces along which the amplitude of the field is constant.

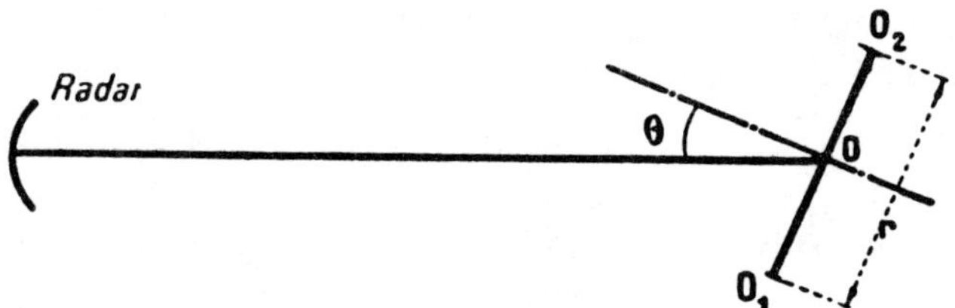

Fig. 5.7 Target consisting of two identical point sources.

where d_1 and d_2 are the distances from the radar to O_1 and the radar to O_2 respectively. The resultant wave is given by

$$E = A\left[\exp\left(-\frac{4\pi j d_1}{\lambda}\right) + \exp\left(-\frac{4\pi j d_2}{\lambda}\right)\right]$$

and hence

$$E = 2A\exp\left[-\frac{2\pi j(d_1 + d_2)}{\lambda}\right]\cos\left[\frac{2\pi(d_1 - d_2)}{\lambda}\right]$$

Then if we denote the distance from the radar to O by D, where

$$D \approx \frac{d_1 + d_2}{2}$$

we obtain

$$E = 2A\exp\left(-4\pi j\frac{D}{\lambda}\right)\cos\left(\frac{4\pi r\sin\theta}{\lambda}\right)$$

When the target is fixed and the radar moves over a sphere with center O, the phase of the wave received by the radar is constant and equal to Φ_0 when

$$-\frac{\pi}{2} < \frac{4\pi r \sin\theta}{\lambda} < \frac{\pi}{2} \quad \text{(except for a factor of } 2k\pi)$$

$$\frac{k\lambda}{2r} - \frac{\lambda}{8r} < \sin\theta = \theta < \frac{\lambda}{8r} + \frac{k\lambda}{2r} \quad (k \text{ integer})$$

and the phase of the wave received by the radar is constant and equal to $\Phi_0 + \pi$ when

$$k\frac{\lambda}{2r} + \frac{\lambda}{8r} < \sin\theta = \theta < \frac{3\lambda}{8r} + \frac{k\lambda}{2r}$$

The amplitude of the received signal varies as

$$\left|\cos\left(\frac{4\pi r \sin\theta}{\lambda}\right)\right| = \left|\cos\left(\frac{4\pi r\theta}{\lambda}\right)\right|$$

Thus the spheres with center O are no longer equiphase surfaces: the phase changes over the surfaces in steps of π radians. In addition, the amplitude of the field received over the spheres is not constant (Fig. 5.8).

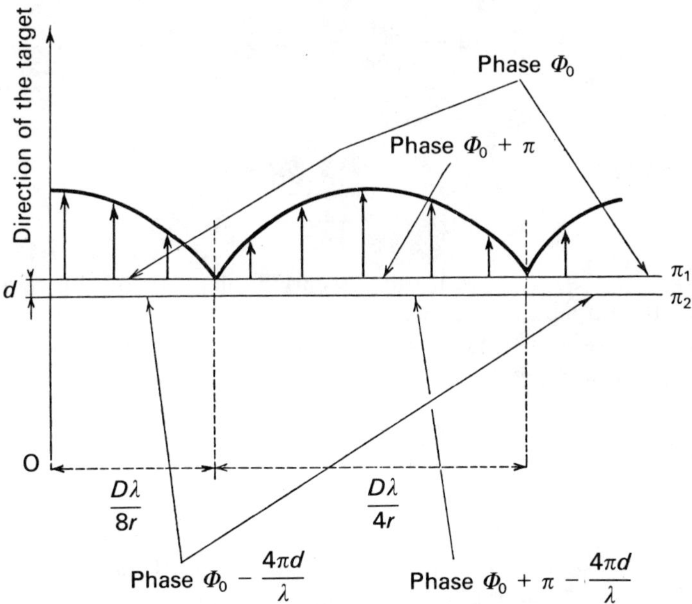

Fig. 5.8 Variation in the phase and amplitude of the field on spherical surfaces with center O.

It is of interest to give orders of magnitude for the parameters in Fig. 5.8. Assume that the target is equivalent to two point sources separated by $r = 5$ m (a twin-jet moving away can often be approximately defined in this way), and that $\lambda = 0.1$ m and $D = 200$ km. It is then found that $D\lambda/4r = 1000$ m and $\lambda/4r = 5$ mrad.

Classical wave collectors (radar antennas) have dimensions of the order of 10 m at the wavelength considered, and therefore it is reasonable to assume that the amplitude of the received field is constant at all points of the wave collector if the target is fixed. Furthermore, if the target is subject to small random movements (this is inevitable), the amplitude of the received field, which is the same at all points of the wave collector, is also random between two limits 0 and $2A$: the receiver does not receive the signal $kS(t - t_0)$, where $S(t)$ is the transmitted signal, but a signal $aS(t - t_0)$, where $|a| = 2A \sin \varphi$ and all values of φ between zero and $\pi/2$ are equiprobable since no value of θ is *a priori* preferred. In other words, a is a random variable whose amplitude distribution is characterized by a probability density $p_1(a)$ such that

$$p_1(a)\,\mathrm{d}a = p(\varphi)\,\mathrm{d}\varphi$$

where

$$\varphi = \arcsin\left(\frac{a}{2A}\right)$$

$$\mathrm{d}\varphi = \frac{\mathrm{d}a}{2A(1 - a^2/4A^2)^{1/2}} = \frac{\mathrm{d}a}{(4A^2 - a^2)^{1/2}}$$

$$p(\varphi) = \frac{2}{\pi}$$

i.e. (Fig. 5.9)

$$p_1(|a|) = \frac{2}{\pi(4A^2 - a^2)^{1/2}}$$

The rate at which a varies with time essentially depends on the velocity of rotation of the target with respect to the radar.

In order to demonstrate the orders of magnitude involved, we consider examples of two important cases.

(a) The target is an airplane flying at $300\,\mathrm{m\,s^{-1}}$ and making a turn with an acceleration of $2g$ ($20\,\mathrm{m\,s^{-2}}$). The rotation rate is therefore $6 \times 10^{-2}\,\mathrm{rad\,s^{-1}}$. Thus if $r = 5$ m and $\lambda = 0.1$ m, the airplane turns through $\lambda/4r$ in 0.08 s. In this case, it can be said, roughly speaking, that a passes from one maximum to the next in a time of the order of 0.1 s. This time is generally appreciably larger than the time for the radar measurement, i.e. it can be assumed that the received signal is constant throughout the radar measurement.

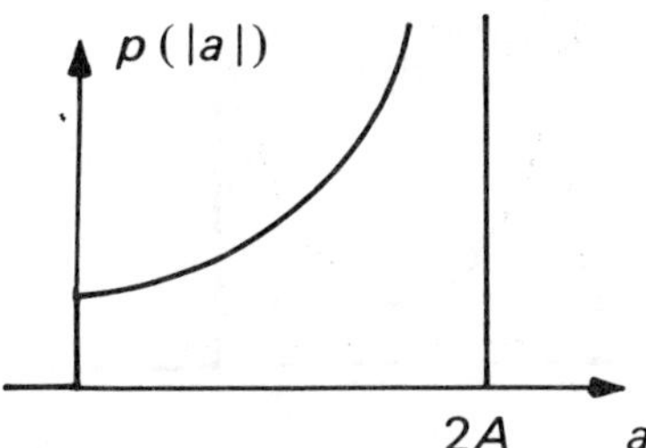

Fig. 5.9 Variation in $p(|a|)$.

(b) The target is an airplane moving perpendicularly to the radar–target direction with a velocity of $300\,\mathrm{m\,s^{-1}}$ at a distance of 5 km. In this case, the velocity of rotation of the line of sight with respect to the target is equal to $300/5 \times 10^3\,\mathrm{rad\,s^{-1}}$, i.e. again $6 \times 10^{-2}\,\mathrm{rad\,s^{-1}}$.

Finally, a characterizes (apart from a factor) the amplitude of the signal which will be transmitted to the radar receiver (it is finally expressed in volts). Nevertheless, the important parameter is the received energy, i.e. the value of a^2 whose mean value (apart from a factor) characterizes the mean value of the power received from the target. The value of a^2 is also (apart from a factor) the radar cross-section of the target, which is therefore a random parameter whose mean value has a physical meaning and which has a probability distribution characterized by its density $p_2(a^2)$, where

$$p_2(a^2)\,\mathrm{d}a^2 = p_1(|a|)\,\mathrm{d}a$$

$$2p_2(a^2)\,a\,\mathrm{d}a = p_1(|a|)\,\mathrm{d}a$$

$$p_2(a^2) = \frac{1}{\pi(4A^2 - a^2)^{1/2}(a^2)^{1/2}}$$

The probability density of the radar cross-section σ_r is therefore given by

$$p_2(\sigma_r) = \frac{1}{\pi(2\Sigma - \sigma_r)^{1/2}\sigma_r^{\;1/2}}$$

where Σ denotes the mean value of σ_r (Fig. 5.10).

In order to complete the study of the "dumbbell" target it should also be noted that the corresponding equiphase surfaces have the staircase form shown in Fig. 5.11 (the equiphase surface is the locus of points where at a given point in time the field has the same phase). Figure 5.12 shows the radiation diagram of such a dumbbell target. It can be seen that it is daisy shaped, whence the name "daisy effect" sometimes used for the fluctuation phenomenon of radar targets.

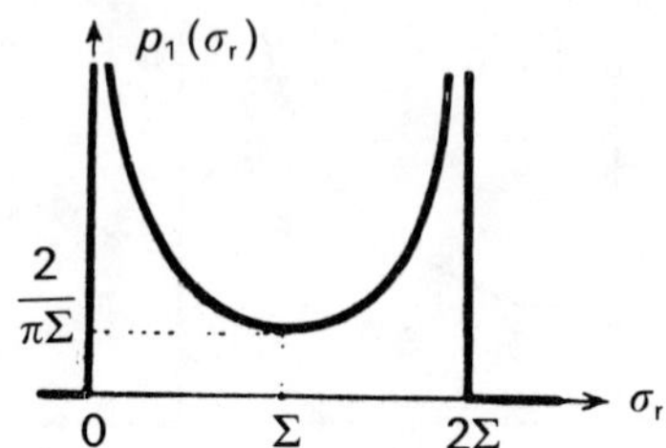

Fig. 5.10 Variation in $p_1(\sigma_r)$.

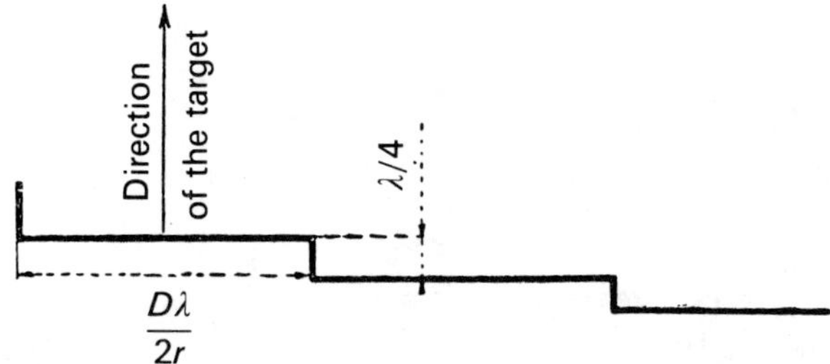

Fig. 5.11 Equiphase surfaces of a dumbbell target.

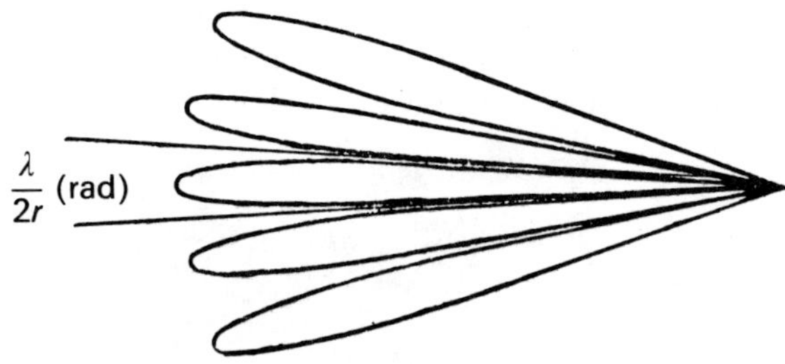

Fig. 5.12 Radiation diagram of a dumbbell target.

5.6 Target consisting of two different point targets

Let us now consider a target consisting of two point sources situated at O_1 and O_2 (the same notation is used as in Section 5.5) but with the source at O_1 having a radar cross-section four times greater than that of the source at O_2. Then the wave from O_1 is given by

$$A \exp\left(-\frac{4\pi j d_1}{\lambda}\right)$$

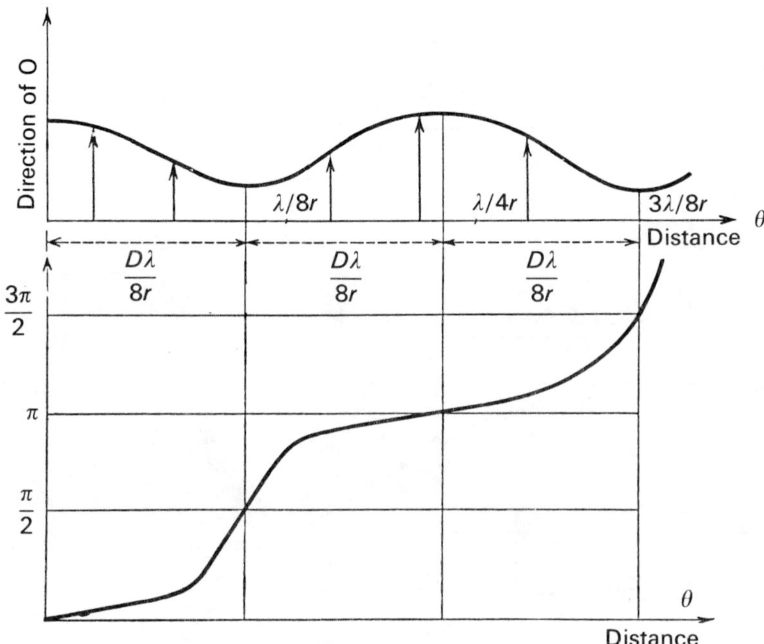

Fig. 5.13 (a) Amplitude and (b) phase of the wave received from two different point targets.

and the wave from O_2 is given by

$$0.5A \exp\left(-\frac{4\pi \mathrm{j} d_2}{\lambda}\right)$$

which gives a resultant wave

$$A\left[\exp\left(-\frac{4\pi \mathrm{j} d_1}{\lambda}\right) + 0.5 \exp\left(-\frac{4\pi \mathrm{j} d_2}{\lambda}\right)\right]$$

When the target is fixed and the radar moves over a sphere with center O, the phase of the received wave varies continuously and the amplitude of the received signal varies as

$$\left[1.25 + \cos\left(\frac{8\pi r\theta}{\lambda}\right)\right]^{1/2}$$

Thus spheres with center O are not equiphase surfaces and the amplitude of the field received over these spheres also varies (Fig. 5.13). Figure 5.14 shows the reradiation diagram of the target.

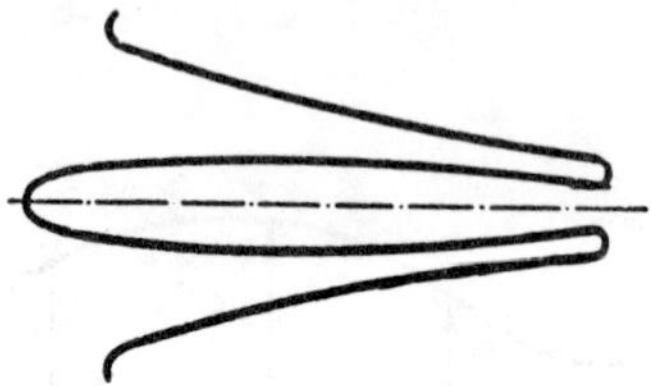

Fig. 5.14 Reradiation diagram for a target consisting of two different point targets.

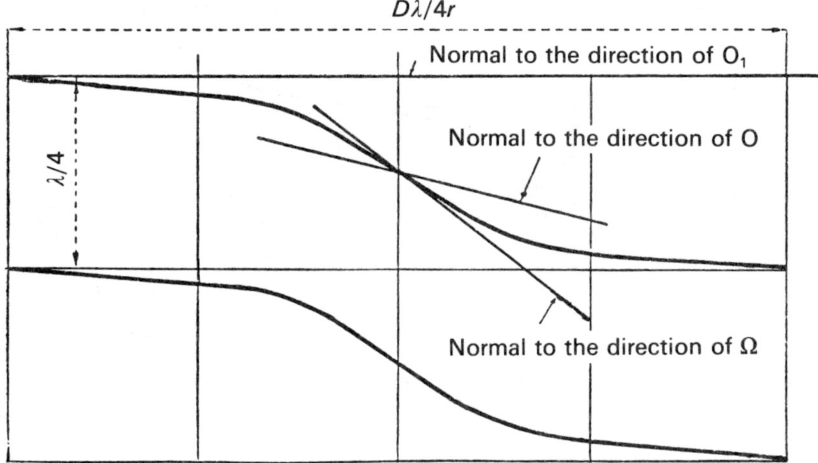

Fig. 5.15 Section of two equiphase surfaces formed by a plane passing through O_1O_2.

Equiphase surfaces, i.e. the loci of points where at a given time the field has the same phase, have nothing in common with the spheres with center O. The section of two of these surfaces formed by a plane passing through O_1O_2 is shown in Fig. 5.15. It can be seen from this figure that an equiphase surface is never normal to the direction of O which is the midpoint of O_1O_2. In half the cases, the normal to the equiphase surface is directed towards a point outside O_1O_2 which may be point Ω symmetrical to O_2 with respect to O_1 (Fig. 5.16). In other cases it lies in the direction of a point lying between O and O_1. It will be shown in Chapter 6 that the radar determines the direction of the normal to the equiphase surface and not the direction of the target. The system behaves as if the target were equivalent to a radiant point which could be located between O_1 and Ω.

Since all targets are subject to small random movements, the following conclusions can be drawn in this particular case:

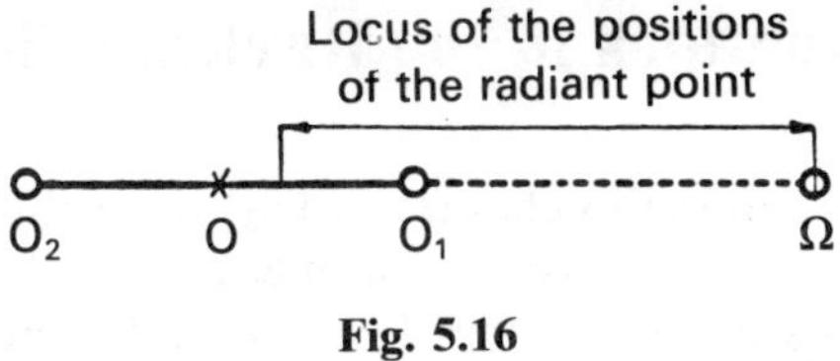

Fig. 5.16

the position of the radiant point has a 50% chance of lying outside the actual target;

the radar cross-section of the target varies between Σ and 9Σ (the ratio of the amplitude of the field at $\theta = 0$ to that at $\theta = \lambda/4r$ is equal to 3, and hence the ratio of the corresponding powers entering the radar is equal to 9);

the size of the radar cross-section decreases as the distance of the radiant point from the target increases.

Remark 5.2

It has already been noted that the target arrangement considered in this section (two similar targets with different radar cross-sections) explains rather well what happens when an airplane flies at low altitude above a reflecting surface (for example, above the sea when the wave radiated by the radar has horizontal polarization) (Fig. 5.17). Let us assume, for example, that the coefficient of reflection from the sea is 0.5. On average, the signal received for trajectory Δ_1 (from the airplane) is twice as large as the signal received for trajectory Δ_2 (coming from the image of the airplane), and therefore on average the radar will determine a direction closer to that of the plane than to that of its image. However, the airplane itself constitutes a complex target with a very distorted reradiation pattern, and it is possible that at the time of measurement its radar cross-section will be small whereas that of the image will be large. In this case, the radar will determine a direction close to the image, and there will be little hope of determining the direction of the airplane.

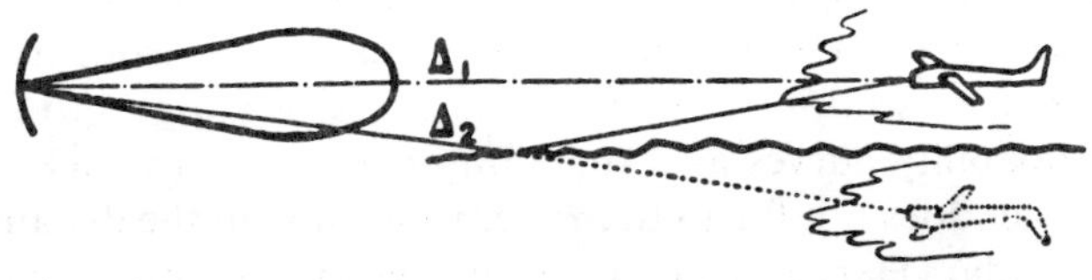

Fig. 5.17 Signal received from an airplane flying at low altitude above a reflecting surface.

5.7 Fluctuations in the Rayleigh distribution

The discussion so far in this chapter clearly shows that a target has a very poorly defined radar cross-section which fluctuates with time, although the rate of fluctuation is generally rather low (the radar cross-section remains almost constant for the duration of the radar measurement). Thus radar cross-section is a random variable characterized by a rather narrow frequency spectrum (or a rather large correlation interval) and by an amplitude distribution of which an example is given in Section 5.5.

Just as when we have a random variable where everything is ignored, it is said that its distribution is gaussian (this is only valid if this variable is the sum of a large number of independent random variables), in the case of radar we define an extremely arbitrary target (or a target undergoing extreme fluctuations) in the following manner. At a given moment the received field always has two components: component 1 in phase with, or with opposite phase to, a given phase reference, and component 2 in phase quadrature with this reference. It is assumed that, for an extremely random target, component 1 has equal probabilities of being positive or negative and has a gaussian distribution and that component 2 has the same distribution. Under these conditions the amplitude A of the received field has a Rayleigh distribution, i.e. the probability density of A is given by

$$p_1(A) = \frac{2A}{A_0^2}\exp\left(-\frac{A^2}{A_0^2}\right) \tag{5.1}$$

where A_0^2 is the mean square value of A. Thus (cf. the calculations of Section 5.5) the probability density of the radar cross-section σ of an extremely arbitrary target can be written as

$$p_2(\sigma) = \frac{1}{\sigma_0}\exp\left(-\frac{\sigma}{\sigma_0}\right) \tag{5.2}$$

where σ_0 is the mean value of σ (σ is proportional to A^2). σ_0 has a physical meaning as it represents the average power radiated by the target in the direction of the radar. It can therefore characterize the target in a valid manner and enables a non-fluctuating target (metal sphere) with a constant radar cross-section σ_0 to be compared with an extremely arbitrary target with a mean radar cross-section σ_0. In particular, it is very useful to compare the detection probability curves as a function of R in the two cases.

If the target does not fluctuate, R depends only on the distance from the target to the radar. Hence, if the distance is given (assuming that the target is in the direction of maximum radiation of the radar), the value of R can be determined. In the case of a very complex target under the same conditions, only the mean value of R, i.e. R_0, can be determined. In practice, the receiver of the ideal radar, regardless of whether it correlates the received signal over a time T with a delayed fraction $kS(t - t_0)$ of the transmitted signal or filters the received signal with an optimum filter with characteristic $k\Phi(-f)$ or $k\Phi(f_0 - f)$, delivers at the output in the absence of the target (at a good distance) a gaussian noise $n(t)$ with zero mean and a variance which is set equal to unity in order to simplify the argument and, in this example, a useful signal whose amplitude is $R^{1/2} = s$. The detection threshold is defined in such a way that, in the absence of a useful signal, the probability that the noise exceeds this level is equal to a given value (the false alarm probability P_f which is equal to 10^{-3} for example). A target is only detected if the amplitude of the useful signal plus noise is larger than this threshold, and the probability of such an occurrence is the detection probability P_f. The variation in P_d as a function of R for a nonfluctuating target is shown in Chapter 3, Fig. 3.4.

If the target fluctuates, it is possible that R (proportional to σ) will be very high at the measurement time and therefore the detection probability will be close to unity, but it is also possible that R will be very small and then there will be little chance of detecting the target. Mathematically, we can write

$$p_2(R) = \frac{1}{R_0}\exp\left(-\frac{R}{R_0}\right)$$

and

$$p_1(s) = \frac{2s}{R_0}\exp\left(-\frac{s^2}{R_0}\right)$$

Then for $P_f = 10^{-3}$ (base threshold equal to approximately 3)

$$P_d = \frac{1}{(2\pi)^{1/2}}\int_0^\infty \frac{2s}{R_0}\exp\left(-\frac{s^2}{R_0}\right)\mathrm{d}s\int_{3-s}^\infty \exp\left(-\frac{n^2}{2}\right)\mathrm{d}n$$

i.e. the detection probability P_d is the mean value of the probability that at a given time $s + n > 3$ (useful signal plus noise is higher than the threshold) or $n > 3 - s$.

Table 5.1 Detection probability as a function of R_0

R_0 (dB)	P_d	
	Complex target	*Nonfluctuating target*
3	0.07	0.05
6	0.18	0.15
9.5	0.40	0.50
12	0.58	0.85
14	0.69	0.98
17	0.83	≈ 1.00

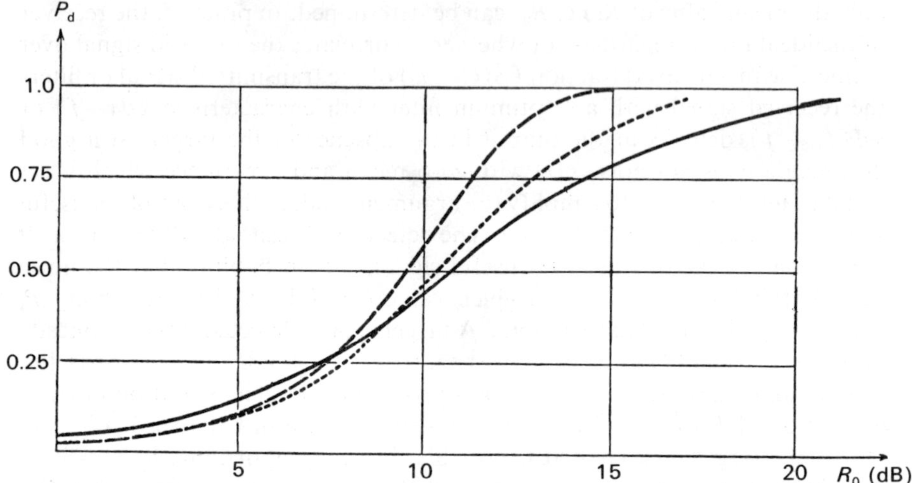

Fig. 5.18 Detection probability P_d versus R_0 for a false alarm probability of 10^{-3}: – – –, nonfluctuating target; - - - -, fluctuating target following a Rayleigh distribution (addition before the threshold is set); ——, fluctuating target following a Rayleigh distribution (normal radar).

Table 5.1 shows the calculated values of the detection probability P_d of an ideal radar for a nonfluctuating target and for a fluctuating target following the Rayleigh distribution as a function of the mean value R_0 of R (obtained by replacing σ by its mean value σ_0 in the radar equations) when the threshold is adjusted to a false alarm probability P_f of about 10^{-3}. These data are also shown by the full curve in Fig. 5.18†.

In order to obtain a detection probability of 90% for a very complex target with a mean radar cross-section σ_0, it is necessary to transmit a power 7–8 dB higher than the power required to detect a nonfluctuating target with

†P_d can generally be written as

$$P_d = P_f + \left\{\exp\left[-\frac{A^2}{2} + R_0\right]\right\}\left[2\pi\left(1 + \frac{2}{R_0}\right)\right]^{-1/2} \int_{-A/(1+2/R_0)^{1/2}}^{\infty} \exp\left(-\frac{v^2}{2}\right) dv$$

where

$$P_f = \frac{1}{(2\pi)^{1/2}} \int_A^{\infty} \exp\left(-\frac{v^2}{2}\right) dv$$

When $P_f < 10^{-2}$ and $P_d > 0.10$, the expression can be simplified to

$$P_d = \frac{\exp[-A^2/(2 + R_0)]}{(1 + 2/R_0)^{1/2}}$$

so that P_d can be calculated for all useful values of P_f and R_0 (see second and third fold-out graphs at the end of the book).

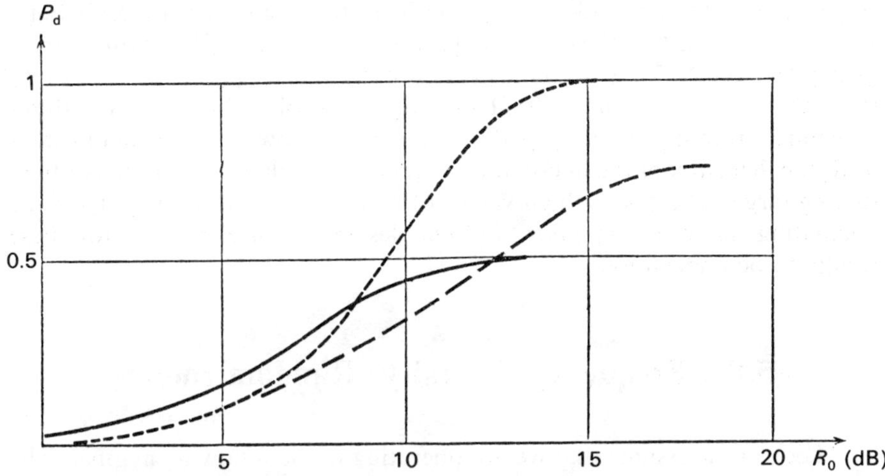

Fig. 5.19 Detection probability P_d versus R_0: ----, nonfluctuating target; – – –, EC target, diversity radar (addition before the base threshold is set); ——, EC target, normal radar.

Table 5.2 P_d as a function of R_0 for an EC target

R_0 (dB)	P_d
3	0.10
6	0.22
9	0.42
12	≈0.5
20	≈0.5

P_d can never exceed 0.5.

the same (average) radar cross-section σ_0 with 90% probability under the same conditions, i.e. the power is multiplied by a factor of 5–6. However, if a detection probability of 50% is acceptable, almost the same power is required regardless of whether the target is a sphere or is very complex.

In order to understand these phenomena better, let us illustrate the preceding abstract mathematics with an example. Let us assume that we are dealing with an extremely complex (EC) target which undergoes very large fluctuations such that the signal $s = R^{1/2}$ has a 50% probability of being zero and a 50% probability of being equal to $(2R_0)^{1/2}$. Thus the mean value of s^2 is equal to R_0 and it is possible to compare such a target with a nonfluctuating target permanently providing $s = R_0{}^{1/2}$. The detection probability corresponds to $2R_0$ for half the time and to the false alarm probability (which is very small) for the remainder of the time. Thus the mean detection probability $P_d(R_0)$ is equal to half the detection probability of the nonfluctuating target giving $2R_0$. Table 5.2 gives the detection probability P_d as a function of R_0 for such a target (see Fig. 5.19). This simple example, which does not differ very much from the target studied in Section 5.5, shows that for small values of R_0 the detection of the fluctuating target is better than that of the nonfluctuating target whereas for high values of R_0 the detection probability of the fluctuating target is very poor, and enables the physical reasons for these results to be understood.

5.8 Frequency diversity. Random radars

Let us now assume that we are operating in the following manner. The transmission power is halved so that we have two transmitters identical in all respects except that their frequency is different. We also have two different receivers whose outputs are respectively a signal s_1 corresponding to $R_0/2$ and embedded in a gaussian noise with standard deviation unity and a signal s_2 corresponding to $R_0/2$ and immersed in a gaussian noise with standard deviation unity. Signals s_1 and s_2 have the same probability distribution (the probability that s_1 or s_2 is zero is 0.5, and the probability that s_1 or s_2 is $R_0{}^{1/2}$ is 0.5) but fluctuate in an independent manner if the difference between the transmission frequencies is sufficiently large† and the two noises are independent. Thus if the signals leaving each receiver are added together, we have a noise with standard deviation $2^{1/2}$ and a signal $s_3 = s_1 + s_2$ or a noise with standard deviation unity and a signal $s = s_1/2^{1/2} + s_2/2^{1/2}$.

Normal operations (setting the threshold so that $P_f = 10^{-3}$) can then be carried out on the system (signal plus noise) and the following results are obtained:

the probability that $s_1 \times 2^{1/2} = 0$ is 0.5;
the probability that $s_1 \times 2^{1/2} = (R_0/2)^{1/2}$ is 0.5;
the probability that $s_2/2^{1/2} = 0$ is 0.5;
the probability that $s_2/2^{1/2} = (R_0/2)^{1/2}$ is 0.5.

† Consider a target whose largest dimension is d. The number of daisy petals in its radiation pattern is of the order of $5d/\lambda$, i.e. $(5d/c)f$, where f is the frequency used and c is the velocity of light. In approximate terms, it can be assumed that if the frequency changes by a quantity Δf so that this number increases by unity, the pattern changes completely. This occurs for $\Delta f \approx c/5d$. This gives the order of magnitude of the frequency jump which must be made to obtain a new independent radar cross-section. For example, if $d = 10\,\mathrm{m}$, $\Delta f \approx 10\,\mathrm{MHz}$.

Table 5.3

R_0 (dB)	P_d
3	0.05
6	0.14
9	0.29
12	0.47
15	0.67
18	≈ 0.75
20	≈ 0.75

This also means that the following hold:

the probability that $s = 0$ is 0.25;
the probability that $s = (R_0/2)^{1/2}$ is 0.5;
the probability that $s = (2R_0)^{1/2}$ is 0.25.

The mean detection probability associated with such a radar is therefore given by half the detection probability of the nonfluctuating target for $R_0/2$ plus a quarter of the detection probability of the nonfluctuating target for $2R_0$. The final results are given in Table 5.3 and Fig. 5.19.

The use of diversity radar (two channels added before the threshold is set) has made it possible to improve the detection probability considerably for the same transmitted power at relatively high values of R_0. Figure 5.20 shows the amplitude distribution of s in classical radar (Fig. 5.20(a)) and diversity radar (Fig. 5.20(b)).

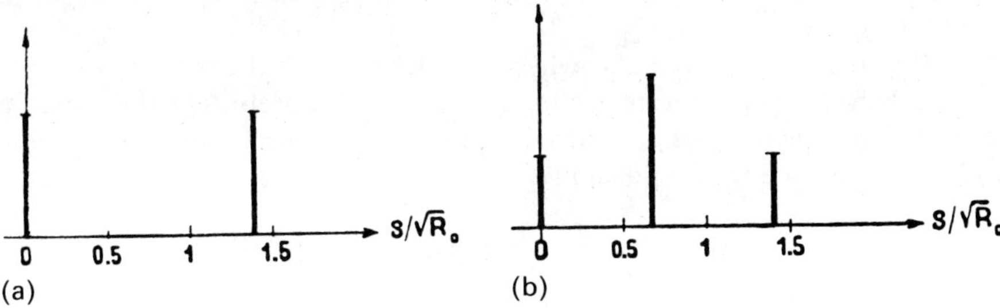

Fig. 5.20 Amplitude distribution of s in (a) classical radar and (b) diversity radar.

If we now return to the extremely complex target, it is logical to expect that the full curve in Fig. 5.18, which represents the variation in the detection probability of an extremely complex target as a function of R_0, will be appreciably modified if a diversity radar is used. This can be seen quite easily if we compare the following amplitude distributions.

(1) The amplitude distribution $p_1(s)$ for the useful signal (the noise is adjusted to a level such that its standard deviation is unity) for a complex target and normal radar:

$$p_1(s) = \frac{2s}{R_0}\exp\left(-\frac{s^2}{R_0}\right)$$

(2) The amplitude distribution $p_1(s)$ for an extremely complex target and a diversity radar with two channels (addition before the threshold is set):

$$p_1(s) = 2\exp\left(-\frac{2s^2}{R_0}\right)\left[2\frac{s}{R_0}\exp\left(-\frac{2s^2}{R_0}\right) + \Theta\left(\frac{s \times \sqrt{2}}{R_0^{1/2}}\right)\left(\frac{\pi}{2}\right)^{1/2}\left(\frac{4s^2}{R_0} - 1\right)\frac{1}{R_0^{1/2}}\right]$$

where†

$$\Theta(x) = \frac{2}{\pi^{1/2}}\int_0^{\infty} \exp(-t^2)\,\mathrm{d}t$$

(3) The amplitude distribution $p_1(s)$ for a nonfluctuating target.

Figure 5.21 shows the curves of $p_1(s)\,R_0^{1/2}$ as a function of $s/R_0^{1/2}$ for these three cases. It can be seen that the amplitude distribution of s for diversity radar approaches the distribution corresponding to a nonfluctuating target.

Calculation of the detection probability for the diversity radar is described by the curves in Fig. 5.21, and the result is shown by the dotted curve in Fig. 5.18. It can be seen that if a detection probability of 0.5 is satisfactory, the diversity radar does not have any advantages. However, if a detection probability of 0.9 is required, use of a two-channel diversity radar makes it possible to save approximately 3 dB of the transmitted power for an extremely complex target, i.e. 50% of the power transmitted when the radar is not a diversity system.

If we use a diversity radar where the number n of channels is large, the system behaves as if we were dealing with a constant useful signal s_0 equal to $n^{1/2}$ times the mean value of s corresponding to R_0/n accompanied by a noise with a standard deviation of unity:

$$s_0 = n^{1/2}\int_0^{+\infty} \frac{2ns^2}{R_0}\exp\left(-\frac{ns^2}{R_0}\right)\mathrm{d}s$$

$$= \left(R_0\,\frac{\pi}{4}\right)^{1/2}$$

Thus the target appears to be nonfluctuating with a constant radar cross section equal to 0.79 times its mean value. Finally, the "fluctuation noise" produces a loss of 1 dB. In practice, if we require detection probabilities

† The derivation of this formula is classical and relatively complicated, and, like the formula itself, is of no intrinsic interest. Only the dotted curve in Fig. 5.21 is of interest.

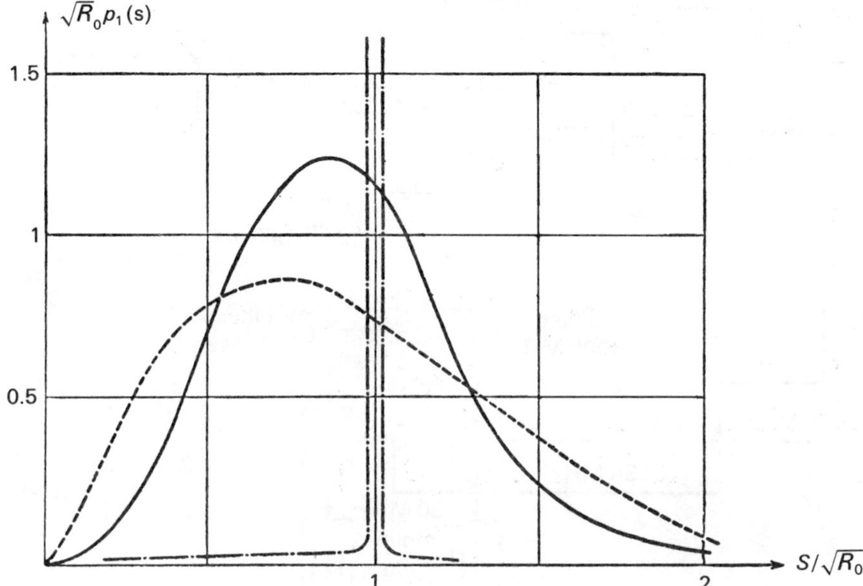

Fig. 5.21 Calculation of the detection probability for diversity radar: –·–, target showing almost no fluctuation; - - - -, complex target with diversity radar (two channels added before the threshold is set); ——, complex target with normal radar.

greater than 0.9 or false alarm probabilities below 10^{-10}, we can assume (to the nearest 0.5 dB) that $n \geqslant 4$ can be taken as large.

Figure 5.22 shows a possible arrangement for a diversity pulse radar with two channels. Transmitter 1 transmits pulses of duration 4 μs with a peak power of 1 MW at 3000 MHz at times 0, 4 ms, 8 ms, 12 ms, ... Transmitter 2 transmits pulses of duration 4 μs with a peak power of 1 MW at 3100 MHz at times 8 μs, 4008 μs, 8008 μs, 12 008 μs, ... Thus the duplexer, the antenna and the entire microwave guide never have to withstand peak powers of more than 1 MW, whereas if transmitters 1 and 2 were to transmit simultaneously the instantaneous peak power in these components could reach 4 MW at times. When this arrangement is used it is necessary to delay the signal received at 3000 MHz by 8 μs before combining the output signals of the two receivers.

5.8.1 Random radars

A similar reduction of the target fluctuation effect is obtained when the carrier frequency is changed n times during the radar measurement (for

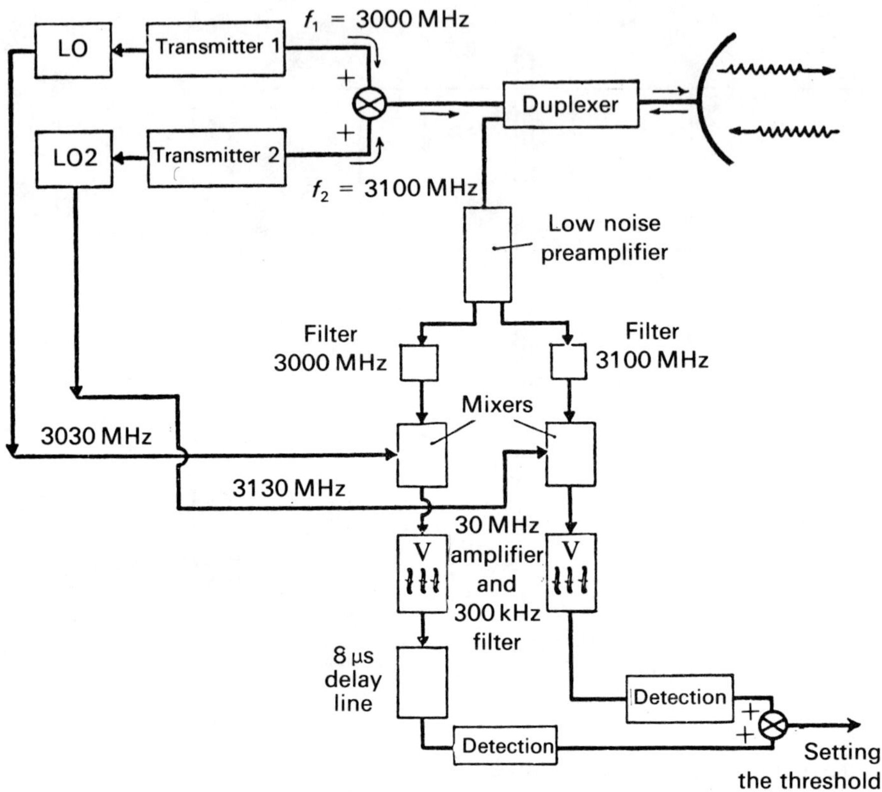

Fig. 5.22 Diversity pulse radar with two channels.

example, by performing each transmission with a carrier frequency selected at random in a given frequency band, thus obtaining a frequency-agile radar). In this case, the equivalent of what would have been obtained with a diversity radar with n channels can be produced. Thus if an agile radar transmits more than four different frequencies during the radar measurement, a target can be considered as nonfluctuating (with a constant radar cross-section which is slightly less than its mean value (1 dB less for an extremely arbitrary fluctuating target and 0–2 dB less in the general case)).

5.9 Conclusions

A mean radar cross-section can be defined for an arbitrary target in free space for a given frequency, presentation and polarization. In practice, the radar cross-section of the target fluctuates on both sides of this mean value, but the fluctuation is usually slow in comparison with the measurement time of the radar (which rarely exceeds a few tens of milliseconds except in tracking radars where the antenna remains permanently pointed in the direction of the target). The direction Δ of the radiant point fluctuates simultaneously, generally around a mean direction Δ_m passing close to the center of gravity of the target, so that Δ is close to Δ_m when the radar cross-section is large but is very far from Δ_m when the radar cross-section is small. The correlation between the direction of the radiant point and the level of the signal received by the radar may also enable angular measurements with large errors to be eliminated. The effect of target fluctuation is minimized when a number of different central frequencies are used during the measurement, and is almost eliminated when this number exceeds 4.

Since the radar cross-section of the target is a random function of time, it has a spectrum. This spectrum normally has a simple pattern in which the amplitude decreases regularly with frequency. However, if the target contains an object such as a propellor, compressor blade etc. rotating at a velocity Ω (in rad s^{-1}), the spectrum contains peaks which can be very high at the fundamental frequency Ω and its harmonics 2Ω, 3Ω, . . . (Fig. 5.23). The reason for this is obvious.

The above statements hold only in infinite free space and for given aspect angle and wave polarization. In practice, at least two constraints must be remembered.

(1) Reflection from the ground may have a marked effect on the behavior of the radar cross-section of a low-flying target.

(2) The presence of the atmosphere may lead to a variation in the intensity of absorption of the energies involved (which must be included in the radar equation).

It should also be remembered that the atmosphere may reflect waves in certain cases. Thus clouds sometimes behave as targets with a very high radar

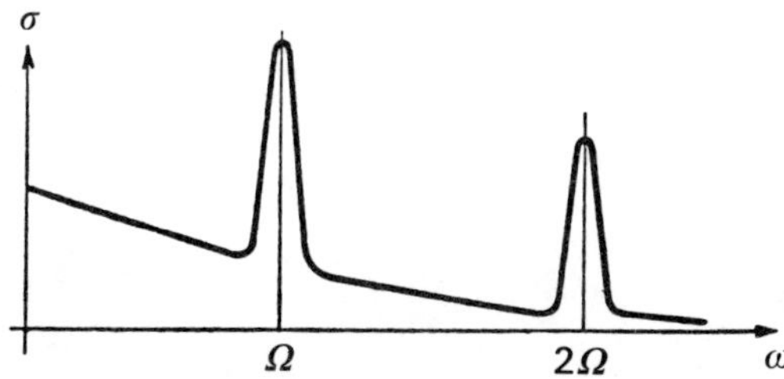

Fig. 5.23 Spectrum of the radar cross-section of a target containing a rotating object.

cross-section in linear polarization so that the targets inside them become invisible. This effect is generally less marked in circular polarization and hence this is usually used in order to "eliminate clouds". When targets are not clouds but metal targets (planes, missiles etc.), it is helpful to know the behavior of a target for the different polarizations used. In fact, little is known about this. However, it appears that in general the mean radar cross-section of a target is almost the same regardless of whether the wave has horizontal, vertical or circular polarization. Radar engineers should also remember that wave propagation is not always rectilinear and they have to understand the phenomena of trophospheric refraction, ionospheric reflection etc.

Remark 5.3

If the radar cross-section of a target varies with the carrier frequency, it is clear that the position of the radiant point also varies with frequency. Thus if the carrier frequency of the radar changes during the time required to measure the radial velocity of the targets, the velocity at which the radiant point is moving over the same time is measured. This velocity can be extremely large. Therefore it is not possible to eliminate fixed echoes with a radar transmission whose central frequency varies over the measurement time (by more than 3×10^{-3} for example).

Chapter 6

Angle Measurement Using Radar

6.1 General comments

Radar is by definition an apparatus for measuring the range of a target (although some radar systems measure only the radial velocity of the targets), but obviously this information alone is not sufficient to locate the target. Direction measurement has been performed, if only in a crude manner, since the early days of radar. In addition, in order to avoid the necessity of transmitting large amounts of energy during the measurement period, the transmission must be concentrated in a preferred direction or a preferred plane.

Historically, therefore, it was necessary to develop radar systems using directional antennas transmitting in a highly preferential manner at a given time in a vertical semiplane and rotating regularly about a vertical axis. This type of system, which is known as two-dimensional search radar, only receives an appreciable signal from the target when the antenna is pointing in its direction and allows a rough measurement of the azimuth of the target to be made. Similarly a quick measure of the altitude of the target can be made using an antenna which also concentrates its transmission in one plane but this time scans about a horizontal axis contained in this plane and perpendicular to the direction of the target. The signal received during scanning with such a height-finding radar is normally a maximum when the target is in the plane of maximum radiation, and hence the elevation angle of the target can be measured and the altitude can be determined (since the radial distance is known).

In tracking radars intended for continuous location of the direction of a single target, the horizontal scanning function and the height-finding function can be obtained rapidly and simultaneously using the same antenna. This antenna, which transmits in a single preferred direction, rotates about a direction Δ (generating a cone of revolution with axis Δ) with the aim of causing Δ to point towards the target. This is normally achieved when the received signal remains constant as the antenna rotates. This system is known as a conical scanning radar. In practice, the movement of the radar beam about Δ is obtained by using a primary source located in front of the circular antenna, in the focal plane but outside the axis Δ. Then the rotation of this source about Δ provides the required scanning.

Since the development of these early systems, methods of measuring angles have been improved by using automatic pulse-counting systems combined with search radars for example or by using monopulse procedures to measure the azimuth or the elevation of the target. The latter procedure can also be used to measure the azimuth and the elevation simultaneously if rough estimates of their values are available.

6.2 Two-dimensional search radar

The reader should consult the paper by Drabovitch [13] as an aid to understanding the concepts discussed in this chapter (particularly Section 6.2.1).

6.2.1 Continuous-wave operation for cooperating targets

We consider a search radar for measuring the azimuth of a cooperating target which transmits a continuous signal with a constant level considerably higher than the level of the signal reflected naturally from the target (skin return), i.e. the target is equipped with a responder. The radar antenna is assumed to rotate about a vertical axis with angular velocity $\Omega \,\mathrm{rad\,s^{-1}}$. Let the voltage gain of this antenna for the elevation at a given time be a function $g(\theta)$ of the azimuth θ. For simplicity assume that the antenna has a regular cross-section and that any section through a horizontal plane is illuminated (by the primary source if we are dealing with a reflector) in the same manner. Then there is a simple relationship between $g(\theta)$ and the illumination $A(v)$ of the antenna which is expressed as a function of the spatial frequency $x/\lambda = v$ of the antenna (Fig. 6.1). $A(v)$ is zero for $|v| > L/\lambda$, where $2L$ is the span of the antenna. In fact, $g(\theta)$ is the Fourier transform of $A(v)$.

When the antenna rotates, the signal received from a target at azimuth θ_0 varies according to

$$Kg(\Omega t - \theta_0)$$

Determination of θ_0 is equivalent to determining the time t_0 when the signal $Kg(\Omega t - \theta_0)$ is a maximum. This problem has already been studied in Chapter 3. The measurement error of the time t_0 is random and gaussian with

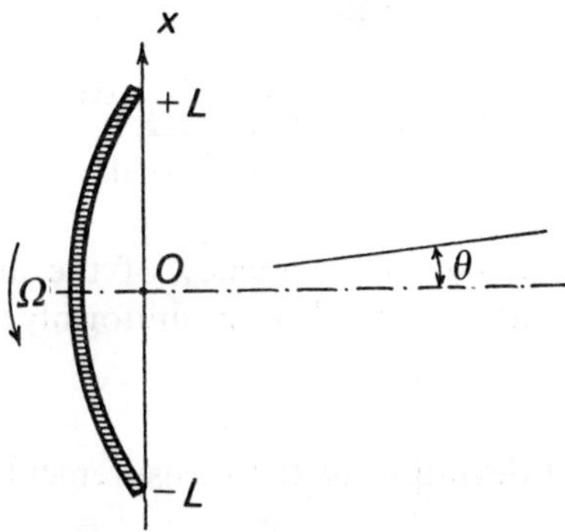

Fig. 6.1 Rotating antenna.

zero mean and standard deviation

$$\frac{1}{2\pi B R^{1/2}}$$

where

$$R = \frac{2E}{N_0}$$

E is the energy of the received signal, N_0 is the spectral density of the accompanying noise and B is the second-order moment of the spectrum of $Kg(\Omega t)$. Drabovitch [13] has shown the following.

(a) The energy E of the signal is a function of the gain in the elevation angle of the antenna in the direction considered (and of course of the responder power and antenna–target distance) and the angular velocity Ω (if the measurement time T is sufficiently large), but it is not a function of the azimuth pattern $g(\theta)$ of the antenna. This follows from the definition of the antenna gain.

(b) The value of B can easily be calculated. Since the Fourier transform of $Kg(\Omega t)$ is equal to $A(f/\Omega)$, apart from a factor, we have

$$B^2 = \frac{\int_{-\infty}^{+\infty} f^2 |A^2(f/\Omega)| \, \mathrm{d}f}{\int_{-\infty}^{+\infty} |A^2(f/\Omega)| \, \mathrm{d}f}$$

$$= \Omega^2 \frac{\int_{-\nu_0}^{+\nu_0} \nu^2 |A^2(\nu)| \, \mathrm{d}\nu}{\int_{-\nu_0}^{+\nu_0} |A^2(\nu)| \, \mathrm{d}\nu} = \Omega^2 B_\nu{}^2$$

where

$$B_\nu^{\,2} = \frac{\int_{-\nu_0}^{+\nu_0} \nu^2 |A^2(\nu)| \, d\nu}{\int_{-\nu_0}^{+\nu_0} |A^2(\nu)| \, d\nu}$$

$\nu_0 = L/\lambda$ is the (spatial) cut-off frequency of the antenna and B_ν is the second-order moment of $|A(\nu)|$. For a uniformly illuminated antenna, $B_\nu = L/\lambda\sqrt{3}$.

Thus the standard deviation of the measurement error of t_0 is

$$\frac{1}{2\pi\Omega B_\nu R^{1/2}}$$

and the standard deviation of the measurement error of θ_0 is

$$\frac{1}{2\pi B_\nu R^{1/2}}$$

This expression is analogous to Woodward's formula in which, under certain conditions, R depends not on the azimuth pattern but on the measurement time and $A(\nu)$ performs exactly the same function for the angular measurement as the Fourier transform $\Phi(f)$ of the transmission signal performs for the range measurement. In particular, the quality of the angular measurement depends only on $|A(\nu)|$ and not on arg $A(\nu)$. Of course, this is only valid to the extent that the operations at reception are carried out properly. It follows from this that, with the reservation made in Section 3.6.3.1, the azimuth discriminating power depends only on $|A(\nu)|$ and that it cannot be better than

$$\frac{1}{2\nu_0} = \frac{\lambda}{2L}$$

6.2.2 Pulse operation on a cooperating target

If the responder transmits a periodic pulse train, it can no longer be assumed that a signal $Kg(\Omega t - \theta_0)$ is being received continuously; rather, the signal is sampled. However, the most probable position of the maximum can theoretically be measured just as well if the sampling frequency is higher than $2\Omega\nu_0 = 2L\Omega/\lambda$ provided that the computer involved is sufficiently sophisticated, which is only rarely the case.

6.2.3 Operation on a noncooperating target

In this case the problem is different because the received signal varies as

$$K|g^2(\Omega t - \theta_0)|$$

and, under certain conditions, the target fluctuates so that K is no longer a constant. The consequences of the fluctuation of K are discussed briefly later. Because of the new shape of the received signal, the autocorrelation function $\varrho_A(\nu)$ of $A(\nu)$ replaces $|A(\nu)|$ as the key curve. One consequence of this is that the extreme limit for resolution can only be improved by $1/4\nu_0$ instead of by $1/2\nu_0$ (i.e. $\lambda/4L$, still with the reservation given in Chapter 3, Section 3.6.3.1). However, the general philosophy is similar (the standard deviation of the measurement error of θ_0 is inversely proportional to the square root of the ratio R).

Remark 6.1

For small values of θ/α the signal received from a target can generally be written in the form

$$y(\theta) = K \exp\left(-\frac{2.8\theta^2}{\alpha^2}\right)$$

where α is the width of the lobe at 3 dB (transmission or reception lobe in the case when the same antenna is used for both transmission and reception). It can therefore be written as a function of time:

$$y(t) = K \exp\left(-\frac{2.8\Omega^2 t^2}{\alpha^2}\right)$$

By using eqn (3.18) (Chapter 3, Section 3.6.4) we find that two signals can be distinguished if the time interval between them is $0.80\alpha/\Omega$, i.e. the angular discriminating power is 0.80α. A search radar can distinguish two targets by their azimuthal difference if this difference is more than three-quarters of the width of the 3 dB azimuth beam when the same antenna is used for transmission and reception. It is obvious that if targets can be distinguished by the differences between their radial velocities or ranges, the fact that they are in the same azimuth is irrelevant. This commonsense statement is not without value.

6.3 General principle of monopulse radars

Figure 6.2 shows the radiation pattern in the vertical plane of maximum radiation of a search radar antenna. A pattern of this type is usually obtained

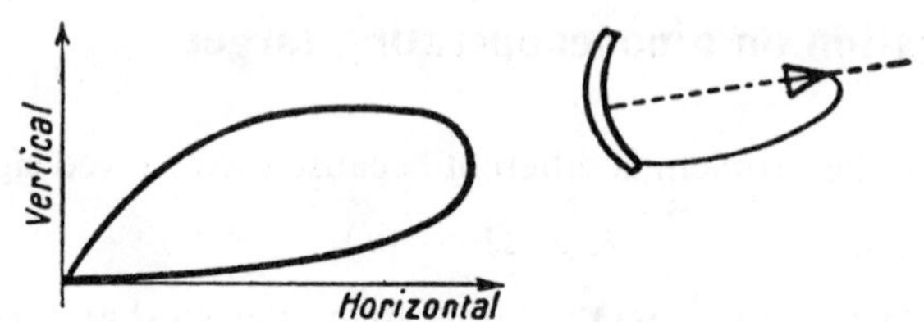

Fig. 6.2 Radiation pattern of a search radar antenna.

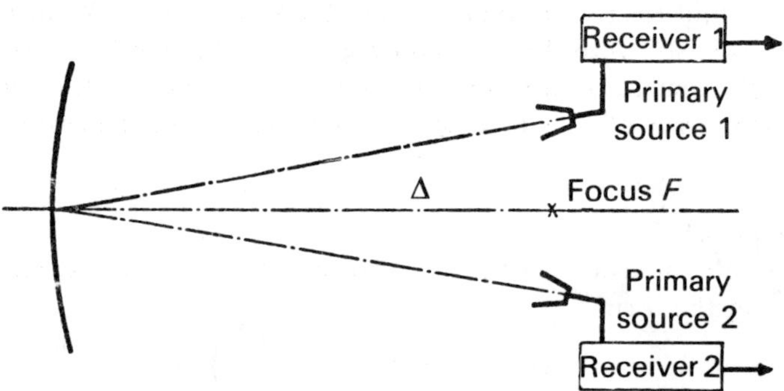

Fig. 6.3 Monopulse radar. The complete figure is symmetric about the line Δ.

by placing a single primary source at the focus of a parabolic reflector (slightly deformed to obtain the "cosecant square").

Let us assume that the antenna consists of a parabolic reflector and two horns located in the same vertical plane on each side of the focus (Fig. 6.3). It is obvious that if the two primary sources transmit identical radiation receivers 1 and 2 will receive two signals of the same amplitude and phase from a target situated on the axis Δ. However, if the target is not on Δ, receivers 1 and 2 will receive signals 1 and 2 respectively which in general do not have the same amplitude and are not in phase. Therefore the elevation angle of the target can be determined by comparing either the phase or the amplitude of the signals received simultaneously at receivers 1 and 2. (The term monopulse follows directly from the fact that the measurement is carried out on signals received simultaneously from the same single transmission signal.) In this case the antenna radiation pattern is as shown in Fig. 6.4. Thus the elevation angle of the target can be measured at each repetition period as the antenna illuminates the target. It is obvious that such a measurement is only possible when the signals received by the two receivers are at a sufficiently high level. If this is not the case it is necessary to compare either the phase or the amplitude of the noise in each receiver; these are normally

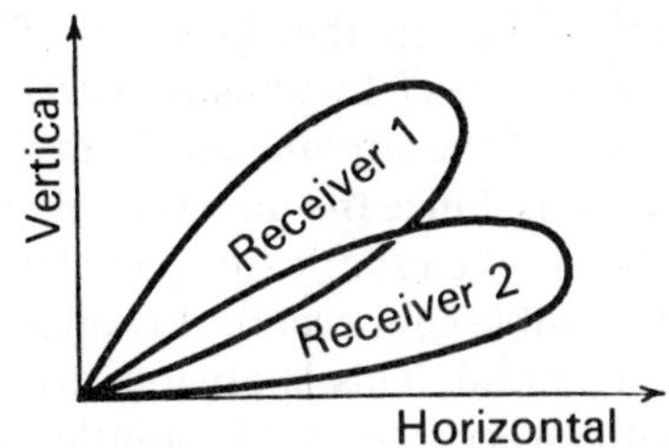

Fig. 6.4 Radiation pattern of a monopulse radar antenna.

only distantly related to the elevation of the target. Therefore, *a priori*, elevation angle measurements obtained using monopulse procedures are only valid when the value of R for each reception channel is sufficiently high, i.e. when the antenna passes over the target.

However, it is possible to instal n horns on the focal line of the reflector so that there are n reception channels. In this case channels 1 and 2, channels 2 and 3, . . . , channels $n - 1$ and n are compared simultaneously. Then, for example, if the received signal with the highest level is in channel 4, the received signal in channel 3 is fairly strong, the received signal in channel 5 is weak and the signals received in the remainder of the channels are very weak, it is obvious that the elevation angle of the target is close to the axis $\Delta_{3,4}$ of beams 3 and 4 and comparison of signals 3 and 4 will provide an accurate value for the elevation of the target.

Let us now assume that we have two targets A and B that cannot be distinguished by their range, their radial velocity or their azimuth and that A is almost on the axis $\Delta_{2,3}$ and B is on the axis $\Delta_{4,5}$. Then receivers 2 and 3 receive an appreciable signal from A and receivers 4 and 5 receive an appreciable signal from B. Thus comparison of 2 and 3 (or interpolation between 2 and 3) and of 4 and 5 makes sense, but comparison of 3 and 4 does not make sense. However, in this case it is very difficult to know which interpolations make sense and which do not. We can also say that the resolution of the elevation angle is poorer than the elevation angle itself between two axes $\Delta_{j-1,j}$ and $\Delta_{j+1,j+2}$. Finally, it is necessary to define a confusion or ambiguity zone in a plane with the four dimensions range, radial velocity, azimuth and elevation, inside which it is not possible to distinguish two targets.

6.4 Angular fluctuation of targets

As the procedures for angular measurement using radar became more accurate, unusual phenomena were observed. For example, in the case of an airplane flying in a straight line in the direction of the radar, the angle measured by the radar was subject to unusual variations: on average, the radar measured the direction of the airplane but from time to time it indicated completely aberrant directions. Since the direction of an airplane with a span of 20 m located 10 000 m from the radar cannot be defined better than to the nearest 2 mrad, it is not possible to measure this direction with an accuracy to better than the nearest 2 mrad. This is obvious from elementary physics; if a quantity is defined with an accuracy E it is pointless to attempt to measure it with an error less than E. However, the large errors observed at times were not expected with expensive high precision radar. When it was noticed that the times when the radar indicated aberrant directions corresponded rather well to the times when the signal received from the target was at a minimum, it was realized that a radar system does not in fact measure the direction of the target but the direction of the normal to the equiphase surface of reradiation from the target, i.e. the direction of the radiant point. This direction can be very different from the target direction, particularly when the received signal is at a minimum (see Chapter 5).

We now return to the measurement of angles using radar. It has been shown in Chapter 5 that although the amplitude of the reradiated field varies along an equiphase surface, this variation is very slow, i.e. it can be assumed that the received field is constant at all points on the antenna (wave collector). Consequently, when a wave collector rotates about a fixed point, the signal received by it is a maximum when all points on the collector are illuminated by the fields in phase, i.e. when the collector is in the plane tangential to the reradiation equiphase surface of the target. Therefore, when the antenna of a search radar rotates about a vertical axis, the maximum received signal is obtained when the azimuth of its principal radiation plane is in the direction of the radiant point (normal to the equiphase surface etc.). Similarly, because the equiphase surfaces are also parallel, if the wave collector encloses a cone of revolution (scanning radar) the received level is constant over the period of revolution (if this period is small with respect to the rate of fluctuation of the target) when the axis of the cone is in the direction of the radiant point.

A monopulse radar with two receivers behaves as if there were two identical collectors whose centers may or may not coincide and which may or may not be parallel. When the plane bisecting these two collectors contains the direction of the radiant point the signals received by the two collectors have the same level and are in phase (Fig. 6.5). However, in the general case when this plane does not contain the direction of the radiant point, the signals received by each collector have different levels and different phases (Fig. 6.6). Thus the direction of the radiant point can be determined either by the equality of the phases of the signals received by the two collectors or by the equality of the amplitudes received by the two collectors. In any event, in the

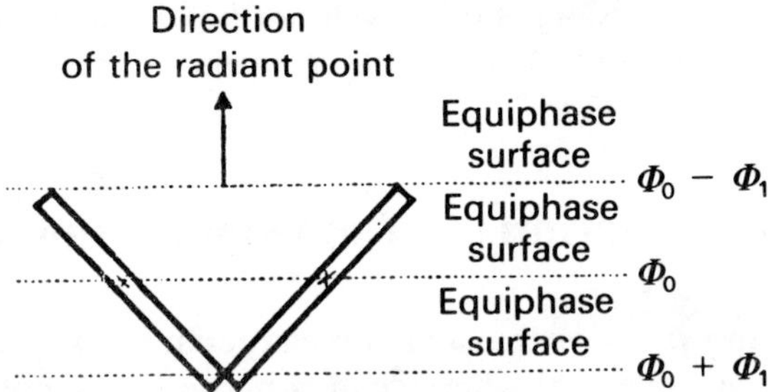

Fig. 6.5 Collectors arrayed such that the plane bisecting them contains the direction of the radiant point.

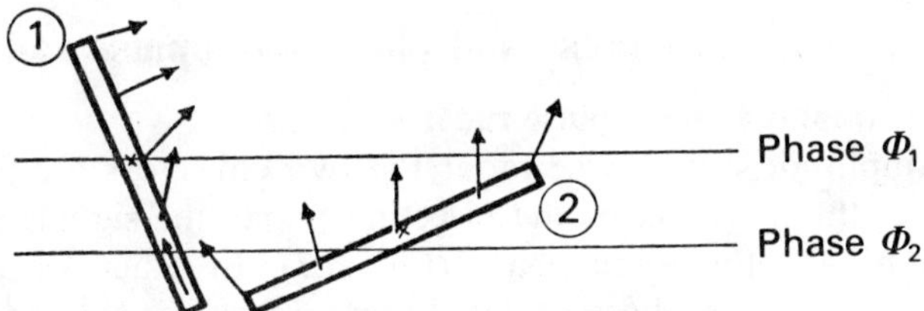

Fig. 6.6 Collectors arrayed such that the plane bisecting them does not contain the direction of the radiant point. In collector 1 the fields have very different phases; their resultant is small and has the phase Φ_1. In collector 2 the fields are almost in phase; their resultant is large and has the phase Φ_2.

general case (the wave collector is small compared with the irregularities of the reradiation equiphase surfaces) a monopulse radar also measures the direction of the radiant point.

6.4.1 Errors in angular measurement using radar

Most errors in angular measurement using radar originate from four different causes:

(1) the fluctuation of the radiant point, which can be reduced by using diversity radar or random radar;

(2) the difference between theory and practice (faults in the antenna patterns, receiver faults etc.);

(3) measurement parameters (widths of the antenna patterns, pulse count criterion etc.);

(4) the noise accompanying the useful information.

In the following sections, fault 1 will not be examined, fault 2 will be examined briefly, and faults 3 and 4 will be examined in detail by means of examples.

6.5 Description of monopulse radars

It has been shown above that a monopulse radar behaves as if there were two identical wave collectors whose centers may or may not coincide and which may or may not be parallel, and that the direction of the radiant point can be determined by the equality of either the phases or the amplitudes of the signals received by the two collectors.

6.5.1 Amplitude monopulse and phase monopulse radars

6.5.1.1 Amplitude monopulse radar

In an amplitude monopulse radar the two collectors are arranged so that, whatever the direction of the bisecting plane, the signals received by each collector have the same phase (Fig. 6.7). In other words the two collectors have the same phase center. Figure 6.8 shows the most common practical arrangement for obtaining two collectors at a particular angle which have approximately the same phase center. This arrangement provides the equivalent of two wave collectors at an angle a/F (where a is the separation of the primary sources and F is the focal length of the parabola) whose phase centers are separated by approximately a (they do not completely coincide). For example in the S band (10 cm) we have $D = 7\,\mathrm{m}$, $F = 5\,\mathrm{m}$, $a = 0.1\,\mathrm{m}$ and $a/F = 1°$, which is the equivalent of two wave collectors with an aperture of 7 m at an angle of 1° whose phase centers are separated by approximately 0.1 m.

6.5.1.2 Phase monopulse radar

In a phase monopulse radar the two collectors are arranged so that, regardless of the orientation of the system, the signals received by each collector have the same amplitude and phases differing by $\Delta\Phi$ (Fig. 6.9).

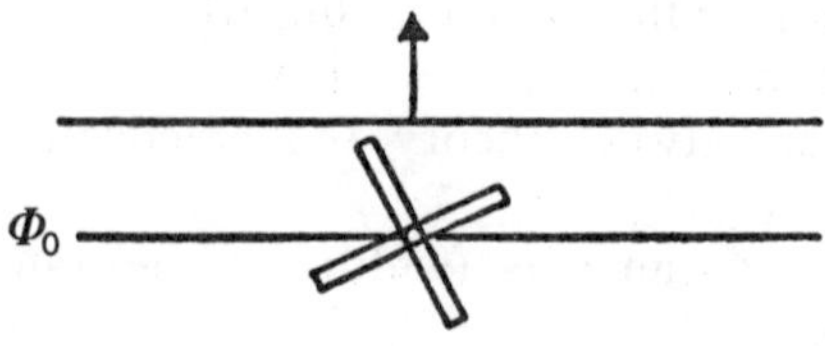

Fig. 6.7 Schematic diagram of an amplitude monopulse radar.

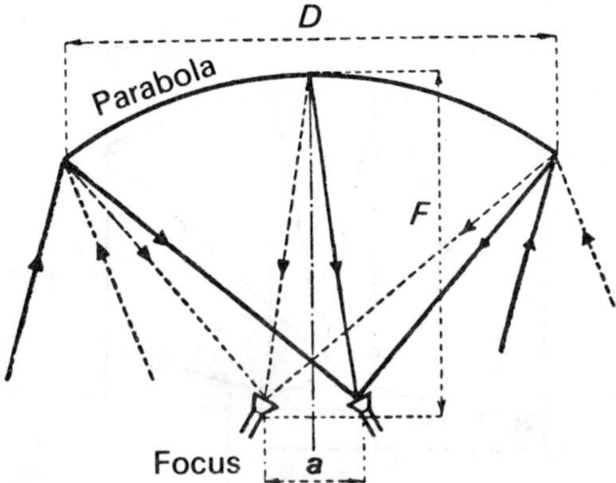

Fig. 6.8 Practical arrangement for obtaining two collectors with the same phase center.

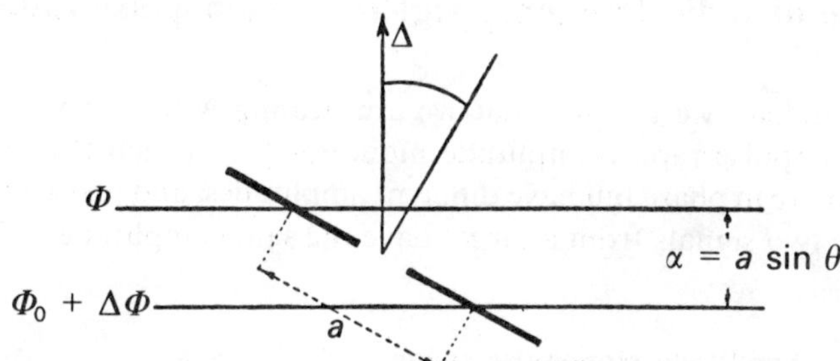

Fig. 6.9 Schematic diagram of a phase monopulse radar.

It can be seen from Fig. 6.9 that

$$\Delta\Phi = \frac{2\pi a \sin\theta}{\lambda}$$

where a is the distance between the phase centers of the two wave collectors. In practice the equivalent of these two wave collectors in parallel is obtained with the arrangement shown in Fig. 6.10. A single reflector is used with two primary sources located at its focus. Each primary source only illuminates part of the reflector, and the two illuminated zones may or may not overlap. For example, in order to obtain two phase centers separated by 2 m and the same lobe aperture as in the previous case (at the same frequency), it is necessary to use an antenna whose aperture is approximately 7 m + 2 m = 9 m.

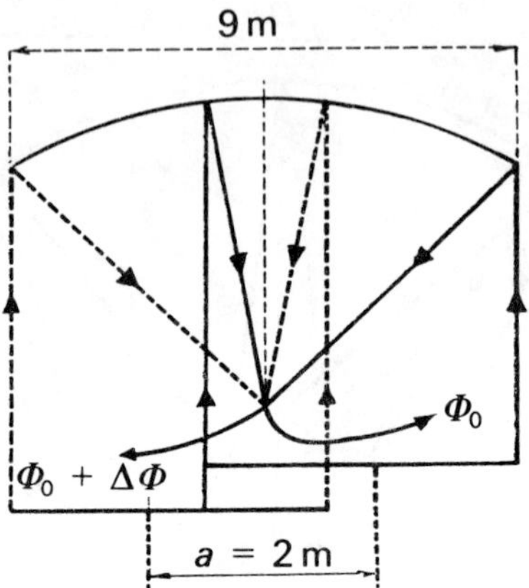

Fig. 6.10 Example of a practical arrangement for a phase monopulse radar.

6.5.2 Use of radio frequency signals in monopulse radars

For simplicity we assume that we are dealing with two very different types of monopulse radar: amplitude monopulse, in which the two signals from a target are in phase but have different amplitudes, and phase monopulse, in which the two signals from a target have the same amplitude but different phases.

6.5.2.1 Amplitude monopulse radar

We denote the expressions giving the amplitudes of the received signals S_1 and S_2 by $G_1(\theta)$ and $G_2(\theta)$ respectively, where θ is the angle which the direction of the target makes with the symmetry plane of the antenna. If the antenna is symmetric, we can write

$$G_1(\theta) = G_2(-\theta)$$

If θ remains constant (a given target direction) but the range or power of the target vary, the levels of both S_1 and S_2 are multiplied by the same factor. The problem consists in obtaining from S_1 and S_2 the best possible signal E which vanishes for $\theta = 0$ and is a linear function of θ. Several solutions of this problem are known.

Amplitude operation (amplitude–amplitude monopulse radar)

The two most common solutions are

$$E = \log\left(\frac{S_1}{S_2}\right)$$

and

$$E = \frac{S_1 - S_2}{S_1 + S_2}$$

As we have already seen, the function $G_1(\theta)$ for many radars can be written

$$G_1(\theta) = G_0 \exp[-k(\theta - \theta_0)^2]$$

A difficulty frequently encountered in producing amplitude monopulse radar is to obtain this distribution with good accuracy. In this case

$$E = \log\left(\frac{S_1}{S_2}\right) = \log\left(\frac{G_1}{G_2}\right)$$

becomes

$$E = k[(\theta + \theta_0)^2 - (\theta - \theta_0)^2] = 4k\theta\theta_0$$

Thus E is a linear function of θ and vanishes for $\theta = 0$. In this expression k characterizes the width of a lobe and $2\theta_0$ the separation of the lobes (Fig. 6.11). As k increases, the lobes become narrower.

In practice, signals S_1 and S_2 are amplified in logarithmic amplifiers which are as identical as possible and then the output signals of these amplifiers are subtracted from each other (in video for example). It can easily be seen that this is where the difficulty of implementation lies. If, for example, we assume that the two amplifiers are truly logarithmic but have different gains, we no longer obtain $E = 4k\theta\theta_0$ but $E = 4k\theta\theta_0 + E_0$ (Fig. 6.12).

If E is given by

$$E = \frac{S_1 - S_2}{S_1 + S_2} = \frac{G_1 - G_2}{G_1 + G_2}$$

we obtain the expression

$$E = \tanh(2k\theta\theta_0)$$

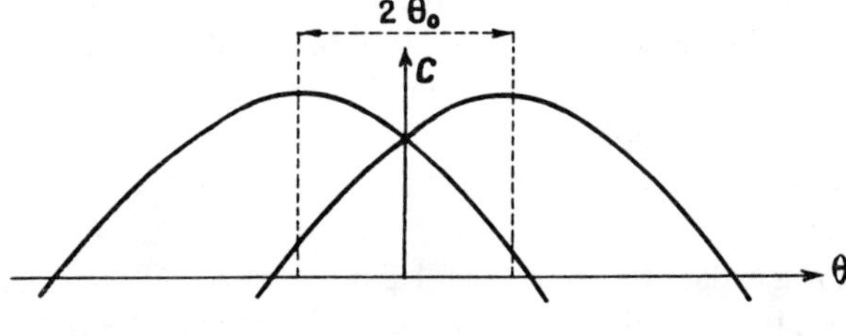

Fig. 6.11

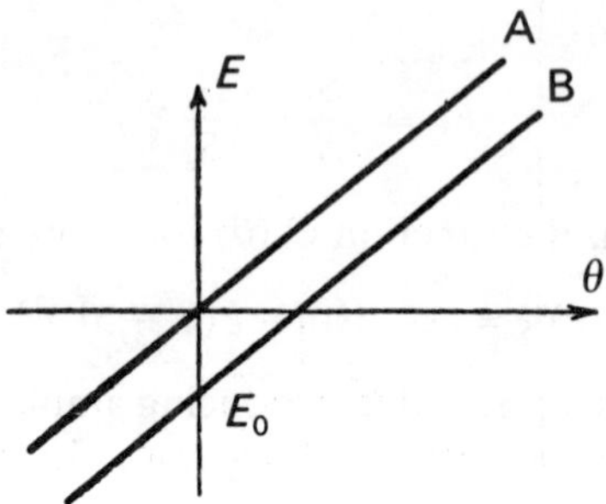

Fig. 6.12 E versus θ for systems with identical logarithmic amplifiers (line A) and logarithmic amplifiers with different gains (line B).

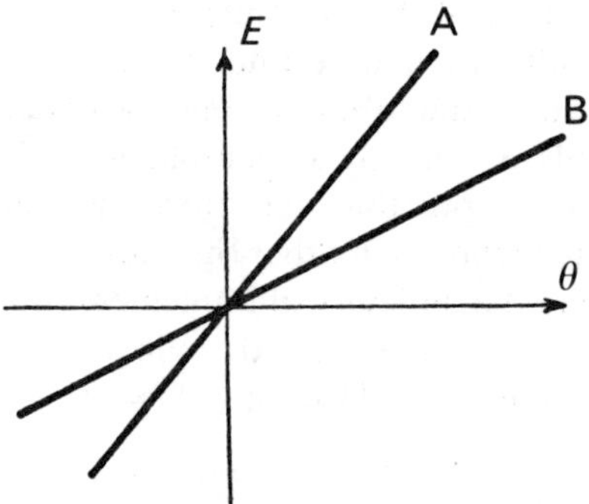

Fig. 6.13 E versus θ for reception chains with the same gain (line A) and different gains (line B).

which is an increasing function of θ whose middle section (θ small) is linear and which vanishes for $\theta = 0$. In practice, the sum $\Sigma = S_1 + S_2$ and the difference $\Delta = S_1 - S_2$ are formed using microwaves (with the help of couplers or magic tees). Σ and Δ are amplified independently, and the quotient is formed at the end of the reception chain (in video for example). What happens if the two reception chains have different gains? In this case kE is measured rather than E. The signal obtained still vanishes for $\theta = 0$ but the slope $\mathrm{d}E/\mathrm{d}\theta$ is no longer correct (Fig. 6.13).

It can therefore be seen that these two procedures should not be used indiscriminately.

In a tracking radar the error signal E must be zero for $\theta = 0$, but a slight variation in the slope $\mathrm{d}E/\mathrm{d}\theta$ is acceptable. In contrast, in a three-dimensional search radar E need not vanish completely for $\theta = 0$, but $\mathrm{d}\theta/\mathrm{d}E$ should not vary with time or as a function of θ. Thus the solution

$$E = \frac{S_1 - S_2}{S_1 + S_2}$$

is generally preferred for tracking radar and the solution

$$E = \log\left(\frac{S_1}{S_2}\right)$$

is generally preferred for three-dimensional search radar.

Phase operation (amplitude–phase monopulse radar)

The main problem encountered in amplitude operation is that different chains with exactly the same gain are required. This drawback can be overcome by using the following procedure.

We form the sum $\Sigma = S_1 + S_2$ and the difference $\Delta = S_1 - S_2$ in the microwave region before any amplification or frequency shift. In this way we obtain two signals in phase (or with opposite phases if $S_1 - S_2 < 0$). We then use 3 dB couplers to form $\Sigma + \mathrm{j}\Delta$ and $\Sigma - \mathrm{j}\Delta$, and in this way obtain two signals which are in phase if $\Delta = 0$ and whose phase difference Φ is given by

$$\Phi = 2 \arctan\left(\frac{\Delta}{\Sigma}\right)$$

if Δ is nonzero (Fig. 6.14). With the assumptions of the previous section we obtain

$$\tan\left(\frac{\Phi}{2}\right) = \tanh(2k\theta\theta_0)$$

The phase difference Φ is measured at an intermediate frequency using standard procedures and provides a measurement of θ. The curve of Φ as a function of θ passes through the origin ($\Phi = 0$ for $\theta = 0$) and is linear near $\theta = 0$. Any difference in the gains of the two reception chains has no effect on the measurement of Φ. However, if one of the chains shifts the phase of

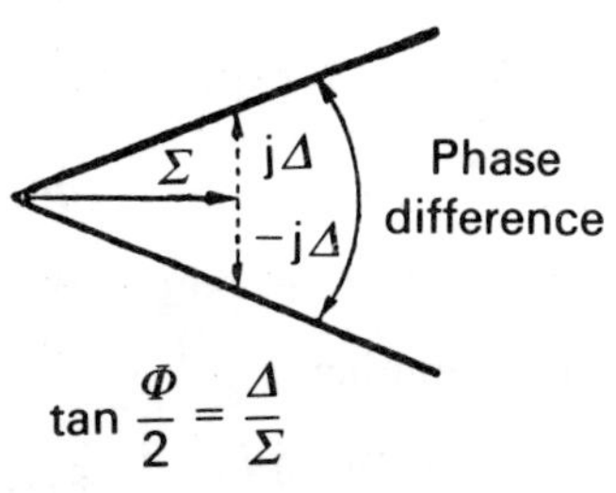

Fig. 6.14

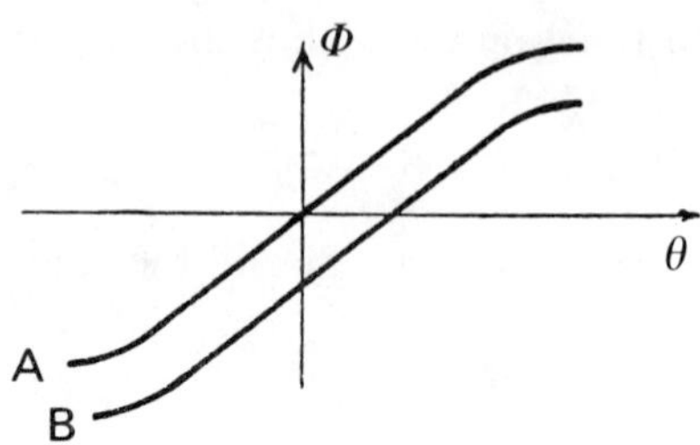

Fig. 6.15 Φ versus θ for amplification giving the correct phase (curve A) and for amplification in which the phase of one signal is shifted with respect to the other (curve B).

one signal with respect to the other, this will have the effect of shifting the curve $\Phi = f(\theta)$ which will no longer pass through the origin (Fig. 6.15). However, it is easier technologically to obtain two correct amplification chains of this type than to obtain two chains with the same gain.

6.5.2.2 Phase monopulse radar

In a phase monopulse radar the phase difference Φ between the signals S_1 and S_2, which have the same amplitude, is measured. This phase difference is proportional to the sine of the off-axis angle:

$$\Phi = k \sin \theta$$

The obvious procedure is to measure Φ. It is not essential for the two reception chains to have the same gain but they must not shift the phase of one signal with respect to that of the other. A radar of this type is known as a monopulse phase–phase radar.

We can also define the difference

$$\Delta = S_1 - S_2$$

and the sum

$$\Sigma = S_1 + S_2$$

Since

$$S_1 = S_2 \exp(-\mathrm{j}\Phi)$$

we obtain

$$\Delta = S_2[1 - \exp(-\mathrm{j}\Phi)]$$

$$\Sigma = S_2[1 + \exp(-\mathrm{j}\Phi)]$$

and Δ/Σ is a pure imaginary number

$$\frac{\Delta}{\Sigma} = \frac{1 - \exp(-\mathrm{j}\Phi)}{1 + \exp(-\mathrm{j}\Phi)} = \mathrm{j} \tan\left(\frac{\Phi}{2}\right)$$

whose modulus is

$$\frac{\Delta}{\Sigma} = \tan\left(\frac{\Phi}{2}\right)$$

Since the modulus of a quotient is equal to the quotient of the moduli, it is possible to determine Φ by measuring the ratio of the amplitudes Δ and Σ of the signals.

In principle, Δ and Σ can be obtained directly in the microwave region and the arguments of Section 6.5.2.1 apply. In particular, it is important that the gains of the two chains used are the same; if they are different there is a risk that the slope of the curve of the error signal as a function of the tracking error is incorrect. Signals S_1 and S_2 must follow equal trajectories prior to the combination which produces Δ and Σ, otherwise a zero error will be made which is serious in the case of a tracking radar.

6.5.3 Effect of receiver noise on the accuracy of angular measurements made using a monopulse radar

6.5.3.1 Phase monopulse radar

In a phase monopulse radar it is necessary to measure the phase difference Φ between two signals which are each received by a wave collector followed by a receiver. This phase difference is related to the angle θ to be measured as follows:

$$\Phi = k\theta$$

We now need to determine the accuracy with which this phase difference can be defined bearing in mind the fact that each of the signals consists of the useful signal accompanied by noise. For simplicity we assume that the useful signal has a constant amplitude S and an angular frequency ω, which is either fixed (classical radar) or varies with time (pulse compression radar), and we determine the accuracy to which the phase of such a signal can be defined when it is accompanied by noise. This assumption is frequently valid in practice.

We consider a signal $S \sin(\omega t)$ with duration T, which has zero phase by definition, accompanied by a gaussian noise occupying a spectrum of width Δf. It should be noted that ω is the instantaneous angular frequency of the signal and can vary as a function of time over the duration T of the measurement. The power of the signal is therefore given by $S^2/2$ and the energy of the signal over the measurement time is given by $S^2T/2$. The noise can be written in the form

$$n(t) \sin[\omega t + \varphi(t)]$$

where $n(t)$ and $\varphi(t)$ are random functions of time. The mean value of

$$n^2(t)\sin^2[\omega t + \varphi(t)]$$

represents the mean power of the noise and the mean value of

$$\frac{n^2(t)\sin^2[\omega t + \varphi(t)]}{\Delta f}$$

represents the spectral density of the noise. It is assumed that the noise is low with respect to the useful signal (R is large) (this assumption is justified by the considerations of Chapter 3). Under these conditions, the combination of the useful signal and noise can be written

$$S\sin(\omega t) + n(t)\sin[\omega t + \varphi(t)] = \sin(\omega t)(S + n\cos\varphi) + \cos(\omega t)\, n\sin\varphi$$

At any given time the apparent phase Φ_a of the combination of the useful signal and noise is not generally zero and can be expressed as

$$\tan\Phi_a = \frac{n\sin\varphi}{S + n\cos\varphi}$$

$$\Phi_a \approx \frac{n(t)\sin[\varphi(t)]}{S}$$

The measurement of Φ_a can be repeated a number of times and the average taken. However, it has already been established that it is only possible to carry out $T\Delta f$ independent measurements. Finally, the measured value of Φ appears as a gaussian random variable whose variance is given by

$$\frac{\overline{n^2\sin^2\varphi}}{S^2 T\Delta f} = \frac{1}{2R}$$

The presence of noise means that, instead of measuring a zero phase, a value differing from zero is found. This error appears as a gaussian variable with zero mean and standard deviation $1/(2R)^{1/2}$, i.e. when we measure the phase of a useful signal accompanied by noise we make a gaussian error whose standard deviation is $1/(2R)^{1/2}$ rad.

Thus, when two signals with the same power accompanied by (independent) noise with the same characteristics are received in each reception channel of a monopulse radar, the error in the measurement of the phase difference between these two signals has a standard deviation of $(1/R)^{1/2}$ rad, where R is the ratio of the energy received during the measurement to the spectral density of the noise for one of the signals. Therefore the error due to noise in the measurement of the angle θ has a standard deviation inversely proportional to $R^{1/2}$. We thus find a similar result to that obtained for the error in the radar measurement of range or radial velocity.

6.5.3.2 Amplitude monopulse radar

As an example we consider the case where θ is obtained by calculating

$$E = \log\left(\frac{S_1}{S_2}\right) = \log S_1 - \log S_2$$

where

$$S_1 = S_0 \exp[-k(\theta - \theta_0)^2]$$

$$S_2 = S_0 \exp[-k(\theta + \theta_0)^2]$$

We assume that S_1 is accompanied by a noise n_1 and S_2 is accompanied by a noise n_2 (these are independent gaussian noises occupying a spectrum of width Δf, where the useful signal lasts for a time T).

Instead of measuring $E = \log S_1 - \log S_2$, we measure

$$E + \mathrm{d}E = \log(S_1 + n_1) - \log(S_2 + n_2)$$

where

$$\mathrm{d}E = \frac{n_1}{S_1} - \frac{n_2}{S_2}$$

We can make $T\Delta f$ independent measurements and the variance of the final error in the measurement of E is given by

$$\frac{1}{2R_1} + \frac{1}{2R_2}$$

If $\theta = 0$, the two received signals S_1 and S_2 are equal on average if they are transmitted to similar receivers (same characteristics for the noise in the two receivers). We therefore have

$$R_1 = R_2 = R_0$$

and the standard deviation of the error in the measurement of E is $1/R^{1/2}$, which is a very similar result to that found for the phase monopulse radar.

6.5.4 Relationship between the monopulse procedures

In the preceding sections we have distinguished between amplitude monopulse radar and phase monopulse radar because historically the two procedures were developed in parallel by teams which were often in competition, each being convinced of the superiority of its own solution. It has since been shown [14] that these two breeds of monopulse radar are not really different but to a first approximation are two different ways of doing the same thing.

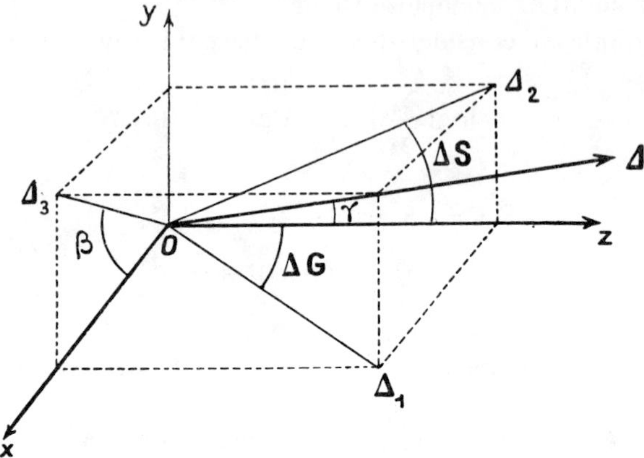

Fig. 6.16 Geometry of a conical scanning radar: Oz, antenna axis; xOy, reference plane of the meridian; Δ, target direction.

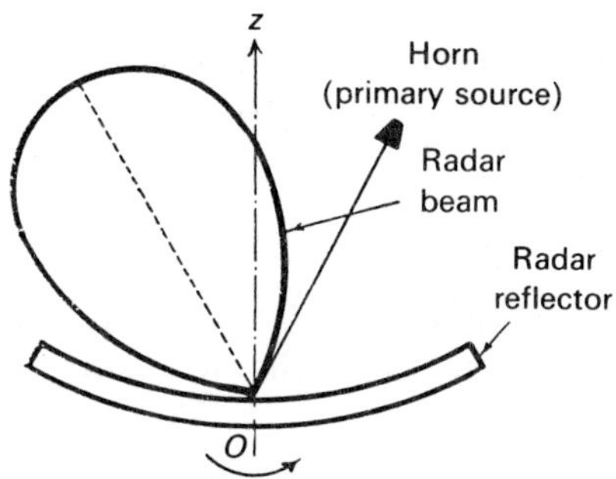

Fig. 6.17 Set-up for measuring γ and β.

6.6 Conical scanning radar

The aim of conical scanning radar, which is generally used for tracking, is to measure the three coordinates of a target with respect to a coordinate system whose reference is the radar antenna. To do this we first measure the range using one of the procedures described already, and at the same time we measure the angle γ between the axis of the antenna (which is an axis of revolution) and the target direction, and the angle β between the antenna axis plane, the target direction and a reference plane passing through the antenna axis (meridian plane) (Fig. 6.16).

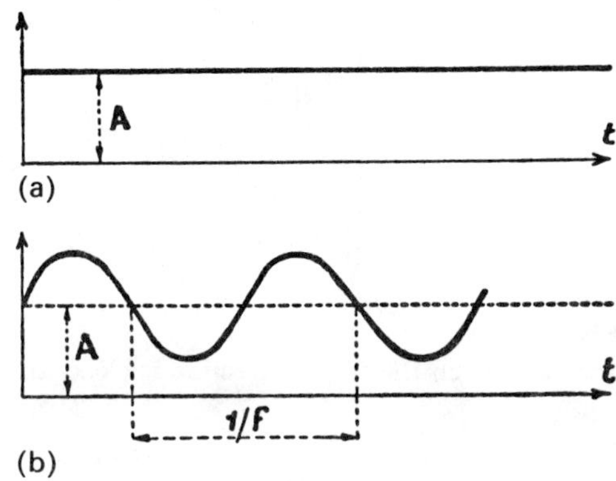

Fig. 6.18 Signal received when the target is located (a) on and (b) off the axis of the reflector.

The principle for measuring β and γ is as follows (Fig. 6.17). A circular antenna is illuminated by a primary source located outside the axis which rotates about this axis with a velocity f rev s^{-1}. Under these conditions the received signal has a constant amplitude if the target is located on the axis of the reflector (Fig. 6.18(a)) but its amplitude varies as a function of time if the target is not on the axis (Fig. 6.18(b)). This is an approximately sinusoidal signal with frequency f whose phase with respect to a reference sinusoidal wave (with frequency f) gives the angle β of the antenna–target axis plane with respect to a reference plane. Furthermore, as the target direction becomes closer to the reflector axis the amplitude of the sinusoidal signal becomes smaller (this is only valid for target directions close to the reflector axis). Thus the received signal is approximately

$$S = A[1 + k\gamma \sin(2\pi ft + \beta)] \tag{6.1}$$

In fact, the received signal is a signal with a radio frequency F (Fig. 6.19) (10 000 MHz for example) and A is a function of the distance from the target to the radar.

However, if the received signal is appropriately detected and a suitable automatic gain control system is used, after amplification we obtain at the output of the receiver a signal represented by eqn (6.1) in which A is a constant. After this signal has been passed through a highpass filter designed to eliminate the continuous component, we obtain a signal

$$S' = K\gamma \sin(2\pi ft + \beta) \tag{6.2}$$

Thus when γ and β vary slowly, the output signal of the receiver appears as a signal with frequency f amplitude modulated by γ and phase modulated by β.

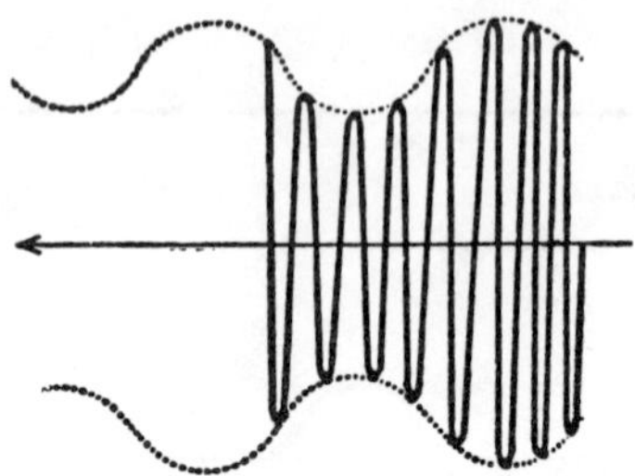

Fig. 6.19 Signal received when the target direction is close to the reflector axis.

We now need to extract the values of γ and β from S'. They are given by

$$\gamma \sin \beta \approx \tan \gamma \sin \beta = \tan \Delta S \approx \Delta S$$
$$\gamma \cos \beta \approx \tan \gamma \cos \beta = \tan \Delta G \approx \Delta G$$

and represent the angle between the plane xOz and the plane $xO\Delta$ and the angle between the plane yOz and the plane $yO\Delta$ respectively (see Fig. 6.16). $\gamma \sin \beta$ and $\gamma \cos \beta$ are generally error signals, and two separate servomechanisms are used in order to eliminate them by classical procedures so as to make Oz coincide with $O\Delta$ when the radar is used for tracking.

We therefore multiply S' by $\sin(2\pi ft)$, i.e. we demodulate S' by a signal in phase with the reference signal, and obtain

$$S' \sin(2\pi ft) = \frac{K\gamma}{2}[\cos \beta - \cos(4\pi ft + \beta)]$$

If this demodulation or multiplication is followed by lowpass filtering which eliminates the frequency $2f$ (this filtering is normally carried out by the servomechanism), we finally obtain

$$\frac{K}{2} \gamma \cos \beta$$

Similarly, multiplication of S' by $\cos(2\pi ft)$ demodulation of S' by a signal in quadrature with the reference) followed by lowpass filtering gives

$$\frac{K}{2} \gamma \sin \beta$$

6.7 Effect of target fluctuation on the accuracy of radar measurements

In addition to the fact that the amplitude fluctuation of targets is inevitably accompanied by fluctuation of the radiant point which results in fluctuation of the angular measurement regardless of the procedure used, amplitude fluctuation itself introduces problems, two examples of which are examined here.

6.7.1 Conical scanning radar

If the target is located on the radar axis ($\beta = 0$ and $\gamma = 0$) but its amplitude fluctuates at the scanning frequency, S in eqn (6.1) varies according to

$$S = S_0 \sin(2\pi f t + \beta)$$

Hence it can be deduced that the target does not lie on the axis. This is also true if the target fluctuates at a frequency close to the scanning frequency while the difference between these two frequencies is less than the pass band of the radar servomechanism. Even if the amplitude fluctuation of S is negligible, the fluctuation of the coordinates β and γ of the target may introduce a similar undesirable effect if the modulation operation is imperfect (which is always the case).

6.7.2 General case

Radar measurements of range, radial velocity and angle are normally distorted by noise, and the corresponding errors are generally gaussian with a standard deviation of $K/R^{1/2}$ where R is the ratio of the energy received during the measurement to the spectral density of the noise. If the target does not fluctuate R is independent of time and is equal to its mean value R_0. Thus we can say that, because of the noise, the measurement has a gaussian error ε with standard deviation $K/R_0{}^{1/2}$ and probability density

$$p_1(\varepsilon) = \frac{1}{K}\left(\frac{R_0}{2\pi}\right)^{1/2} \exp\left(-\frac{\varepsilon^2 R_0}{2K^2}\right)$$

Let us now assume that the target is fluctuating. For example, there is a 50% chance that $R = 0.1R_0$ and a 50% chance that $R = 1.9R_0$ so that the mean value of R is equal to R_0. The probability density of ε is then

$$p_2(\varepsilon) = \frac{0.5}{K}\left(\frac{R_0}{2\pi}\right)^{1/2}\left[0.32 \exp\left(-\frac{\varepsilon^2 R_0}{20K^2}\right) + 1.38 \exp\left(-\frac{0.95\varepsilon^2 R_0}{K^2}\right)\right]$$

and it is no longer gaussian.

Numerical examples of the two cases are given below.

(1) Nonfluctuating R:
the probability that $|\varepsilon| < 0.5K/R_0^{1/2}$ is 0.38
the probability that $|\varepsilon| < K/R_0^{1/2}$ is 0.68
the probability that $|\varepsilon| < 2K/R_0^{1/2}$ is 0.95
the probability that $|\varepsilon| < 3K/R_0^{1/2}$ is 0.997

(2) There is a 50% chance that $R = 0.1R_0$ and a 50% chance that $R = 1.9R_0$:
the probability that $|\varepsilon| < 0.5K/R_0^{1/2}$ is 0.31
the probability that $|\varepsilon| < K/R_0^{1/2}$ is 0.53
the probability that $|\varepsilon| < 2K/R_0^{1/2}$ is 0.73
the probability that $|\varepsilon| < 3K/R_0^{1/2}$ is 0.83

In other words, in the example chosen here the fact that the target fluctuates has multiplied errors with low probability by 15%, errors with 70% probability by 2 and errors with 95% probability by more than 5.

In the case of a very complex target which can be assumed to fluctuate according to a Rayleigh distribution, it can be shown that the probability density of the error has a Student distribution† and that

the probability that $\varepsilon < 0.5K/R_0^{1/2}$ is 0.34
the probability that $\varepsilon < K/R_0^{1/2}$ is 0.58
the probability that $\varepsilon < 2K/R_0^{1/2}$ is 0.81
the probability that $\varepsilon < 3K/R_0^{1/2}$ is 0.90

It is therefore incorrect to say that the error due to noise in the measurement of a parameter of a fluctuating target is gaussian. The errors for a fluctuating target are worse than those for a nonfluctuating target with the same radar cross-section and become increasingly worse as the probability increases.

† It is found that the probability density can be written as

$$p_3(\varepsilon) = \frac{1}{K(2\pi R_0)^{1/2}} \int_0^\infty \exp\left(-\frac{R}{R_0}\right) R^{1/2} \exp\left(-\frac{\varepsilon^2 R}{2K}\right) \mathrm{d}R$$

$$p_3(\varepsilon) = \frac{R_0^{1/2}}{K} \frac{1}{(2 + R_0\varepsilon^2/K^2)^{3/2}}$$

This calculation is due to L. Gerardin (Thomson-CSF) who was one of the first to study the effect of target fluctuation on the accuracy of radar measurement.

6.7.3 Accuracy of a radar system performing several consecutive measurements

The accuracy of radar measurements can sometimes be improved by carrying out n consecutive measurements. However, the improvement obtained depends on the conditions existing during the measurements.

If the target does not fluctuate during the measurement period and if all n measurements are identical, i.e. if R is the same for all measurements, the errors are divided by $n^{1/2}$ according to well-known statistical theory.

If the target does not fluctuate during the measurement period but, as is the case in a two-dimensional search radar, R varies during the sequence of n measurements, the above result obviously no longer holds. Thus in a search radar, measurements of range, radial velocity or angle can only be averaged over a number of measurements a little less than $\alpha f/V$, where α is the 3 dB width of the azimuth beam, f is the recurrence frequency of the radar and V is the velocity of rotation of the antenna. Under these conditions, the errors are divided by approximately $(\alpha f/2V)^{1/2}$.

If the target fluctuates from one measurement to the next, as is the case in random radar, it can no longer be assumed that there are n measurements each with a gaussian error. Thus the final variance will not be given by the sum of the elementary variances divided by n^2. For example, if the target fluctuates following a Rayleigh distribution, each measurement has an error with a Student distribution and averaging the measurements produces little improvement in their accuracy.

6.8 Fluctuation of the radiant point. Angular target signature

We explained in Section 6.4 that rather unusual results may be obtained in the presence of a target which is not a point because the normal to the equiphase surface does not generally lie in the direction of the center of the target but in the direction of a migratory radiant point and the radar measures the direction of this normal.

When the collector, which is assumed to be planar with a span $2L$, is located in the reradiation field of a target, the field $E(x)$ at a point on the collector with abscissa x $(-L < x < L)$ is given by

$$E(x) = E_0 + E_1 \sin\left(2\pi \frac{x}{2L}\right) + E_1' \cos\left(2\pi \frac{x}{2L}\right) + E_2 \sin\left(2\pi \frac{x}{L}\right) + E_2' \cos\left(2\pi \frac{x}{L}\right) + \dots$$

If E_0 and its variation with time are known it is possible to measure two independent target parameters. The same result is obtained if both E_0 and E_1 are known at a single time. However, the measurement of more parameters than these provides more information about the target.

For example, a dumbbell target is characterized by four independent parameters: the amplitude of the signal received from the sphere at one end of the dumbbell, the amplitude of the signal from the sphere at the other end and the azimuths of the two spheres (it is sufficient to work in the horizontal plane). If

$$E_0 = \frac{1}{2L}\int_{-L}^{+L} E(x)\,\mathrm{d}x$$

is known, the amplitude of the signal received from the two spheres can be determined. If

$$E_1 = \frac{1}{L}\int_{-L}^{+L} E(x)\sin\left(2\pi\frac{x}{2L}\right)\mathrm{d}x$$

is known, the direction of the radiant point can be determined. In a monopulse amplitude radar E_0 is given (very approximately) by the sum signal and E_1 is given (very approximately) by the difference signal. However, we can also measure†

$$E_1' = \frac{1}{L}\int_{-L}^{+L} \bar{E}(x)\cos\left(2\pi\frac{x}{2L}\right)\mathrm{d}x$$

which is sometimes called a deviation signal, and

$$E_2 = \frac{1}{L}\int_{-L}^{+L} \bar{E}(x)\sin\left(2\pi\frac{x}{L}\right)\mathrm{d}x$$

This information is sufficient to characterize the dumbbell target completely.

This procedure enables us to improve target characterization (to obtain a target signature) and possibly to improve the resolving power of the antenna (with the reservation given in Chapter 3, Section 3.6.3.1). In order to extract valid information from the measurement of E_2, E_2', . . . , an adequate measurement margin (signal-to-noise ratio) is required, and hence it can be said that the resolution depends on the signal-to-noise ratio. Thus we are departing further and further from the initial concepts of resolution in optics.

† To the best of our knowledge S. Drabovitch (Thomson-CSF) was the first to suggest this measurement and to devise a method of performing it by using multimode sources.

Chapter 7

Data Processing of Radar Information. Radar Coverage

7.1 Coherent integration

For simplicity let us assume that the radar makes a total of n identical measurements, i.e. the antenna gain in the target direction does not vary during these measurements. Each of these elementary measurements lasts for a time T' and corresponds to a value of R. Thus in a single measurement we receive a useful signal with amplitude $R^{1/2}$ accompanied by a gaussian noise with a standard deviation of unity. If the total measurement time T is less than $1/f_D$, where f_D is the uncertainty in the Doppler frequency of the target or the Doppler frequency of the radiant point of the target which is assumed to have zero velocity in the absence of any information to the contrary, it is possible to sum the n received signals to obtain a useful signal with amplitude $nR^{1/2}$ accompanied by a gaussian noise with standard deviation $n^{1/2}$. Thus the radar behaves as if it had received a useful signal with amplitude $(nR)^{1/2}$ accompanied by a gaussian noise with a standard deviation of unity. The radar also behaves as if the received energy under consideration were the total energy received over the total measurement time. Therefore the results given in Chapter 4 are still valid, and it is not necessary to know whether the measurement was made continuously over a time T or whether it was made over n time intervals of duration T' with $nT' < T$.

If we are dealing with a classical pulse radar, it is assumed that the frequency difference between the local oscillator and the transmitter is almost constant during the measurement, i.e. any variation is less than to the nearest $1/T$, and that summing is carried out correctly before detection. It is then said that the radar is performing coherent integration and classical radars satisfying the first of these conditions are called coherent. No additional conditions are assumed for correlation radars such as those described in Chapter 4, Section 4.1.

7.2 Integration after detection

Let us assume that the radar is coherent, that the antenna gain in the target direction does not vary during the measurement time T and that the target does not fluctuate during T. We also assume that the target has a zero radial velocity, although it has a Doppler frequency f_D (an error f_D is made in our knowledge of the Doppler frequency). For simplicity, we assume that the radar transmits a signal of constant amplitude and frequency f during each elementary measurement of duration T' and that $T' \ll 1/f_D$.

The first measurement is made between $t = 0$ and $t = T'$.
The second measurement is made between $t = t_1$ and $t = t_1 + T'$.

$\vdots$

The kth measurement is made between $t = t_{k-1}$ and $t = t_{k-1} + T'$.
The nth measurement is made between $t = t_{n-1}$ and $t = t_{n-1} + T'$.

The output signal of the radar is therefore, apart from a factor, the integral over the duration of an elementary measurement of

$$\cos(2\pi f t)\{A\cos[2\pi(f + f_D)t] + \text{noise}\}$$

where f is, if necessary, a slowly changing variable (we ignore the case, which occurs rarely in practice, of amplitude modulation within the signal). Thus we obtain a useful signal

$$\int_{t_{k-1}}^{t_{k-1}+T'} A\cos[2\pi(f + f_D)]\,t\cos(2\pi f t)\,\mathrm{d}t \quad = \quad B\cos(2\pi f_D t_{k-1})$$

accompanied by output noise. The useful signal can therefore be written as

$$R^{1/2}\cos(2\pi f_D t_{k-1})$$

which gives $R^{1/2}$ if $f_D = 0$ and is accompanied by a gaussian noise with a standard deviation of unity (Fig. 7.1).

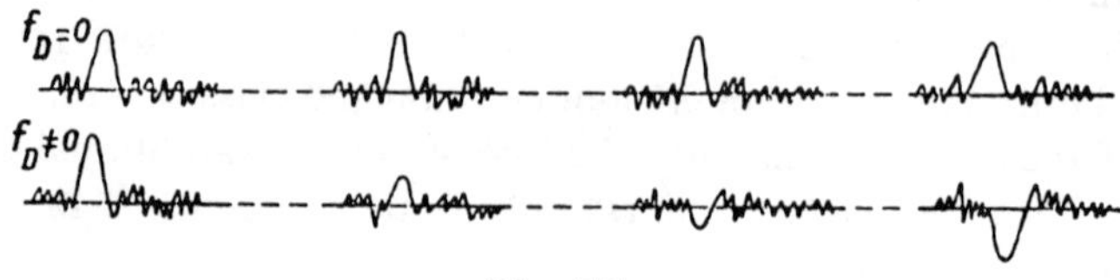

Fig. 7.1

If the total duration of the n elementary measurements is large compared with $1/f_D$ ($T > 1/f_D$), about half of the n useful signals will be positive and about half will be negative so that their sum will be approximately zero, although the addition of the noise components will give a total noise with a standard deviation of $n^{1/2}$. Thus if the total duration of the measurement is greater than $1/f_D$ no advantage is gained by performing coherent integration on the n elementary measurements contained in the time T. In this case, it is customary to "detect" the elementary signals and then sum them. This is known as noncoherent integration after detection.

The detection operation is performed on the signal with intermediate frequency (to avoid a loss of 3 dB). This means that instead of working on the combination

$$R^{1/2}\cos(2\pi f_D t_{k-1})\,C(t) + b_{k-1}(t)$$

where $\overline{b_{k-1}{}^2} = 1$, at the kth measurement ($C(t)$ is unity at the signal position and zero elsewhere), the receiver supplies

$$R^{1/2}\cos[2\pi(f_i + f_D)\,t_{k-1}]\,C(t) + d_{k-1}(t)\cos[2\pi f_i t + \varphi_{k-1}(t)]$$

where f_i is the intermediate frequency and

$$n_{k-1}(t) = d_{k-1}(t)\cos[2\pi f_i t + \varphi_{k-1}(t)]$$

is a gaussian noise with zero mean and power equal to unity which implies that $d_{k-1}(t)$ has a Rayleigh distribution (see Chapter 1, Section 1.8). This is the signal which is detected. The detection characteristic is not simple in general, but for simplicity it is assumed to obey a square law (this is not too far removed from the behavior observed in practice). It is therefore assumed that the elementary signals are squared (the high frequency components are eliminated) and then added.

After square-law detection and elimination of the high-frequency components, we obtain

$$RC^2(t) + 2C(t)\,R^{1/2}d_{k-1}(t)\cos[2\pi f_D t_{k-1} - \varphi_{k-1}(t)] + d_{k-1}{}^2(t)$$

where $RC^2(t)$ is equal to R at the signal position and is zero elsewhere,

$$2C(t)\,R^{1/2}d_{k-1}(t)\cos[2\pi f_D t_{k-1} - \varphi_{k-1}(t)]$$

is a gaussian random term at the signal position with zero mean and standard deviation $2R^{1/2}$ and $d_{k-1}{}^2(t)$ is a random variable with mean value 2 which is not gaussian (its distribution is of the form $p(v) = \frac{1}{2}\exp(-v/2)$ for $v > 0$, where $v = d_{k-1}{}^2$). Thus after square-law detection, addition of n signals and elimination of the continuous component (mean value of $d_{k-1}{}^2(t)$), we obtain a term nR, a random gaussian term with zero mean and standard deviation $2(nR)^{1/2}$ and a random nongaussian term $D(t)$ with zero mean at the signal position, and the term $D(t)$ alone elsewhere.

The analysis of $D(t)$ is simplified if it is assumed that the number n of integrated measurements is sufficiently large for $D(t)$ to remain gaussian. (In practice, this assumption is valid for $n > 10$ if $P_f = 10^{-3}$, for $n > 20$ if $P_f = 10^{-5}$ and for $n > 40$ if $P_f = 10^{-10}$, where P_f is the false alarm probability.) Its variance is then given by

$$n\overline{[d_{k-1}{}^2(t) - 2]^2} = n[\overline{d_{k-1}{}^4(t)} - 4]$$

Since

$$\overline{d_{k-1}{}^4(t)}\,\overline{\{\cos[2\pi f_i t + \varphi_{k-1}(t)]\}^4} = \overline{n_{k-1}{}^4(t)} = 3$$

$$\overline{d_{k-1}{}^4(t)}\,\frac{3}{8} = 3 \Rightarrow \overline{d_{k-1}{}^4(t)} = 8$$

the variance of $D(t)$ is equal to $4n$. Thus if n is sufficiently large the system behaves as if we have a term $Rn^{1/2}/2$ and a gaussian random term with zero mean and a standard deviation of $R^{1/2}$ if R is large and unity if R is small (generally $(R + 1)^{1/2}$) at the signal position, and a random gaussian term with zero mean and a standard deviation of unity elsewhere. Thus, for $P_f = 10^{-3}$ we set $K \approx 3.1$, for $P_f = 10^{-5}$ we set $K \approx 4.4$ and for $P_f = 10^{-10}$ we set $K \approx 6.3$, where K is the threshold level.

Thus the detection probability is the probability that a random gaussian term with zero mean and a standard deviation of $(R + 1)^{1/2}$ is larger than

$$K - \frac{Rn^{1/2}}{2}$$

We therefore obtain a detection probability of 0.5 when

$$K - \frac{Rn^{1/2}}{2} = 0$$

or

$$R = \frac{2K}{n^{1/2}} \tag{7.1}$$

It was shown in Chapter 3 that, if the threshold is set before detection and only one measurement is made, the detection probability is the probability that a gaussian random term with zero mean and a standard deviation of unity is larger than $K - R^{1/2}$. Therefore the detection probability of 0.5 is obtained for $K = R^{1/2}$, i.e. $R = K^2$. This is a factor of $K/2$ greater than the value obtained from eqn (7.1) with $n = 1$ ($K/2$ has values of 1.55, 2.14 and 3.18 for false alarm probabilities of 10^{-3}, 10^{-5} and 10^{-10} respectively). The following theorem is therefore proved.

Theorem 7.1

When a sufficiently large number n of elementary measurements are integrated after square-law detection, the energy required per elementary

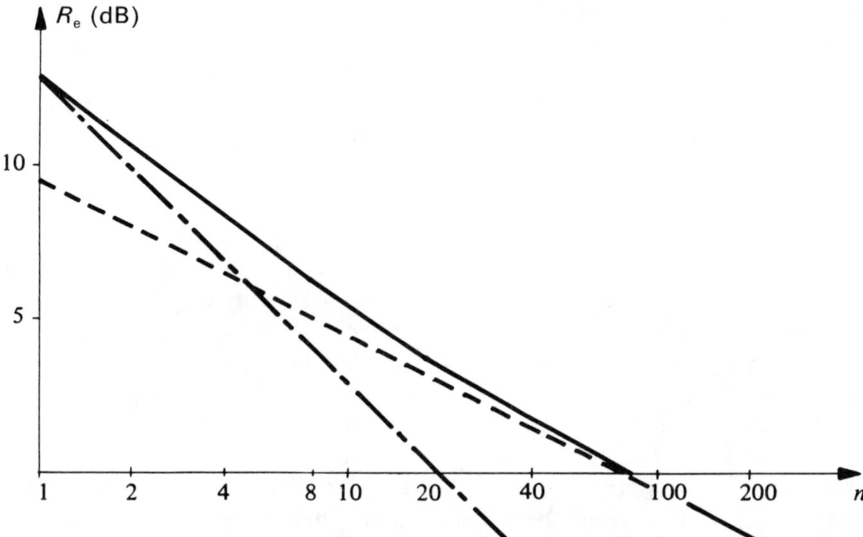

Fig. 7.2 Plot of R_e versus n for $P_d = 0.5$ and $P_f = 10^{-5}$: - · -, coherent integration of n pulses; – – –, asymptotic curve corresponding to theorem 7.1; ——, noncoherent integration (after detection) of n pulses.

measurement to obtain a detection probability P_d of 0.5 decreases in inverse proportion to $n^{1/2}$. It is equal to the energy that would be required if there was only one measurement divided by $(K/2)n^{1/2}$.

Calculations performed for $P_d = 0.9$ and P_f in the range 10^{-3}–10^{-10} show that (to the nearest decibel) this result also holds for $P_d = 0.9$ and for P_d in the range 0.5–0.9. If n is small, we can no longer assume that the distribution of D is gaussian. The calculation is therefore more complicated although it is still feasible.

Typical results are shown in Fig. 7.2 where the value R_e of R required per elementary measurement is plotted as a function of n for $P_d = 0.5$ and $P_f = 10^{-5}$. The solid curve and the chain curve represent the noncoherent integration (after detection) and the coherent integration, respectively, of n pulses. The broken curve is the asymptotic curve corresponding to Theorem 7.1. If n is small, noncoherent integration (after detection) of n pulses is approximately equivalent to coherent integration.

Remark 7.1

It can easily be seen that classical coherent radar satisfies the preceding distributions regardless of the size of the error in the knowledge of the

Doppler effect of the target. If the target is fluctuating but does not fluctuate during the total measurement period, the preceding distributions are also satisfied and the arguments made remain valid. If the target fluctuates frequently during the total measurement period, the gain obtained by integration is augmented by an additional gain due to the radar's functioning as a diversity radar or an agile radar (see Chapter 5).

In summary, the fact that the radar is not coherent or that the Doppler frequency is poorly known prevents coherent integration from being carried out. Coherent integration makes it possible to divide the power to be transmitted by the number of elementary measurements. When coherent integration is not possible, detection (at intermediate frequency) must be carried out before the elementary measurements are integrated. If this detection is square-law and a sufficient number of elementary measurements are made, integration after detection makes it possible to divide the power to be transmitted by two to three times the square root of the number of elementary measurements. This result becomes invalid if the false alarm probability specified is too high ($>10^{-3}$) or too low ($<10^{-10}$) or if the detection probability specified is too high (>0.9) or too low (<0.5).

7.3 Integration in a search radar

Let us consider a search radar rotating with a velocity V using the same antenna for transmission and reception and carrying out brief measurements (over a few microseconds) at a repetition frequency f (of the order of 100–1000 Hz). Then the radar makes $f\alpha/V$ measurements during the dwell time of the beam over the target, where α is the 3 dB azimuth beamwidth. For example, if $f = 250$ Hz, $\alpha = 1°$ and $V = 36 \text{ deg s}^{-1}$, $f\alpha/V = 7$.

When the antenna rotates measurements are made every $1/f$ s, and when the direction of maximum radiation is not close to the target direction the measurements are made with small values of R. However, when the direction of maximum radiation is close to the target direction the measurements are made with large values of R which has a maximum R_M. Then, for each rotation of the antenna, the value of R is only larger than $R_M/4$ for $f\alpha/V$ consecutive measurements (if the target does not fluctuate).

7.3.1 Coherent integration

Let us assume under these conditions that the radar is coherent, that the target is not subject to the Doppler effect (or has a negligible or known

Doppler effect) and that its fluctuation during the beam dwell time is negligible, and that coherent integration is carried out on n pulses (including the most powerful pulse).

If $f\alpha/V$ is sufficiently large and n is small (e.g. $f\alpha/V = 10$ and $n = 3$), the three measurements used correspond to almost the same value R_M of R and the system behaves (see Section 7.1) as if R_M had been multipled by n, i.e. as if we had made a single measurement with a value of R equal to nR_M.

However, if n is not small compared with $f\alpha/V$ (e.g. $f\alpha/V = 10$ and $n = 100$), coherent integration will produce the following: the addition of n gaussian noises with a standard deviation of unity, giving a noise with standard deviation $n^{1/2}$, and the addition of n useful signals of which some are large (close to ${R_M}^{1/2}$) and the remainder are small. If the signal received from a target is written as (see Chapter 6, Section 6.2.3)

$$R = {R_M}^{1/2} \exp\left(- \frac{2.8\theta^2}{\alpha^2} \right)$$

it is found, once all the calculations have been made (as soon as $\alpha f/V$ is of the order of a few times unity), that the addition of n useful signals gives

$$1.06 \frac{\alpha f}{V} {R_M}^{1/2} \Theta(X)$$

where

$$X = \frac{0.83n}{\alpha f/V}$$

Thus the system behaves as if we were dealing with a gaussian noise with a standard deviation of unity and a useful signal with amplitude

$$1.06 \frac{\alpha f}{V} \left(\frac{R_M}{n} \right)^{1/2} \Theta(X)$$

For $n = 1$, the useful signal is equal to $R^{1/2}$. For n small compared with $\alpha f/V$, the useful signal is equal to $(nR)^{1/2}$. If n is very large, the useful signal tends to zero as expected. The final result of this calculation is that the useful signal is a maximum when $n \approx \alpha f/V$.

Logically, therefore, it is necessary to integrate $\alpha f/V$ measurements and then the system behaves as if we had made a single measurement with

$$\alpha = 0.65 \frac{\alpha f}{V} R_M$$

7.3.2 Integration after square-law detection

Let us now assume, as is the case in a large number of applications, that the radar cannot carry out coherent integration and consequently integration is carried out on n measurements after square-law detection. Let us also assume that the target does not fluctuate while the n measurements are being performed, as is generally the case. Under these conditions, the calculations described in Sections 7.2 and 7.3.1 give the following results for detection probabilities of the order of 0.5 and false alarm probabilities of the order of 10^{-3}–10^{-10}, which is generally the case when integration is carried out after square-law detection in noncoherent or coherent radar in the presence of targets with unknown radial velocity: integration of a number n of measurements (pulses) which is small compared with $\alpha f/V$ has the effect of multiplying R_M by $K/2n^{1/2}$, if n is sufficient; the optimum number of measurements (pulses) which must be integrated is of the order of $\alpha f/V$ (rigorously, it is approximately $0.8\alpha f/V$, but the difference between $n \approx \alpha f/V$ and $n \approx 0.8\alpha f/V$ is not large). Under these conditions the system behaves as if a single measurement were being made with

$$R = 0.7 \frac{K}{2}\left(\frac{\alpha f}{V}\right)^{1/2} R_M$$

The integration has the effect of multiplying R_M by $0.7\,(K/2)(\alpha f/V)^{1/2}$.†

Note. According to the assumptions made in Section 7.2, this result is only valid if $\alpha f/V$ is sufficiently large.

7.4 Use of digital extractors

The integration procedures which have just been analyzed involve first summing a given number of signals obtained at a given distance at each elementary measurement (often after detection) and then producing a threshold from the sum obtained. This can now be done almost perfectly at reasonable cost using digital technology because the price of memories is very low. However, until recently the high price of these memories meant that only a single level was quantified in a "digital extractor", the principles of which are explained in this section.

† For any n, the integration has the effect of multiplying R by

$$\frac{\alpha f}{V} \frac{1.5}{n^{1/2}} \Theta(\gamma)$$

where

$$\gamma = \frac{1.18n}{\alpha f/V}$$

The coefficient 0.7 represents the loss due to the fact that the radar beam is gaussian and not rectangular (this loss is equal to 1.5 dB and is called the beamshape loss).

For example, let us assume that we are dealing with a radar performing three consecutive (identical) measurements and that a computer then applies the following rules. In order for a target to be declared present it is necessary and sufficient that at least one elementary measurement gives a positive result (signal higher than the threshold level). A detection probability of 0.5 corresponds to a detection probability per measurement of $1 - (1 - 0.5)^{1/3} = 0.2$, and a false alarm probability of 10^{-5} corresponds to an elementary false alarm probability of $1 - (1 - 10^{-5})^{1/3} = 3.3 \times 10^{-6}$. The value of R corresponding to $P_d = 0.2$ and $P_f = 3.3 \times 10^{-6}$ is 11.3 dB. It is therefore necessary to have $R = 11.3$ dB per elementary measurement, whereas the curve of Fig. 7.2 shows that a perfect noncoherent integration requires only $R = 8.1$ dB per elementary measurement. The procedure described therefore causes a loss of approximately 4 dB.

Let us now assume that the detection criterion is modified as follows. In order for a target to be declared present each of the three elementary measurements must give a positive result. Therefore, for a final detection probability of 0.5 each elementary measurement must have a detection probability of $(0.5)^{1/3} = 0.8$, and for an overall false alarm probability of 10^{-5} the elementary false alarm probability must be $(10^{-5})^{1/3} = 0.0215$. This pair of values ($P_d = 0.8$ and $P_f = 0.0215$) corresponds to $R = 9.2$ dB, i.e. this time the detection criterion produces a loss of only about 1 dB.

It can therefore be seen that the use of detection criteria which enable the integration to be performed by logical operations (use of digital computers known as extractors) can provide almost the same results as perfect analog integration if these criteria are suitably chosen.

Let us consider, for example, a search radar operating with a useful spectrum of width approximately Δf. Then the range resolution is of the order of $1/\Delta f$ s (or $150 \times 10^{-6}/\Delta f$ m), where Δf is in hertz (see Chapter 3). On average the duration of the noise signals is also $1/\Delta f$, and therefore it is necessary to quantify range in cells of length $1/\Delta f$. We therefore proceed in the following manner. The signals received on each scan are, after detection, brought to a threshold level so that a given elementary false alarm probability P_{fe} is obtained. This means that, in the absence of the target and after setting the threshold, there is a probability P_{fe} of finding a signal above the threshold in a range cell.

Let us assume that the computer declares a target present if a signal above threshold is found in the same range cell in each of three successive scans. The probability that a false alarm will be found at a particular distance (in a given range cell) is equal to P_{fe}^3. Thus the false alarm probability is P_{fe}^3, and the most probable number of false alarms per antenna revolution is given by

$$P_{fe}^3 Q \frac{360f}{3v}$$

where Q is the number of range cells, f is the repetition frequency and v is the velocity of rotation of the antenna (in deg s^{-1}). For example, if $\Delta f = 500\,\text{kHz}$, so that $1/\Delta f = 2\,\mu\text{s} \rightarrow 300\,\text{m}$, $Q = 2000$ (useful range, 600 km) and $P_{fe} = 10^{-5}$, the most probable number of false alarms per antenna rotation is 17.

A pulse-counting criterion of this type is effectively used in the simple computers known as interference suppressors. More generally, in such an interference suppressor a target is declared if a signal above threshold is found in the same range cell in at least N successive scans (here $N = 3$). For example, if five such signals are found in succession, it is reasonable to assume that the target direction is the direction of maximum antenna radiation when the third signal is received. This gives an approximate measurement of the target direction.

More complex and better counting criteria can be used when more advanced computers (extractors) are available. An example of such a criterion chracterized by two parameters N and n is given below. The sequence of results obtained in a given range cell is called a pulse train. It consists of a sequence of present signals (above the threshold) or absent signals (below the threshold). Then the pulse train corresponds to a target when there are at least N present pulses which are not separated by more than n consecutive absent pulses. If this condition is satisfied, by definition the pulse train begins when there are no more than n consecutive absent pulses and finishes when there are n consecutive absent pulses. The direction of the target is taken as the direction of the maximum antenna radiation in the middle of the pulse train. The following arrangement illustrates this definition for $N = 3$ and $n = 1$ (the 0s indicate absent pulses and the 1s indicate present pulses):

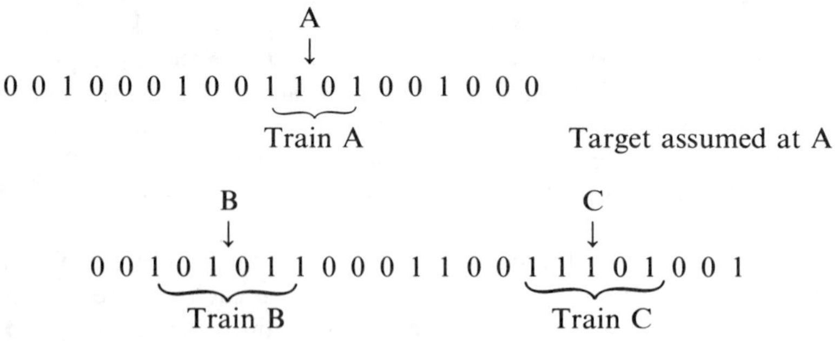

It can be proved that, if a false alarm probability of the order of 10^{-3}–10^{-5} is imposed with such a criterion, the parameters N and n (and P_{fe} which fixes the threshold) cannot be chosen arbitrarily. This is logical if we wish to obtain the best result. For example, if $\alpha f/V = 4$, we must take $P_{fe}^2 \approx 10^{-2}$, $N = 2$ and $n = 2$ or 3. However, if $\alpha f/V = 8$; we must take $P_{fe} \approx 10^{-1}$–10^{-2}, $N = 3$ and $n = 3$ or 4. The results are obtained by making the realistic assumption that the lobe is gaussian. If the lobe is assumed to be rectangular, which is not realistic, completely different results are found which are not valid.

Two important results can also be obtained.

(1) A detection criterion of this type gives almost the same result (to the nearest 2 dB approximately) as an integration (after detection of a number of pulses close to $\alpha f/V$) when appropriate values of N and n are chosen.

(2) When the detection probability obtained is greater than 0.1 or 0.2 (i.e. in all cases of interest), the error on the azimuth of the target (which is taken as the azimuth of the middle of the pulse train) never exceeds V/f for $\alpha f/V < 10$. For example, if $\alpha = 1°$, $f = 250$ Hz and $V = 36\ \mathrm{deg\,s^{-1}}$, $\alpha f/V = 7$ and $V/f = 0.14°$. Therefore the error on the measurement of the azimuth is always less than 0.14°.

Remark 7.2

Obviously errors on the azimuth of the target caused by fluctuation of the radiant point have to be taken into account elsewhere.

7.5 Some variations on the radar equation

7.5.1 Tracking radar

The ratio R for a single pulse is given by the expression

$$R = \frac{2E_p G\sigma S_R}{(4\pi D^2)^2}\frac{1}{kT_N L}$$

where G is the antenna gain (on the axis), S_R is the radar cross-section of the receiver ($S_R = G\lambda^2/4\pi$), E_p is the energy transmitted per pulse, L takes account of various losses and can be ignored, D is the range of the target, σ is the radar cross-section of the target and T_N is the noise temperature of the receiving system. In fact, N pulses are received during the measurement, i.e. during the time constant of the servomechanism of the radar turret. In the best case (N small or coherent integration), the required value R_{min} of R can be divided by N and we obtain

$$R_{min} = \frac{2(NE_p)\,G\sigma S_R}{(4\pi)^2 D^4 k T_N L}$$

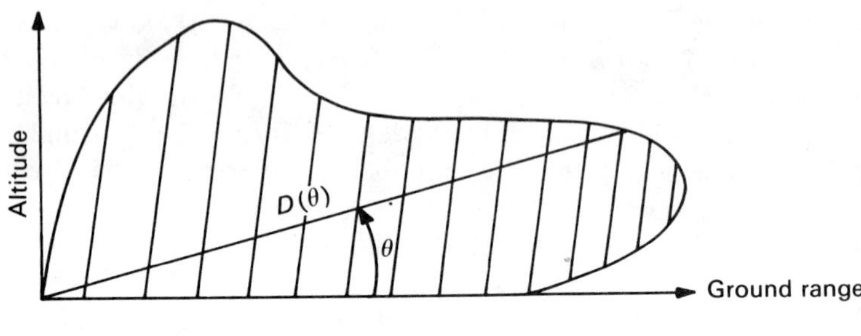

Fig. 7.3

Therefore it is the energy transmitted during the time constant of the turret servomechanism which is important. If λ is multiplied by τ at a given S_R, NE_p must be multiplied by τ^2. If S_R is multiplied by β at fixed λ, NE_p must be divided by β^2.

7.5.2 Two-dimensional surveillance radar

The search radar must provide coverage of the shaded zone in Fig. 7.3 which has area Σ. The value of R for a target at an elevation angle θ and a pulse with transmitted energy E_p is given by

$$R = \frac{2E_p G^2(\theta)\lambda^2\sigma}{(4\pi)^3 D^4(\theta) k T_N L}$$

where $G(\theta)$ is the antenna gain at the elevation angle θ. $G(\theta)$ must vary with θ for R to remain constant irrespective of θ, and hence we can write

$$G^2(\theta) = K^2 D^4(\theta)$$

where

$$K^2 = \frac{(4\pi)^3 R k T_N L}{2E_p \lambda^2 \sigma}$$

Thus we obtain

$$2K\Sigma = K\int_0^{\pi/2} D^2(\theta)\,d\theta = \int_0^{\pi/2} G(\theta)\,d\theta = \frac{4\pi}{\alpha}$$

where α is the width (say at 3 dB) of the azimuth beam, and

$$E_p = 8\pi k T_N L \frac{\Sigma^2\alpha^2}{\lambda^2\sigma} R = 8\pi k T_N L \frac{\Sigma^2}{D_H{}^2\sigma} R$$

where D_H is the horizontal span of the antenna ($\alpha \approx \lambda/D_H$). As the antenna illuminates the target $N = \alpha f_r/V$ pulses are received, where f_r is the repetition

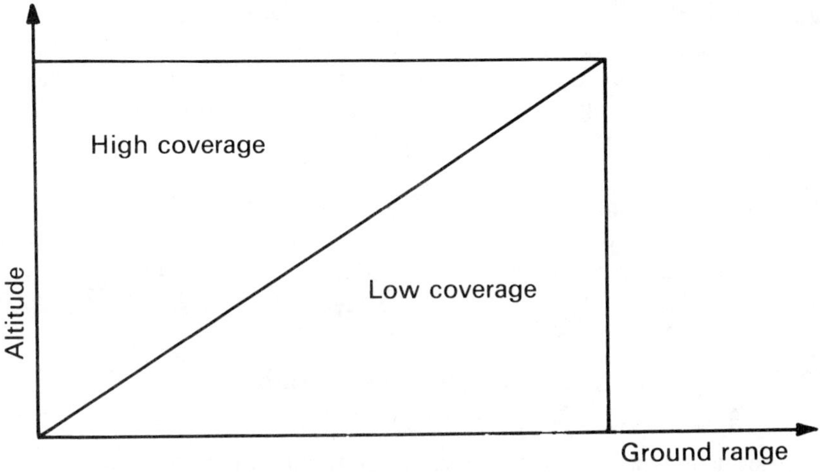

Fig. 7.4

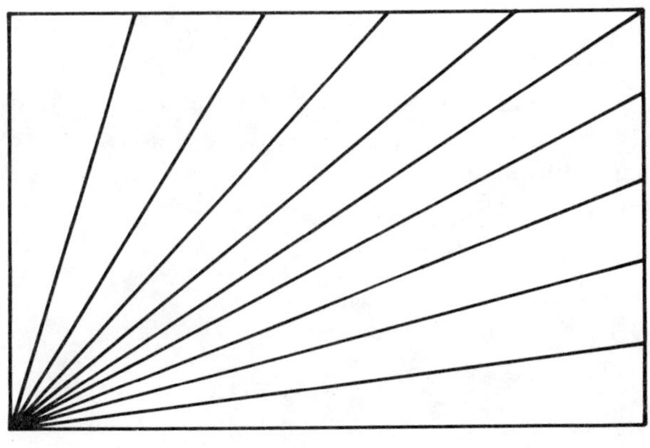

Fig. 7.5

frequency and V is the velocity of rotation of the antenna ($V = 2\pi/T_0$ where T_0 is the period of revolution of the antenna). It is assumed that there is coherent integration or that N is small, and the loss due to the lobe effect is neglected. Therefore the energy to be transmitted can be divided by N, which gives

$$\frac{\alpha f_r E_p}{V} = 2\pi k T_N L R \frac{\Sigma^2}{{D_H}^2 \sigma}$$

$$P_m T_0 = 16\pi^2 k T_N L R \frac{\Sigma^2}{{D_H}^2 \sigma \alpha}$$

where P_m is the mean power transmitted. If λ is multiplied by τ (with D_H constant), α will be multiplied by τ and $P_m T_0$ will be divided by τ. In order to have the same pattern for the elevation the antenna height, and therefore its area, must be multiplied by τ. If D_H, and therefore the antenna area, is multiplied by δ (with λ constant), α will be divided by δ and $P_m T_0$ will also be divided by δ. Hence we obtain the following conclusion: the product of the mean power, the antenna area, the radar cross-section of the target and the period of revolution of the antenna is determined by the coverage required, and (to a first approximation) is independent of the wavelength and the span of the antenna.

7.5.3 Three-dimensional surveillance radar

Let us now assume that the coverage required is rectangular as is often the case (Fig. 7.4). As shown in Fig. 7.4 this coverage can profitably be divided in two, because

$$2\left(\frac{\Sigma}{2}\right)^2 = \frac{\Sigma^2}{2}$$

Each half can be further divided into n equal areas (Fig. 7.5) and hence $P_m T_0$ can be divided by n. However, the height, and therefore the area, of the antenna is then multiplied by n. The conclusion given at the end of Section 7.5.2 still applies.

Chapter 8

Applications of Electronic Scanning Antennas to Radar

8.1 Introduction

A conventional radar (which does not use electronic scanning) is such that any modification of the antenna radiation pattern is obtained by mechanical movements that generally involve rotation of the entire antenna about one or more axes, movement of all or part of the reflector or movement of the primary source. This has three main consequences: any modification of the radiation is slow; it is not possible to modify the radiation pattern significantly; and it is not possible in practice to control the pattern in the secondary radiation directions (characterized by the antenna sidelobes). The direct result of this is that it is not possible to change the principal radiation direction rapidly. Therefore in changing from one direction of interest to another that is quite different it is necessary to pass systematically through a large number of other directions of no interest in which the energy transmitted by the radar is lost. A second result is that if a system with a number of very different patterns is required in order to provide different functions, it is essential to use as many radars as there are functions. Finally, if the parasitic energy (due to jamming or reflection from parasitic echoes) returns to the radar receiver via sidelobes, it is necessary to tolerate the effects passively.

It is said (incorrectly) that a radar antenna has "electronic scanning" if the radiation pattern can be modified without the need for mechanical movements and if therefore the time required to move from one pattern to another is very short with respect to the time (of the order of 1 ms) taken by the electromagnetic signal to go from the radar to the target and back.

An electronic scanning antenna provides the following facilities.

(1) The energy transmitted by the radar can be "time-shared" by sending most of the energy in difficult directions (where jamming or clutter is

significant) and by only sending a small amount of energy in easy directions (clear conditions);

(2) The same radar can be used for different functions (low altitude surveillance, high altitude surveillance, tracking different targets etc.), and new detection methods, such as sequential detection† which often saves energy, can also be applied.

(3) The sidelobes of the antenna pattern can be permanently controlled in order to obtain the best results against jamming by systematically reducing the sidelobe radiation in the direction of the jammers while accepting a larger value in other directions.

It is obvious that it is only possible to obtain the above advantages if the radar management is capable of sufficiently rapid and intelligent reactions as the situation changes. Thus radar management must be carried out using a digital computer since the reaction time of a human operator is much too long in terms of what is required (of the order of 1 ms). However, computer control implies at least two things: a control program must be designed that is "complete and perfect", and that contains the best nonambiguous responses to all situations that are likely to be encountered by the radar (at least to the extent that perfection is possible); the transmitter, the receiver and the signal processing used by the radar must have very stable performances corresponding to what is specified in the management program so that the computer is not faced with an unforeseen situation due to interference, jamming, clutter or simply malfunctioning of the radar.

8.2 Basic principles of electronic scanning. Simplifications

Electronic scanning is based on the following general principle. Since there is a Fourier transform relationship between the antenna pattern at infinity and the illumination of the radiating system (Huygens–Fresnel theorem), it is clear that control of the amplitude and the wave phase at all points of the radiating system makes it possible to control the antenna pattern completely. Thus the antenna pattern will be modified if the amplitude and phase of the wave at the surface of the radiating system are changed.

†Sequential detection consists in making the first detection with a sufficiently high false alarm probability and making a second confirmation detection only on the alarms obtained in this way.

8.2.1 First simplification

Using the Huygens–Fresnel theorem is in fact an error (or an approximation) because it is only valid when a single frequency is radiated whereas a radar antenna radiates a spectrum which on average becomes wider with increasing time. The Huygens–Fresnel theorem is only valid if the large dimension D of the radiating system is appreciably smaller than $c/\Delta f$, where c is the velocity of light and Δf is the width of the spectrum at a given time:

$$D \ll \frac{c}{\Delta f}$$

This relationship can easily be explained.

Let us consider a planar radiating system of length D and a target located a large distance away in a direction at 45° to the axis. The path from the radar to the target depends on which end of the radiating system is used as a starting point and the difference between the two paths is $D\sqrt{2}/2$. Obviously this difference must be negligible compared with the resolving power $c/2\Delta f$ of the radar, and hence we obtain

$$D\sqrt{2} \ll \frac{c}{\Delta f}$$

For example, if $\Delta f = 3\,\text{MHz}$ and $D = 3\,\text{m}$, $c/\Delta f = 100\,\text{m}$. However, if $\Delta f = 100\,\text{MHz}$ and $D = 30\,\text{m}$, $c/\Delta f = 3\,\text{m}$. In the first case it is possible to use the Huygens–Fresnel theorem and to control the antenna pattern by controlling the phase of the wave along the radiating system. In the second case the Huygens–Fresnel theorem does not apply, and in order to control the pattern it is necessary to control the delay between the transmitted (or received) waves at different points of the radiating system. In practice, combinations of delay devices and phase shifters are used in this case.

8.2.2 Second simplification

An overall evaluation of the cost–performance factor shows that it is not profitable to control the amplitude of the wave for the radiating system (as is sometimes done in sonar). Therefore only the phase is controlled.

8.2.3 Third simplification

The radiation pattern is assumed to be "formed" or "at infinity" if the distance of the target from the radiating system is much greater than D^2/λ,

Table 8.1

S (m^2)	λ (m)	S/λ (m)
100	0.2	500
10	0.1	100
1	0.03	30

i.e. to a first approximation at a distance much greater than S/λ if the antenna cross-section is circular or square or a similar shape (which is generally the case) with area S. Some numerical examples are given in Table 8.1.

The relationship between the illumination of the radiating system and the antenna pattern is such that it is useless (if not impossible) to control the phase at each point of the radiating system. It is strictly sufficient to control the phase every half-wavelength. There are two methods of doing this.

(1) When jamming and/or clutter are not serious problems, grating lobes, which are similar to the spill-over lobes found in classical structures, are acceptable. In this case the spacing between two adjacent phase-controlled radiating elements can reach

$$d_2 = \frac{\lambda}{2\sin(\theta_{max})}$$

where θ_{max} is the maximum offset angle of the antenna with respect to the axis of radiation and in a planar system is obtained when all the phases are zero.

(2) Grating lobes are unacceptable and the spacing between two adjacent radiating elements must be restricted to

$$d_1 = \frac{\lambda}{1 + \sin(\theta_{max})}$$

Table 8.2 gives examples of the values of d_2 and d_1 for various values of θ_{max}. This shows why the spacing is close to 0.6λ for most military applications, whereas a spacing larger than 2λ is used in some specific applications. Generally speaking, $d = 0.6\lambda$ means that the total number N of controlled phase shifters is

$$N \approx \frac{3S}{\lambda^2}$$

Table 8.2

θ_{max} (deg)	d_2	d_1
10	2.8λ	0.85λ
30	λ	0.61λ
60	0.58λ	0.54λ

where S is the area of the radiating network. For example, if $S = 1000\,\text{m}^2$ and $\lambda = 0.2\,\text{m}$, there are 75 000 radiating elements, and if $S = 1\,\text{m}^2$ and $\lambda = 0.05\,\text{m}$, there are 1200 radiating elements. There is also a relationship between the maximum gain G of the antenna and the number N of phase shifters:

$$G \approx 3N$$

or (in decibels)

$$G_{\text{dB}} \approx 5 + 10 \log_{10} N$$

8.2.4 Fourth simplification

Obviously there is no point in having infinitely accurate phase control. In practice the phase of the radiating elements is almost always quantized. For example, three-bit quantization means that the phase in each controlled radiating element can only be equal to (in radians)

$$0 \quad \pi/4 \quad \pi/2 \quad 3\pi/4 \quad \pi \quad 5\pi/4 \quad 3\pi/2 \quad 7\pi/4$$

It is clear that the phase is defined modulo 2π rad. The use of such a quantization means that phase errors are introduced between the ideal and actual distributions. These phase errors introduce faults whose order of magnitude is indicated below (for intelligent approximation (truncation) distributions).

First, the phase errors introduce a gain loss which is easy to evaluate. If the phase is quantized with p bits, the maximum error introduced is $\pm\pi/2^p$ rad and the mean square value of the error is (uniform distribution)

$$\frac{\pi^2}{4^p} \times \frac{1}{3}$$

which has an effective value of $\pi/(2^p\sqrt{3})$. The maximum gain that would be obtained by summing N vectors with amplitude a_i that are in phase is equivalent to the sum of N vectors of amplitude a_i and phase $\delta\varphi_i$, where $\delta\varphi_i$ is the phase error at each radiating element, or the sum of N vectors with amplitude $a_i \cos \delta\varphi_i$ that are in phase:

$$\begin{aligned}
G_{\max} &= \sum a_i \\
G_{\text{real}} &= \sum a_i \cos(\delta\varphi_i) = \sum a_i \left(1 - \frac{\delta\varphi_i^{\,2}}{2}\right) \\
&= \sum a_i - \frac{1}{2} \sum a_i \delta\varphi_i^{\,2} = G_{\max} - \frac{1}{2} \sum a_i \delta\varphi_i^{\,2} \\
&= G_{\max} - \frac{1}{2}\frac{\pi^2}{4^p} \times \frac{1}{3} \sum a_i = G_{\max}\left(1 - \frac{1}{2}\frac{\pi^2}{4^p} \times \frac{1}{3}\right) \\
G_{\text{real}}^{\,2} &= G_{\max}^{\,2}\left(1 - \frac{\pi^2}{4^p} \times \frac{1}{3}\right)
\end{aligned}$$

Hence the gain loss (in dB) is given by

$$\Delta G = -10 \log_{10}\left(1 - \frac{\pi^2}{4^p \times 3}\right) = 4.34 \frac{\pi^2}{4^p \times 3} = \frac{14}{4^p}$$

Thus if $p = 2$ bits the gain loss is 0.9 dB, if $p = 3$ bits the gain loss is 0.23 dB and if $p = 4$ bits the gain loss is 0.06 dB.

Second, the phase error modifies the radiated beams and introduces a corresponding angular measurement error. A calculation similar to that for the gain loss, but slightly more complex, gives the following expression for the standard deviation of the angular error:

$$\frac{1.4}{2^p} \frac{\theta_{3\,\mathrm{dB}}}{N^{1/2}}$$

where $\theta_{3\,\mathrm{dB}}$ is the beam width. For example, if $p = 2$ bits and $N = 2000$, the standard deviation of the angular error is $0.008\theta_{3\,\mathrm{dB}}$.

Finally, and most importantly, there is a phase error due to the quantization produced by the grating lobes. The mean level of these grating lobes can be defined in at least two ways.

The mean level relative to the maximum gain is given (in dB) by the approximate expression

$$L = 10 \log_{10} N + 6(p - 1)$$

(the exact value depends on the illumination distribution in terms of the amplitude of the radiating array). Again, the proof is similar to that given above for the gain loss. If $N = 2000$ and $p = 2$ bits, the mean level of the grating lobes due to quantization is 40 dB below the maximum gain. If $N = 1000$ and $p = 6$ bits, the mean level is 60 dB below the maximum gain.

The absolute level (in dB) compared with the isotropic level is given approximately by

$$S = 6(p - 2)$$

(the exact value depends on the illumination distribution, as above). Then, if $p = 2$ the level is near the isotropic level, and if $p = 6$ the level is 25 dB below the isotropic level.

The above formulae give a good idea of the quantization level necessary for particular applications and radar types. $L = 40$ dB is suitable for many terrestrial radars. $L = 50$ dB (or better) is often required of an airborne radar where the small sidelobes are often the only protection against certain types of clutter.

8.3 General description of an electronic scanning antenna

Remark 8.1
In the past, dispersive structures have been used in some radars in order to obtain a linear variation of the phase along the antenna. The distribution of the phase variation (and therefore the principal radiation direction) was modified by changing the transmission frequency. We shall not consider this case below as it does not have the advantages provided by electronic scanning using phase shifters.

Figure 8.1 shows a general block diagram of an electronic scanning antenna for radar.

8.3.1 Radiating elements

The radiating elements (the transmitting and/or receiving elements), of which there are usually N, are of different types, generally dipoles or spirals (see Figs 8.7 and 8.9 below). They are regularly or quasiregularly distributed on the surface of the radiating array with a very small spacing between sources as discussed in Section 8.2.3. This very small spacing produces significant coupling between the radiating sources, and this coupling has two effects.

The first effect, which is very disagreeable, is called blind direction. The coupling between the primary sources means that it is almost impossible to transmit or receive appreciable signals in certain directions (generally far from the antenna axis). If the antenna beam is pointed in these directions, most of the transmitted power is reflected back to the transmitter by the radiating array itself. The second effect, which is advantageous, is obtained for certain types of coupling and makes it possible to keep the antenna gain approximately constant in the direction of maximum radiation for quite considerable deflection angles, whereas normally the maximum gain decreases slowly as the maximum radiation direction moves further away from the array axis.

Unfortunately, it is usually very difficult to predict the coupling effects with any accuracy. The only method available for analyzing the coupling between sources is experimentation using part of the radiating array under actual working conditions. At present calculations can only be used for modifications or extensions of a system which has already been analyzed using experimental data.

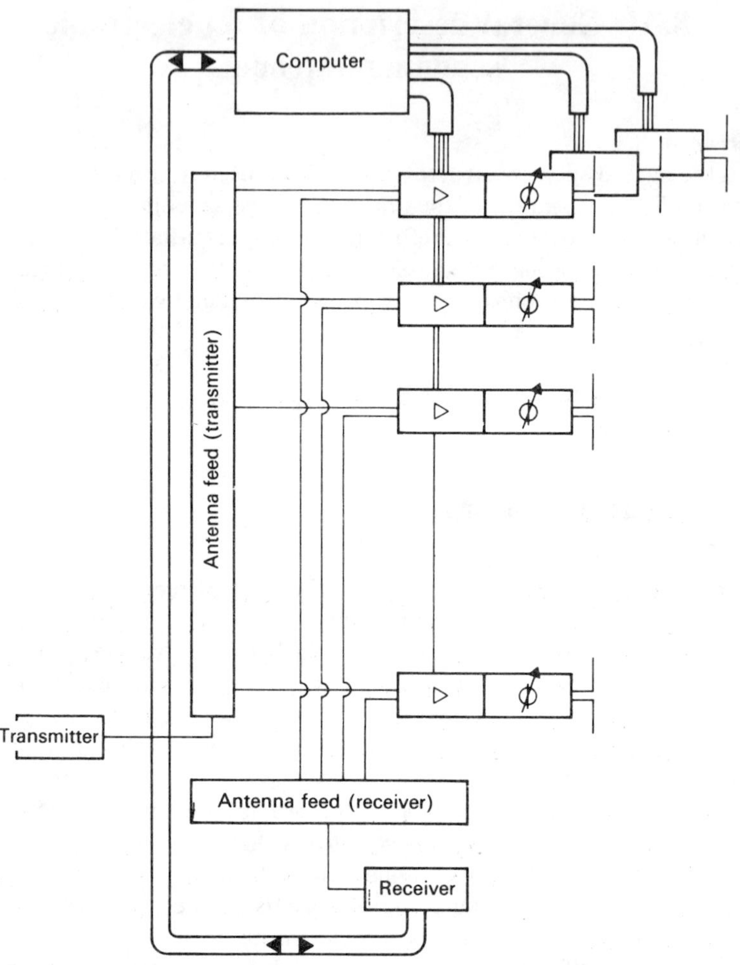

Fig. 8.1 Block diagram of an electronic scanning radar.

8.3.2 Phase shifters

There are essentially two types of phase shifters which use either ferrite elements or pin diode systems. It is generally more convenient to use ferrite phase shifters for short wavelengths (less than 1 cm) and diode phase shifters for long wavelengths (more than 10 cm), and the appropriate phase shifter can be chosen on the basis of the following comments.

Fig. 8.2 4 bit phase shifter. (Courtesy of Thomson-CSF.)

Ferrite phase shifters are heavy, bulky and sensitive to variations in temperature, but they can withstand high power and introduce relatively low losses (of the order of 1 dB for transmission and reception). The pin diode phase shifters are light, compact and fairly insensitive to variations in temperature, but they do not withstand high power very well and introduce relatively high losses (of the order of 2 dB for transmission and reception). However, the losses introduced by phase shifters are often compensated, at least partially, by improved antenna efficiency (a higher gain for a given antenna area).

There are two ways of producing diode phase shifters. Switching phase shifters make the signal take paths of different lengths depending on the control system. They are preferred for large phase shifts (π or $\pi/2$). Perturbation phase shifters produce variable impedances on the transmission line depending on the control system and are preferred for small phase shifts ($\pi/8$ or $\pi/4$). Diode phase shifters are often produced on substrates with high dielectric constants (e.g. 9) and hence their volume can be substantially reduced.

Figure 8.2 shows a 4 bit phase shifter using switching at both ends for large phase shifts and perturbation at the center for small phase shifts. The control amplifiers do not pose any particular problems, but if they are not carefully designed they may lead to considerable and prohibitive consumption.

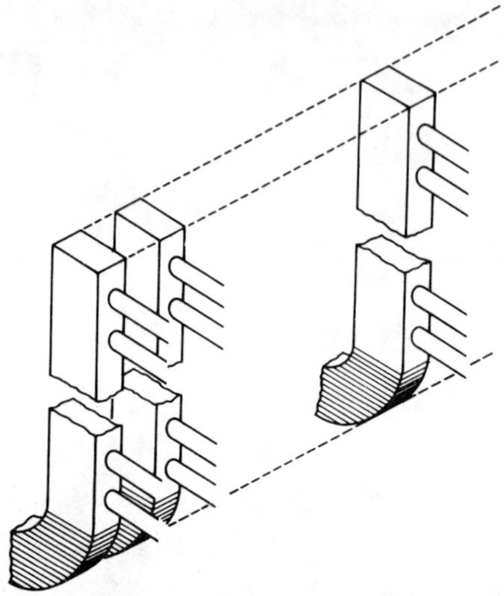

Fig. 8.3 Feed system for a phase shifter.

8.3.3 Feed systems

Many feed systems for phase shifters have been developed, and many of these are very heavy and bulky structures which introduce substantial losses. The systems most frequently used in practice are shown in Figs 8.3–8.5. In the arrangement in Fig. 8.3, which exists in a large number of versions, a column of phase shifters is supplied by the same waveguide (carrying microwave energy) via couplings whose value depends on the decrease in the energy along the guide and on attempts to reduce the illumination at the ends of the antenna with respect to the center. Figures 8.4 and 8.5 show radio frequency feeds.

Figure 8.4 shows a lens array. A primary source illuminates an array of sources that collect its energy, and the phase shifters introduce phase shifts that are the sum of the phase shifts transforming the spherical wave into a planar wave and the phase shifts required to obtain the desired patterns. Hardly any of the rear of the antenna is available for power amplifiers, computers etc.

Figure 8.5 shows a "reflectarray" system. The primary source, which is offset in this case, feeds the sources which capture its energy. This energy

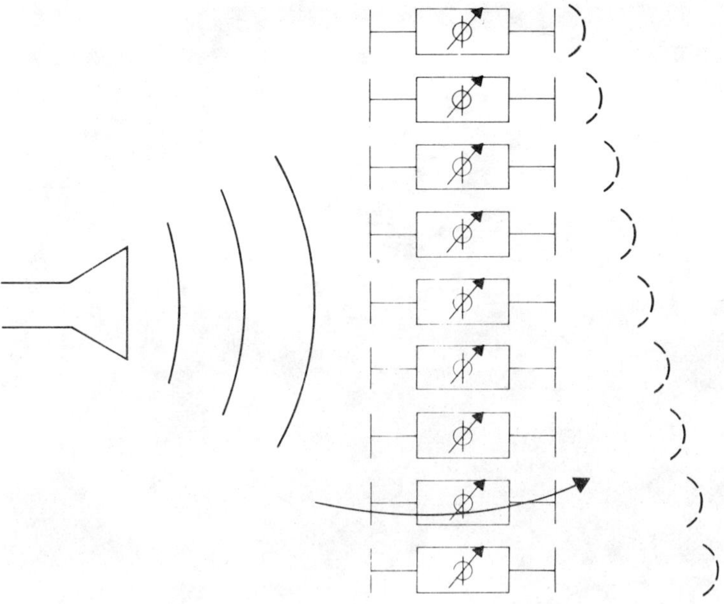

Fig. 8.4 Lens array.

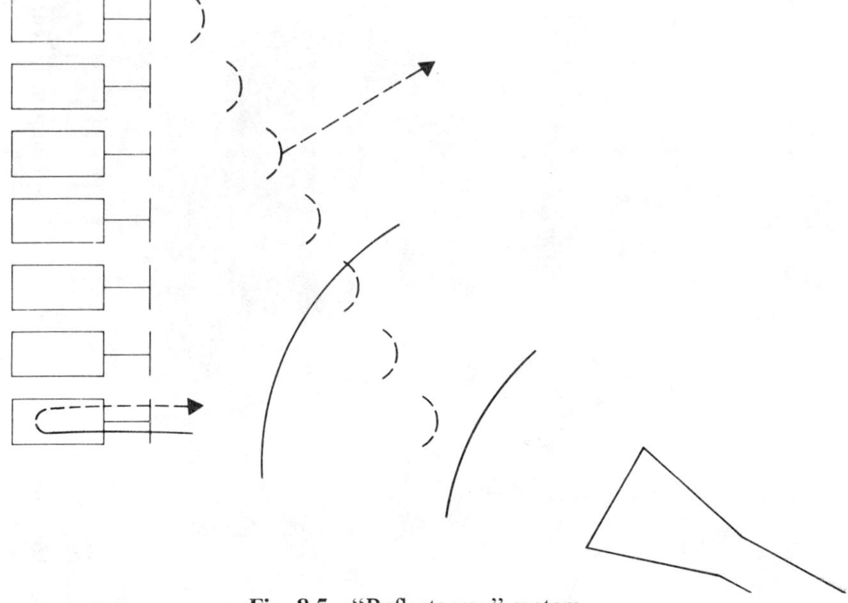

Fig. 8.5 "Reflectarray" system.

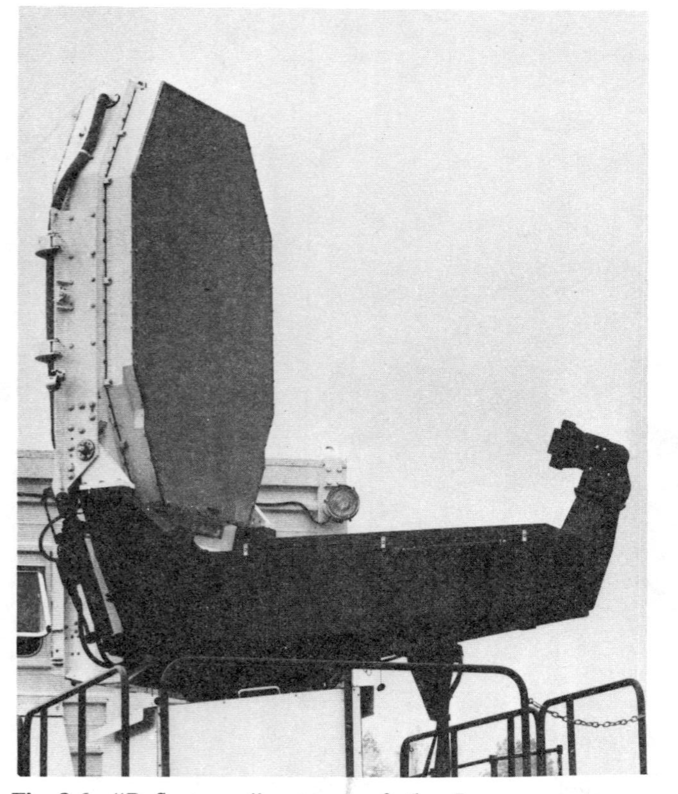

Fig. 8.6 "Reflectarray" system of the Louqsor antenna. (Courtesy of Thomson-CSF.)

Fig. 8.7 Louqsor antenna without its protective radome.

passes through the phase shifter and is reflected and retransmitted by the same sources. Figure 8.6 shows such an antenna with its array-protecting radome and Fig. 8.7 shows the same antenna without its radome. The rear of the antenna is available, but there is the risk of undesirable reflections at the array (specular reflection) and the system is a little less flexible than the lens array.

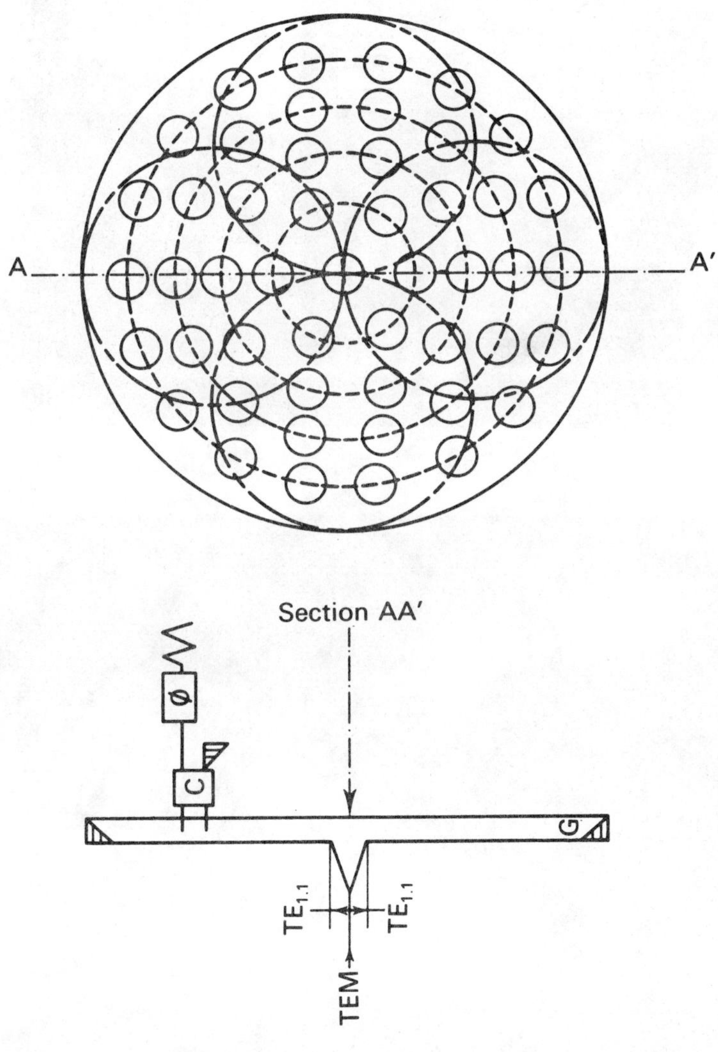

Fig. 8.8

Fig. 8.9 Electronic scanning radar antenna. (Courtesy of Thomson-CSF.)

Fig. 8.10 Rear of the antenna shown in Fig. 8.9. (Courtesy of Thomson-CSF.)

Combinations of the arrangements shown in Figs 8.3–8.5 can also be used, and Fig. 8.8 shows an example of a combination of Figs 8.3 and 8.4 (Fig. 8.9 shows the front of the antenna and Fig. 8.10 shows the rear). Another method of proceeding is to use as many transmitters as sources, with each transmitter containing part of an associated receiver. This is known as the active array solution. The problem (which is not as simple as it appears) is to make all these transmitter–receivers coherent.

8.4 Types of patterns for electronic scanning radars

The use of an antenna with electronic scanning enables very different patterns to be obtained without moving mechanical parts. The main modifications used are shown in Figs. 8.11–8.18. In each case an array antenna consisting of 63 radiating elements with a regular spacing of 0.5λ is assumed to be in the plane of the figure. The antenna is uniformly illuminated and the phase shifters are 4 bit.

8.4.1 Centered pencil beam

If the phases are all zero, the antenna provides a pattern consisting of a pencil beam in the antenna axis with a 1.8° aperture at 3 dB surrounded by sidelobes of which the first two are 13.5 dB below the main lobe.

8.4.2 Offset beam

Figure 8.11 shows the phase distribution corresponding to a pencil beam almost identical with that described above but offset by 7.2°. It also shows the similarity to the offset obtained using a prism. Figure 8.12 shows the original pencil beam (all phases zero) (full curve) and the offset beam corresponding to the pattern in Fig. 8.11 (broken curve).

8.4.3 Broadened beam

Figure 8.13 shows a phase distribution corresponding to a very broad beam (with an aperture of about $\pm 30°$). It also shows the similarity to the broadening obtained using a divergent lens. Figure 8.14 shows the original pencil beam as a full curve and the broadened beam corresponding to the pattern in Fig. 8.13 as a broken curve.

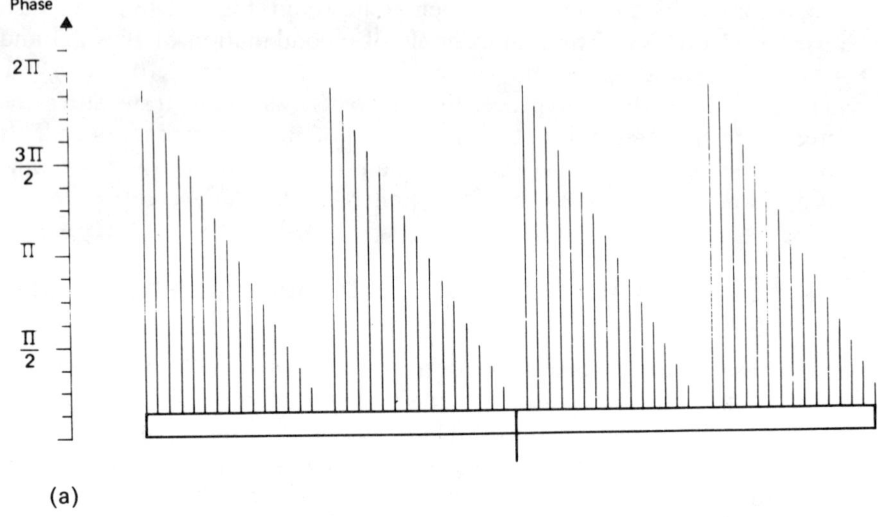

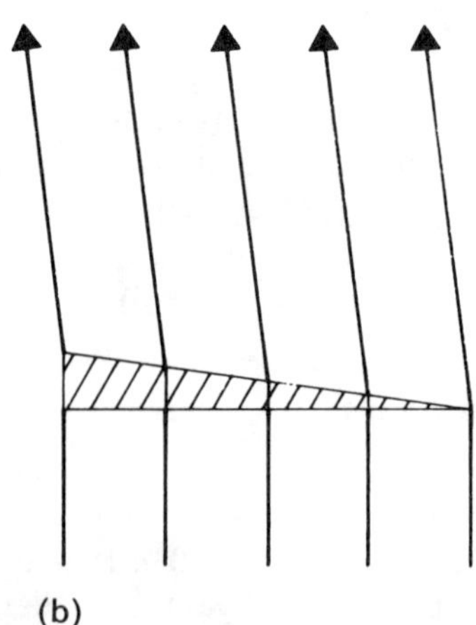

Fig. 8.11 (a) Phase distribution for an offset pencil beam; (b) offset obtained using a prism.

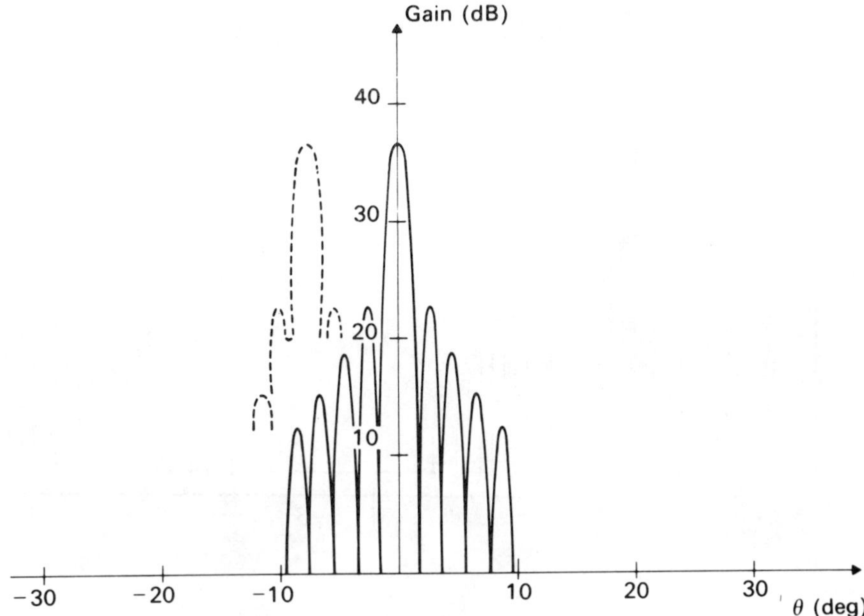

Fig. 8.12 Original pencil beam (——) and the offset beam corresponding to Fig. 8.11 (- - -).

8.4.4 Monopulse difference beam

Figure 8.15 shows the phase distribution corresponding to a pencil beam with a gap of the type used in a monopulse difference antenna. Figure 8.16 shows the original pencil beam as a full curve and the pencil beam with a gap corresponding to the pattern of Fig. 8.15 by a broken curve.

8.4.5 Antijamming antenna

Figure 8.17 shows a phase distribution little different from that giving the original pencil beam and Fig. 8.18 (broken curve) shows the resultant radiation pattern. It can be seen that the structure of the main lobe has been only slightly modified, although the pattern is lowered at the level of the third sidelobe on the right. It is clear that in the presence of a useful target situated on the axis and with the jammer situated in the direction of the center of the third sidelobe, the useful signal-to-jammer ratio will be much improved (53 dB) without the need for moderating the operation of the radar in other respects.

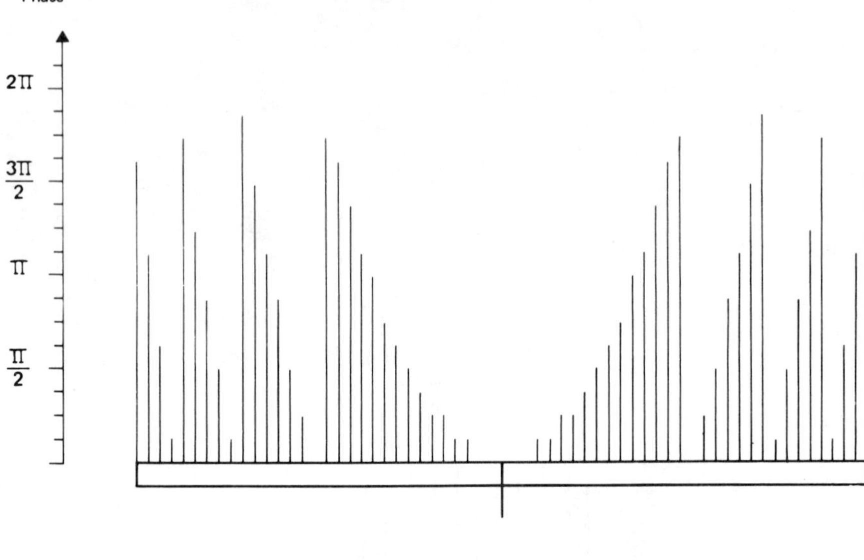

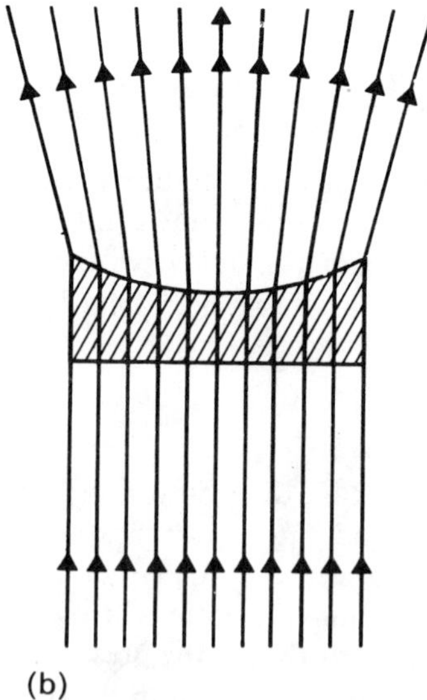

Fig. 8.13 (a) Phase distribution for a broad beam; (b) broadening obtained using a divergent lens.

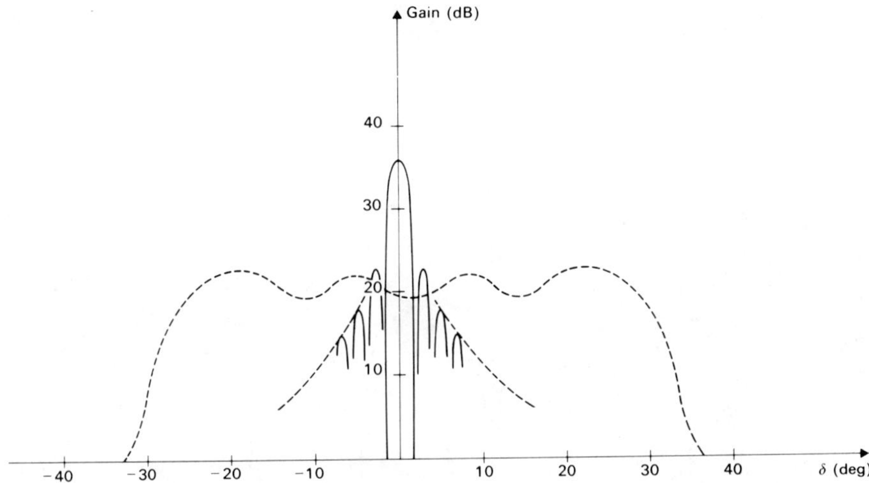

Fig. 8.14 Beam broadening: ——, original pencil beam; – – –, broadened beam.

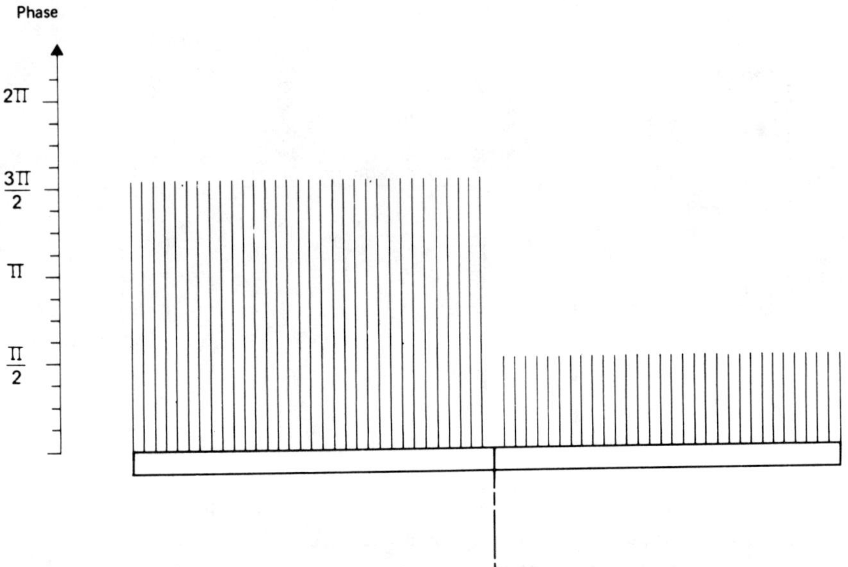

Fig. 8.15 Phase distribution for a pencil beam with a gap.

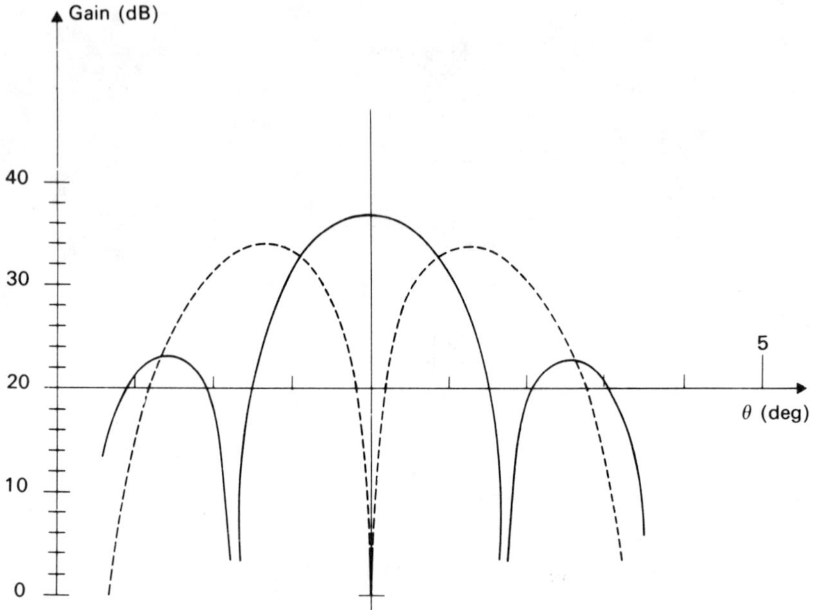

Fig. 8.16 Original pencil beam (——) and the pencil beam with a gap (– – –).

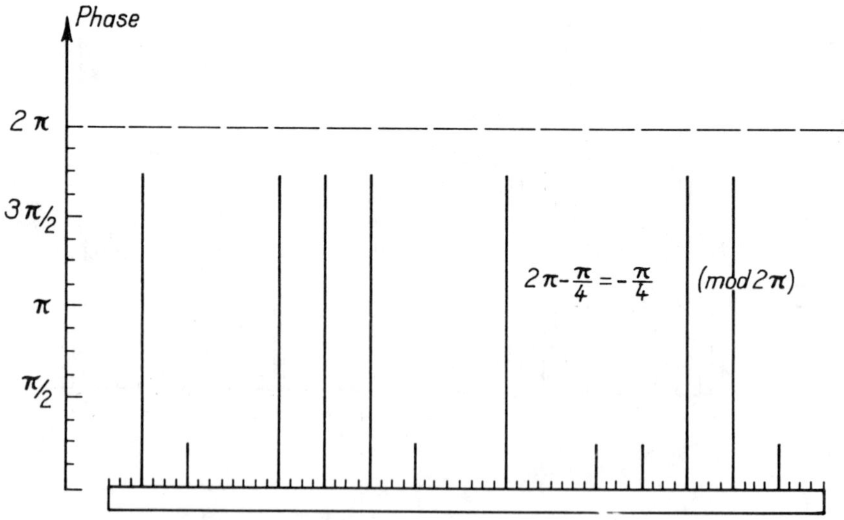

Fig. 8.17 Phase distribution of an antijamming antenna.

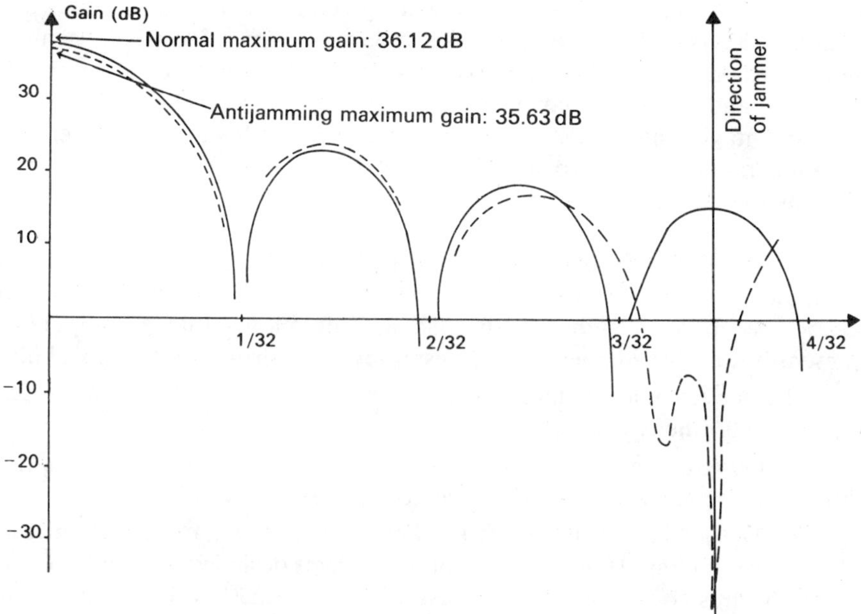

Fig. 8.18 Radiation pattern of an antijamming antenna: ——, normal pattern; - - -, antijamming pattern.

8.5 Final remarks

The capabilities of electronic scanning have been surveyed briefly above. However, it is necessary to add a few additional remarks to complete the picture.

First, the advantages of electronic scanning are limited to particular applications because in practice it is not known how to scan angles significantly higher than 100° (± 50°) without mechanical movement. Of course, this is not a limitation for conventional radars which do not scan anything without mechanical movement, but this limitation together with the higher cost of electronic scanning radar certainly reduces its area of application. In this respect, a parallel can be made with other novel radar techniques such as pulse compression which appeared about 30 years ago. Among other advantages pulse compression can enable larger ranges, a more accurate measurement of range and a significant reduction of the effects of parasitic echoes. However, an amplifier transmitter which is more expensive than a magnetron transmitter is required. This constraint has not removed pulse compression from the scene but has reduced its area of application.

Finally, we can summarize the advantages and disadvantages of electronic scanning radars compared with classical radars as follows. Electronic scanning radars provide much higher performance (an advantage which is difficult to quantify) and a single radar of this type can replace several classical radars (an advantage which is less difficult to quantify). However, an electronic scanning radar is more expensive than a classical radar. There are four main reasons for the higher cost of electronic scanning.

(1) An electronic scanning radar must have almost complete automatic computer control. The timescale is such that the advantages listed above cannot be obtained with human control. This implies that a computer is present but it also implies good transmitter–receiver performance stability.

(2) An electronic scanning antenna is much more complex than a classical antenna, all other things being equal.

(3) Designers have less experience with electronic scanning techniques than with the techniques used in classical radars.

(4) Designers and users often want to include all the capabilities of electronic scanning. Hence the cost of the systems designed is completely out of reach, thus confirming the old saying that "the best is the enemy of the good". However, this is not really a fault of electronic scanning.

Appendices

Appendix A Introduction to Hilbert spaces

A.1 Definition of an abelian group

A set G is called an abelian group when it has an internal law of composition which has all the properties which addition has for real numbers. For this reason this law is often called the addition law. In other words

$$\forall a, b \in G \qquad \exists c \in G \qquad \text{such that } c = a + b$$

This law of composition is associative which means that

$$a + (b + c) = (a + b) + c = a + b + c$$

There exists a neutral element $e \in G$ such that

$$e + a = a + e = a \qquad \forall a \in G$$

$$\forall a \in G \qquad \exists a' \in G \qquad \text{such that } a' + a = a + a' = e$$

(we often write $a' = -a$).

The law of composition (addition) is commutative (otherwise we are dealing with a non-abelian group) which means that $\forall a, b \in G$

$$a + b = b + a$$

For example, the set of vectors (from the same origin) of a plane provided with classical vectorial addition is clearly an abelian group. The set of polynomials of degree n is an abelian group.

A.2 Definition of a ring

A set A is called a ring if it has the following properties.

It has a group structure (with a law which we shall call addition) whose neutral element is generally called zero and is denoted 0.

It has a second law of internal composition, which we shall call multiplication, such that

$$\forall a, b \in A \qquad \exists c \in A \qquad \text{such that } c = a \times b$$

Multiplication is associative and doubly distributive with respect to addition. Associative means that

$$\forall a, b, c \in A \qquad a \times (b \times c) = (a \times b) \times c = a \times b \times c$$

Distributive means that

$$a \times (b + c) = a \times b + a \times c$$

(distributive to the right) and that

$$(b + c) \times a = b \times a + c \times a$$

(distributive to the left). If the multiplication is commutative, the ring is called commutative.

Example A.1

The set of vectors of a plane is not normally a ring because the multiplication of two vectors of the plane (such as is defined by the classical vector product) provides a vector which is not in the same plane and therefore does not belong to the set.

Similarly, the set of polynomials of degree n does not form a ring (the product of two polynomials of degree n is not generally of degree n).

The set of positive or negative integer numbers constitutes a ring (called the ring of relative integers Z) when it is provided with classical addition and multiplication. It should be noted that this set has the structure of an abelian group for addition but not for multiplication because an element of Z does not generally have an inverse (fourth property of a group). It will be seen that this prevents the set Z from being a field.

A.3 Definition of a field

A ring is a field if a set (less the zero element of the additive group) constitutes a group for multiplication. In other words, every element has an inverse belonging to the set. It has already been seen that the ring Z of relative integers is not a field.

Example A.2

The set of rational numbers (denoted Q) is a field.

The set of real numbers (denoted R) is a field of which Q is a subfield (R is a superfield of Q).

The set of complex numbers is a field (denoted C) of which R is a subfield.

The set of residual classes modulo m (denoted Z/m) is a field if m is a prime number.

The set Z/m consists of the elements $\{0\ 1\ 2\ 3\ \ldots\ m - 1\}$. The laws of internal composition are addition modulo m defined as follows.

The result of the addition of two numbers of the set is the remainder of the division by m of the sum of these two numbers: $4 + (m - 2) = 2$.

The second law of internal composition is multiplication modulo m such that the result of the multiplication of two numbers in this set is the remainder of the division by m of the product of these two numbers: if $m = 11$, $4 \times 5 = 9$.

If these two laws hold, Z/m is a field. In fact, it is obvious that it is a ring and for it to be a field it is necessary that every element (except 0) has an inverse which is proved as follows: (a) the inverse of 1 is 1; (b) let another element $a \in Z/m$, since m is prime the highest common factor of a and m is 1. Now, it is known in arithmetic that the highest common factor D of two numbers a and m satisfies the relationship

$$D = xa + ym \quad \text{with} \quad x, y \in Z$$

Here, $1 = ax + ym$. It is therefore clear that $ax = 1$ modulo m and that x (or the remainder of the division of x by m) constitutes the inverse of a. The set Z/m constitutes a Galois field.

We now apply this approach to the case $m = 7$:

$$Z/m = \{0\ 1\ 2\ 3\ 4\ 5\ 6\}$$

The inverse of 1 is 1, the inverse of 2 is 4 ($4 \times 2 = 1$ modulo 7), the inverse of 3 is 5 and the inverse of 6 is 6.

A.4 Definition of a vector space

A set E forms a vector space if on the one hand it is an abelian group and if on the other hand

$$\forall x \in E \quad \text{and} \quad \forall \alpha \in K$$

where K is a field, it is possible, using a law of external composition, to define a "product" of x and α, denoted αx, such that

$$\alpha x \in E \tag{A.1}$$

$$\alpha(\beta x) = (\alpha\beta)x \qquad \alpha, \beta \in K \tag{A.2}$$

$$\begin{aligned} 1.x &= x \\ 0.x &= 0 \end{aligned} \tag{A.3}$$

$$\alpha(x + y) = \alpha x + \alpha y \qquad x, y \in E \tag{A.4}$$

$$(\alpha + \beta)x = \alpha x + \beta x \qquad \alpha, \beta \in K \tag{A.5}$$

The symbols 0 and 1 on the left-hand side of (A.3) are the neutral elements of addition and multiplication respectively in the field K, whereas the symbol 0 on the right-hand side is the neutral element of the abelian group constituted by E. It is regrettable that the same symbol 0 is often used for two different elements. It is said that we are dealing with a vector space over the field K (which is generally an R, C or Galois field). The elements of E are often called vectors.

Definition A.1

F is a vector subspace of E if

$$\forall x_i, x_j \in F \qquad x_i + x_j \in F$$

$$\alpha x_i \in F \qquad \forall \alpha \in K$$

Definition A.2

A linear combination of vectors is defined by

$$\sum_{i=1}^{i=n} \alpha_i x_i$$

where $\{x_i\} \in E$ and $\{\alpha_i\} \in K$.

Definition A.3

A set of vectors $\{x_i\}$ is linearly independent if

$$\sum \alpha_i x_i = 0 \Rightarrow \text{all } \alpha_i = 0$$

Definition A.4

A set of vectors forms a basis of the set E if every vector of E can be expressed in the form of a unique linear combination of basis vectors. There can be several bases, but it can be proved that they have the same number of vectors. This number is the "dimension" of the vector space.

Example A.3

The set of vectors (in the classical sense of the word) from the same origin in a plane provided with vector addition is a vector space over the field R. Collinear vectors form a vector subspace. Any two noncollinear vectors are independent. Any two noncollinear vectors always form a basis for the space, which therefore has two dimensions.

Example A.4

The set of polynomials of order n provided with classical addition constitutes a vector space over the field R. The set of polynomials of degree $n - 1$ is a subspace of the preceding set.

A.5 Hilbert spaces

Definition A.5

A somewhat simplified definition which is sufficient for our purposes is that a vector space E over the field R (or a subfield of R) is a Hilbert space if it is provided with a scalar product defined as a mapping in R of a pair of vectors of E

$$(x_i, x_j) \in R \qquad \forall x_i, x_j \in E$$

such that the scalar product has the following properties. It is commutative:

$$(x_i, x_j) = (x_j, x_i)$$

It is linear:

$$(\alpha x_i + \beta x_j, x_k) = \alpha(x_i, x_k) + \beta(x_j, x_k)$$

$$\forall \alpha, \beta \in R \text{ and } \forall x_i, x_j, x_k \in E$$

It is such that

$$(x_i, x_i) > 0$$

and

$$(x_i, x_i) = 0 \Leftrightarrow x_i = 0 \qquad \forall x_i \in E$$

All these properties are clearly satisfied by the classical scalar product of two vectors (from the same origin). The set of vectors from the same origin in the same plane therefore constitutes a Hilbert space. By extending the terminology used for classical vectors to all Hilbert spaces, we can make the following statements:

The length of a vector x is equal to $(x, x)^{1/2}$.
Two vectors x_i and x_j are orthogonal if $(x_i, x_j) = 0$.

We find in Hilbert spaces nearly all the properties of classical vectors (from the same origin) and therefore those of the associated matrix calculus.

A.5.1 Some properties of Hilbert spaces

Consider a set of classical vectors with n dimensions (e.g. $n = 3$). We define the vector X by

$$X = \begin{vmatrix} x_1 \\ x_2 \\ x_3 \end{vmatrix}$$

and the vector Y by

$$Y = \begin{vmatrix} y_1 \\ y_2 \\ y_3 \end{vmatrix}$$

where $\{x_1 x_2 x_3\}$ and $\{y_1 y_2 y_3\}$ are the three coordinates of the ends of X and Y in a trirectangular trihedron. The scalar product (X, Y) is defined by

$$(X, Y) = x_1 y_1 + x_2 y_2 + x_3 y_3$$

which can also be written as

$$(X, Y) = X^{\mathrm{T}}.Y$$

where X^{T} is the transpose of matrix Y.

A linear transformation of a vector X to a vector Y can be defined as a transformation represented by

$$Y = M.X$$

where M is a matrix with n (3) rows and n (3) columns. It can also be said that the relationship $Y = MX$ is a linear mapping of (X) to (Y) which can be denoted

$$Y = f(X)$$

It can then be said that this mapping performs a homomorphism of X to Y (it can also be said that the matrix M is the operator of the linear mapping). The transformation is called linear because

$$M.\alpha X = \alpha(M.X) \text{ or } f(\alpha X) = \alpha f(X) \qquad \forall \alpha \in R$$

and because $f(X_1 + X_2) = f(X_1) + f(X_2)$.

For example, the transformation defined by

$$M = \begin{vmatrix} 1 & 0 & 0 \\ 0 & 1 & 0 \\ 0 & 0 & -1 \end{vmatrix}$$

transforms a vector into a symmetric vector with respect to a plane of coordinates (it is therefore an isomorphism).

A linear transformation is called symmetric if

$$(X, f(Y)) = (f(X), Y)$$

in other words if the scalar product of X and the transpose of Y equals the scalar product of Y and the transpose of X, i.e.

$$X^{\mathrm{T}}.M.Y = Y^{\mathrm{T}}.M.X$$

This is equivalent to saying that the matrix M is symmetric ($M^{\mathrm{T}} = M$). This is clearly the case in the example above where the linear transformation is symmetric.

When the linear transformation is symmetric, it is known from matrix calculus that n vectors exist which are preserved in the transformation apart from an elongation, i.e.

$$\begin{aligned} f(X_1) &= \lambda_1 X_1 \\ f(X_2) &= \lambda_2 X_2 \\ &\vdots \\ f(X_n) &= \lambda_n X_n \end{aligned}$$

where $\{\lambda_i\} \in R$. It can be proved that these vectors, which are known as eigenvectors (the λ_i are the associated eigenvalues), are orthogonal to each other, i.e.

$$(X_i, X_j) = 0 \quad \text{if } i \neq j$$

In the above example, the three vectors

$$X_1 = \begin{vmatrix} 1 \\ 0 \\ 0 \end{vmatrix} \quad X_2 = \begin{vmatrix} 0 \\ 1 \\ 0 \end{vmatrix} \quad X_3 = \begin{vmatrix} 0 \\ 0 \\ 1 \end{vmatrix}$$

represent three eigenvectors with $\lambda_1 = 1$, $\lambda_2 = 1$ and $\lambda_3 = -1$. They are truly orthogonal.

It is also possible to find a set of n eigenvectors whose length is equal to unity. For the vector space under consideration, they constitute a "basis of n orthonormal eigenvectors" (ortho because they are perpendicular 2×2 and normal because their length is equal to unity). It is easy to prove that these vectors ($V_1, V_2, \ldots, V_n$) are independent and constitute a basis for the vector space.

Any vector X can be put in the form

$$X = \sum a_i V_i$$

where $a_i = (X, V_i)$ (the scalar product of X and V_i or the projection (orthogonal) of X on V_i).

All the properties given above can be generalized to Hilbert spaces.

A.5.2 Example of a Hilbert space

Consider the set E of functions of time $S(t)$ defined by

$$S(t) = \sum_{i=0}^{i=\varphi-1} U_i F_i(t)$$

where

$$\{U_i\} \in R$$

i and φ are positive integers and

$$F_i(t) \triangleq \frac{(\Delta f)^{1/2} \sin[\pi(t\Delta f - i)]}{\pi(t\Delta f - i)}$$

(the symbol $\triangleq$ means equal by definition). It can be seen that $F_i(t)$ is a maximum for $t = i/\Delta f$, and the value of the maximum is $(\Delta f)^{1/2}$. The Fourier transform $\Phi_i(f)$ of $F_i(t)$ is zero for $|f| > \Delta f/2$ and is given by

$$\Phi_i(f) = \frac{\exp(-2\pi \mathrm{j})(i/\Delta f)f}{(\Delta f)^{1/2}}$$

for $|f| < \Delta f/2$. Therefore

$$|\Phi_i(f)|^2 = \frac{1}{\Delta f} \quad \text{for} \quad |f| < \frac{\Delta f}{2}$$

and it follows from this that (Parseval's theorem)

$$\int_{-\infty}^{+\infty} F_i^2(t)\,\mathrm{d}t = \int_{-\Delta f/2}^{+\Delta f/2} \frac{\mathrm{d}f}{\Delta f} = 1$$

and that (Wiener–Kintchine theorem)

$$\int_{-\infty}^{+\infty} F_i(t)\,F_j(t)\,\mathrm{d}t \equiv \int_{-\infty}^{+\infty} F_i(t)\,F_i\left(t - \frac{i-j}{\Delta f}\right)\mathrm{d}t = \frac{\sin\{\pi\Delta f[(i-j)/\Delta f]\}}{\pi\Delta f[(i-j)/\Delta f]} = 0$$

The set E of functions $S(t)$ which obey classical addition clearly constitutes a group and also a vector space since $\forall\alpha \in R$, $\alpha S(t) \in E$. It is possible to define a scalar product of $S_1(t)$ and $S_2(t)$ by

$$(S_1, S_2) = \int_{-\infty}^{+\infty} S_1(t)\,S_2(t)\,\mathrm{d}t$$

With this scalar product, the set E becomes a Hilbert space. It is clear that the set of vectors $F_1(t)$ is a basis of orthonormal vectors, which implies that these vectors are independent (the coefficients U_i are the "coordinates" of

$S(t)$ in this basis). Following classical procedures, it is possible to define the energy of $S(t)$ by

$$\int_{-\infty}^{+\infty} S^2(t)\,\mathrm{d}t$$

i.e. by the square of the "length" of $S(t)$ which is known to be equal to the sum of the squares of the "coordinates" of $S(t)$. Thus

$$\text{energy of } S(t) \quad = \quad \sum U_i^2$$

A.5.3 Terminology

The statement that the set of vectors with a common origin O in a plane is a vector space with two dimensions can be restated as follows.

A point on a plane (the end of a vector with origin O) depends on two independent parameters, for example its x and y coordinates in a cartesian system (an infinite number of such systems can be derived from one another by rotation about O). However, these two independent parameters can be chosen quite differently; for example in a polar representation they can be chosen as the length ϱ of the vector and the polar angle θ which is assumed to lie between zero and 2π). There is obviously a reciprocal relationship between a pair (x, y) and a pair (ϱ, θ).

Similarly, a function $S(t)$ such as that defined in Section A.5.2 depends on φ independent parameters, for example the φ coordinates U_i of this function in the orthonormal basis (or the orthonormal framework) defined by the φ vectors $F_i(t)$. Generally, an n-dimensional vector in Hilbert space depends on n independent parameters, for example the n scalar products of the vector with each of the vectors of an orthonormal basis of the space. The extension of this is that even if the Hilbert space has an infinite number of dimensions, the scalar products of a vector of this space with each of the vectors of an orthonormal basis of this space are independent.

Appendix B
Table of the function $\Theta(x)$

$$0.5[1 - \Theta(x/2^{1/2})] = \frac{1}{(2\pi)^{1/2}} \int_x^{+\infty} \exp\left(-\frac{v^2}{2}\right) dv$$

x	$0.5[1 - \Theta(x/2^{1/2})]$	x	$0.5[1 - \Theta(x/2^{1/2})]$
0.25	0.40	3.00	1.4×10^{-3}
0.50	0.31	3.09	1.00×10^{-3}
0.52	0.30	3.50	2.3×10^{-4}
0.75	0.226	3.72	1.00×10^{-4}
0.84	0.200	4.00	3.2×10^{-5}
1.00	0.158	4.27	1.00×10^{-5}
1.28	0.100	4.76	1.0×10^{-6}
1.50	0.067	5.00	2.9×10^{-7}
2.00	2.28×10^{-2}	5.62	1.0×10^{-8}
2.33	1.00×10^{-2}	6.00	1.0×10^{-9}
2.50	6.2×10^{-3}	6.36	1.0×10^{-10}

Exercises

1 Detection probabilities

(1) We need to detect a nonfluctuating target using a radar system. At the receiver output we find a gaussian noise with zero mean and a standard deviation of unity in the absence of a target, and a useful signal with amplitude $s = R_0^{1/2}$ plus the noise in the presence of a target at long range. Plot the curve of variations in the detection probability P_d as a function of R_0 for a false alarm probability P_f of 5×10^{-5}.

(2) It is now assumed that the target fluctuates so that the useful signal $s = R^{1/2}$ produced at the output of the receiver has one chance out of three of being zero, one chance out of three of being equal to $R_0^{1/2}$ and one chance out of three of being equal to $(2R_0)^{1/2}$. Plot the curve of the variations in P_d as a function of the mean value of the signal-to-noise ratio for a false alarm probability P_f of 5×10^{-5}.

(3) The total power is now assumed to be transmitted by two transmitters which are identical at all points but operate on different frequencies so that the echo areas presented by the target at these two frequencies are entirely independent. We also have two receivers which produce at their outputs a signal s_1 corresponding to $R/2$ immersed in a gaussian noise with a standard deviation of unity and a signal s_2 corresponding to $R/2$ immersed in a gaussian noise with a standard deviation of unity. The target fluctuates so that the probability distributions of s_1 and s_2 are identical with the distribution considered above, i.e. s_1 and s_2 each have, independently of each other, one chance out of three of being zero, one chance out of three of being equal to $(R_0/2)^{1/2}$ and one chance out of three of being equal to $R_0^{1/2}$. Plot the curve of the detection probability as a function of R_0 for a false alarm probability of 5×10^{-5}, given that the noises produced at the output of each receiver are

independent and that the two signals s_1 and s_2 are added before the threshold is set.

(4) What is the detection probability $P_d(R_0)$ for the same false alarm probability under the same assumptions of target fluctuation if a very large number n of identical transmitter–target–receiver paths are used, the total transmitted power remains the same and the echo areas are independent.

What conclusions can be drawn?

2 Use of different types of radar

A radar operates in C band ($\lambda = 5.5$ cm) with a revolving antenna with a 50 dB gain. We consider the range D_0 of the radar for a detection probability of 90% and a false alarm probability of 10^{-5}. The target has a radar cross-section of 1 m^2 and is assumed not to fluctuate.

(1) It is required that the standard deviation of the error in the measurement of the radar–target distance at the range D_0 is less than 150 m. If the transmitted spectrum is bell-shaped, what is the minimum theoretical value for its width at 3 dB?

(2) What is the order of magnitude of the range resolving power?

(3) The radar is intended for surveillance of a zone of radius 300 km centered at the radar. Its repetition frequency is 200 Hz and the period of rotation of the antenna is 6 s. If the useful signal consists of a single pulse, give the number of false alarms produced at each rotation of the antenna.

(4) There is no intentional jamming. The noise figure of the receiving system is 10 dB. Transmission, reception and atmospheric losses are neglected. What is the minimum value of the energy which must be transmitted during the measurement in order to obtain a range of 300 km?

(5) If the transmitted useful signal consists of a single pulse which is not frequency modulated and the error in the range measurement is ten times larger than the theoretical error, give the order of magnitude of the duration of the pulse which must be transmitted in order to obtain the same accuracy for the range measurement. What is the corresponding range resolution (assume the theoretical formula)? What is the peak power of the transmission

signal? What is the width of the pass band of the FM amplifiers of the receiver?

(6) It is now assumed that the target carries a jammer which transmits to the radar a jamming power of 100 W uniformly distributed in a 500 MHz band by means of an antenna with a gain of 3 dB. What value of the peak power must be transmitted with a pulse of the same duration as in (5) in order to obtain a range of 300 km? What is the value of the signal-to-noise ratio corresponding to a target at the limit of the range at the output of the FM amplifier if the image frequency is filtered out? By how many decibels would this value be reduced if the image frequency was not filtered out? Why is a greater difference often found in practice?

(7) It is now assumed that the radar uses pulse compression, and thus transmits a pulse 100 times longer than that defined in (5) (linear modulation of the transmitted frequency). Plot the axis of the ambiguity diagram (in the distance–radial velocity plane). What is the peak power transmitted under the conditions of (4) and (6)? What is the signal-to-noise ratio before and after the compressor filter for a target at a range of 300 km.

(8) In the remainder of this problem it is assumed that the radar is a correlation radar using a transmission signal phase modulated randomly by using a signal with a given frequency whose phase changes randomly by zero or π (equal probability of zero or π) at successive time intervals θ. Calculate the width of the spectrum of such a signal. In order to do this, we determine the following: (a) the autocorrelation function of a signal with constant amplitude whose sign changes or does not change randomly every θ seconds; (b) the Fourier transform of this autocorrelation function (spectrum)†. It is assumed that the signal transmitted by the radar has a spectrum which is simply shifted in frequency with respect to the preceding signal. What value must we take for θ in order to obtain the required range accuracy (150 m)? What is the range resolution of this radar?

(9) What is the minimum duration of the useful signal if the radial velocity resolution of the radar is 25 m s^{-1}? This value will be adopted for the duration of the transmitted signal. In this case, the ambiguity diagram is almost an ellipse whose major axes coincide with the coordinate axes. Draw this diagram. What peak powers should be transmitted in the presence of the jamming specified in (6)?

† If $f(t) = f(-t)$ and if $f(\infty) = 0$, the Fourier transform of $f(t)$ is equal to twice the real part of $F(j\omega)$ if $F(p)$ is the Laplace transform of $f(t)\,u(t)$, where $u(t)$ is the unit level.

(10) The integration time of the correlators is assumed to be equal to the duration of the transmitted signal. What is the signal-to-noise ratio before and after the correlator for a target at 300 km?

(11) If the same antenna is used at transmission and reception, what maximum value can be accepted for the duty cycle of the radar (ratio of mean power to peak power) (the recovery time of the duplexer is assumed to be negligible). What is the corresponding mean transmitted power if this value is used? How many decibels will be lost on targets situated at 150 km, 75 km and 30 km? Plot the curve giving the range error made with this radar as a function of range when it varies between zero and 300 km (under the jamming conditions defined in (6)).

3 Determination of the coverage of a territory by radar

(1) Problem to be solved

It is assumed that a territory runs the risk of being attacked by about 100 enemy airplanes capable of flying at altitudes between 4000 and 20 000 m. These enemy airplanes carry jammers which attempt to interfere with the detection network installed. In order to detect these airplanes, it is planned to instal a radar infrastructure on the territory to be protected. This must provide total coverage of the territory between 4000 and 20 000 m in the presence of jamming. Different solutions can be envisaged in order to arrive at this result: in particular, the range of the selected radar will determine the total number of radar systems to be installed and thus the total cost of the operation. It is possible to envisage a large number of radar systems with low power or a few with high power. Different types of radar can also be envisaged (pulse radar or pulse compression radar). Finally, the problem is to select the optimum infrastructure for providing coverage, i.e. the least expensive solution which provides protection for the territory.

(2) Technical and operational characteristics

Characteristics of the enemy raid

It is estimated that there are 100 enemy airplanes with a regular spacing of 20 km between them (one airplane in 400 km^2). Their velocity varies between Mach 0.6 and Mach 2. Their mean radar cross-section is assumed to be 10 m^2. The airplanes are flying in raid formation at a maximum altitude of 20 000 m. In the raid, jammer and silent airplanes can be found in any proportion and a limited number of the two types of airplanes can be found at low altitude. These last two assumptions are intended to simplify the calculations.

Threat of jamming

The airplanes flying at high altitude may be carrying jammers with a mean power of 200 W per airplane. The jammer transmits by means of an omnidirectional antenna (seen from the radar) whose gain is 3 dB. The jamming is assumed to be noise with constant spectral density. The frequency band jammed by each plane is matched to the frequency band of the radars.

Information to be provided by the radars

All the jamming or silent targets must be detected with a detection probability of 90% and a false alarm probability of 10^{-3}. The accuracy required is ± 1 km, two targets separated by 5 km must be discriminated, and the information renewal rate should be 6 per minute. It is not necessary to measure the altitude. A vertical "coverage hole" with a radius of approximately 35 km (angle of sight of 30°) is allowed at every radar location.

Technical limitations

There are the following restrictions on the antennas: a maximum of two antennas per radar; a maximum area of 50 m^2 per antenna; antenna length should be less than 12 m and antenna height should be less than 6 m; a maximum of 10 beams in a multibeam antenna; a maximum of two tubes can be placed in parallel under an antenna; a loss of at least 2 dB on transmission (in the microwave circuitry) and 2 dB in the data processing in the presence of jamming (receivers with constant false alarm) should be allowed for in the calculations; the power tubes are as specified in (4) below; the antenna sidelobes cannot be assumed to be less than 35 dB below the main lobe.

(3) Feasibility study

The feasibility study should provide the technical characteristics of the chosen solution. In particular, the following should be specified.

(a) antenna characteristics: dimensions, number of beams and the angular dimensions of each beam.

(b) transmitter characteristics: tube(s) used and mode of use.

(c) radar performance: coverage without jamming and coverage in the presence of the threat of jamming. In the latter case an analysis will be made of the range of the jamming airplanes at high altitude and the range of the low altitude airplanes (in the presence of jamming). Jamming penetrating via the sidelobes should be taken into account.

(d) Financial considerations: the reasons for the choice of each parameter must be given.

(4) Additional information

Tubes

Two types of tube can be used: tube A with a maximum peak power of 10 MW and a mean power of 5 kW or tube B with a maximum peak power of 1 MW and a mean power of 20 kW. These tubes cover the frequency band 4500–5000 MHz.

Formulae for determining the cost

In order to determine the cost of the selected solutions, the following empirical formulae are used where S is the total area of the antenna(s) (in m^2), W is the total mean power of the radar (in kW), N is the number of beams (in stacked-beam arrangement) and P is the cost (in millions of dollars). For pulse radars

$$P = 1 + 0.10W + (0.015S + 0.13N)(0.8 + 0.06W)$$

For pulse compression radars

$$P = 1 + 0.05W + (0.015S + 0.13N)(0.8 + 0.02W)$$

(5) Implementation of the study

Examination of the problem shows that the radar antenna should be specified as quickly as possible. The antenna is essentially characterized by the geometric conditions for the accuracy and coverage required.

Characterization of the antenna

Examination of radar coverage diagrams shows that the radar does not need a range (in the presence of jamming) of more than 200 km to provide coverage at 4000 m. Three cases can therefore be considered: (a) radars with a range of 200 km, (b) radars with a range of 140 km and (c) radars with a range of 100 km. Determine the antenna dimensions and the number of beams required for each case.

Characterization of the radar

Either a pulse radar or a pulse compression radar can be used for each of the above cases (i.e. either tube A or tube B can be used). Determine the power necessary for each of these options for cases (a), (b) and (c). When calculating this power, it is important to remember to check the sidelobe jamming and the range of the silent airplanes if necessary.

Choice of the solution

Estimate the cost of each option using the cost formulae given above. The choice of solution is made by trading off the cost against the complexity.

References

1 Angot, A. *Ancillary Mathematics for the Use of Electronic and Telecommunication Engineers*, Editions de la Revue d'Optique, Paris, 4th ed., 1961.

2 Woodward, P.M. *Probability and Information Theory, with Applications to Radar*, Pergamon Press, Oxford, 1953.

3 Carpentier, M.H. Le filtrage dans le détection. *Onde Electrique*, **51**, 827–836, November 1971.

4 Gamow, G. *Mr Tompkins in Wonderland*, Cambridge University Press, Cambridge, 1939.

5 Guilhen, R. and Carpentier, M.H. *French Patent 176949*, 6 December 1968.

6 Guilhen, R. and Carpentier, M.H. *US Patent 3,657,738*, 1969.

7 Chin, I.E. and Cook, C.E. The mathematics of pulse compression, *Sperry Engineering Review*, **12**(3), 11–16, 1959.

8 Carpentier, M.H. and Adamsbaum, A. *French Patent 1,313,075*, 1959; *US Patent 3,123,452*, 1960.

9 Key, E.L., Fowle, E.N. and Haggarty, R.D. A method of designing signals of large time–bandwidth products, presented at the *IRE National Convention*, 22 March 1961.

10 Klander, J.R., Price, A.C., Darlington, S. and Albersheim, W.J. The theory and design of chirp radars, *Bell System Technical Journal*, **39**(4), 745–808, 1960.

11 Cook, C.E. Transmitter phase modulation errors and pulse compression waveform distortion, *Microwave Journal*, **6**(5), 63–69, 1963

12 Skolnik, M.I. *Introduction to Radar Systems*, McGraw-Hill, New York, p. 141, 1962.

13 Drabovitch, S. Applications aux antennes de la théorie du signal, *Onde Electrique*, **14**, 458, May 1965.

14 Carpentier, M.H. *Radars: Theories Modernes*, Dunod, Paris 1962; *Radars: Concepts Nouveaux*, Dunod, Paris, 1966; *Radars: Bases Modernes*, Masson, Paris, 1977, 1981.

Bibliography

Carpentier, M.H. Le filtrage dans la détection, *Onde Electrique*, 827–836, **51**, November 1971.

Drabovitch, S. Applications aux antennes de la théorie du signal, *Onde Electrique*, **14**, 458, May 1965.

Hall, W.M. Pulse radar performance, *Proceedings of the IRE*, **44**, 224, 1956.

Key, E.L., Fowle, E.N. and Haggarty, R.D. A method of designing signals of large time-bandwidth products, presented at the *IRE National Convention*, 22 March 1961.

Klauder, J.R. Design of high-resolution radar signals, *Bell System Technical Journal*, **39**(4), 809–820, 1960.

Lawson, J.L. and Uhlenbeck, G.E. *Threshold signals*, MIT Radiation Laboratory Series, **24**, McGraw-Hill, New York, 1950.

Marcum, J.I. *A Statistical Theory of Target Detection by Pulsed Radar*, RAND Corporation Reports, RM–754, 1947; RM–753, 1948.

Oswald, J. Sur la limitation spectrale et la fréquence instantanée des signaux, *IRE Transactions on Information Theory*, September 1962.

Penin, F. *Cours du Radar*, Ecole Nationale Supérieure de l'Aéronautique, Toulouse, 1970.

Rice, S.O. Statistical properties of a sine wave plus random noise, *Bell System Technical Journal*, **27**, 109–157, 1948.

Ridenour, L.N. (editor) *Radar system engineering*, MIT Radiation Laboratory Series, **1**, McGraw-Hill, New York, 1947.

Shannon, C.E. A mathematical theory of communications, *Bell System Technical Journal*, **27**, 379–423, 623–656, July 1948.

Skenderoff, C. Rôle de la fonction d'ambiguité dans les radars, *Onde Electrique*, May 1965.

Thourel, L. *Cours Professé à l'ENSA*, Ecole Nationale Supérieure de l'Aéronautique, Toulouse.

Turin, G.L. An introduction to matched filters, *IRE Transactions on Information Theory*, **6**, 311, 1960.

Ville, J. Theory and applications of the notion of complex signals, *Câbles et Transmission*, **2**, 61–74, 1948 (in French). (English translation: *Rand Report*, **T-92**, August 1958.)

Woodward, P.M. *Probability and Information Theory, with Applications to Radar*, Pergamon Press, Oxford, 1953.

Index

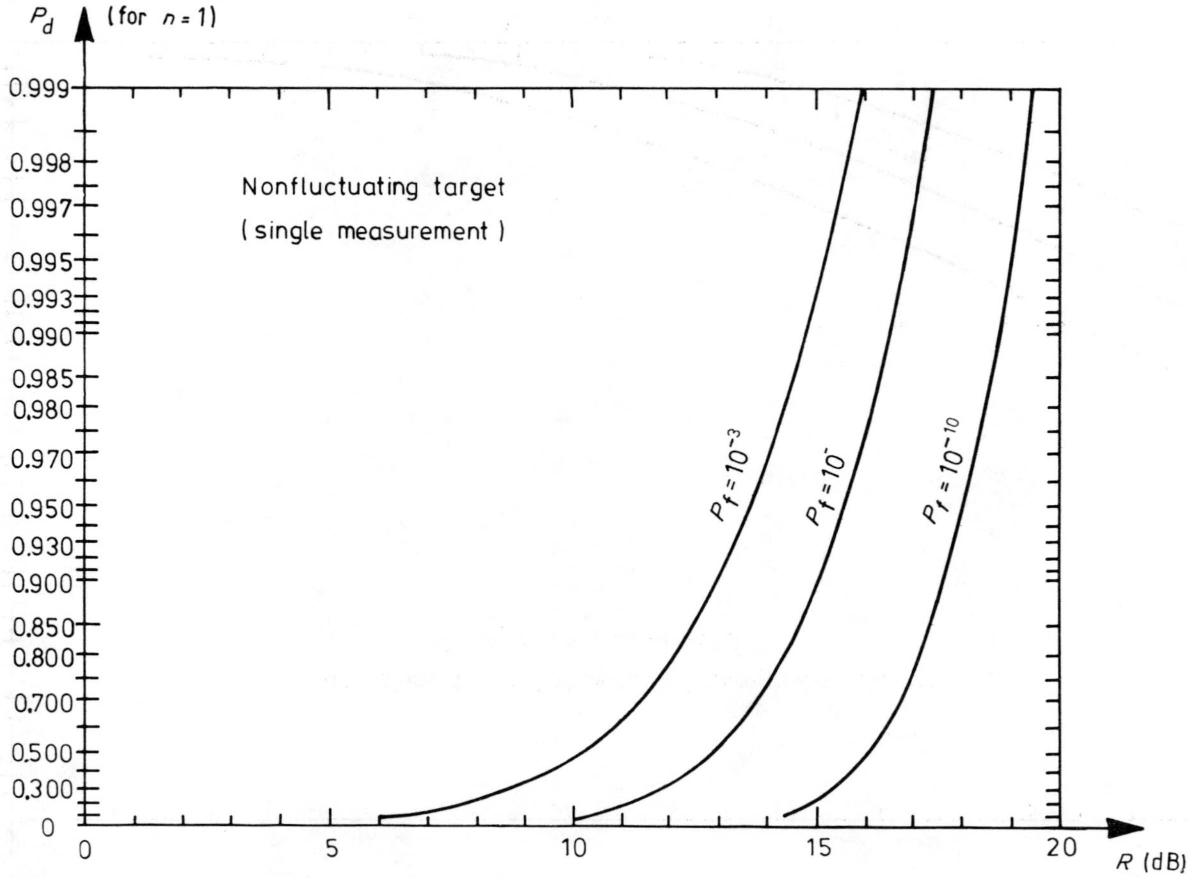
P_d (for $n = 1$)
0.999
0.998
0.997
0.995
0.993
0.990
0.985
0.980
0.970
0.950
0.930
0.900
0.850
0.800
0.700
0.500
0.300
0
Nonfluctuating target
(single measurement)
$P_f = 10^{-3}$
$P_f = 10$
$P_f = 10^{-10}$
0
5
10
15
20
R (dB)

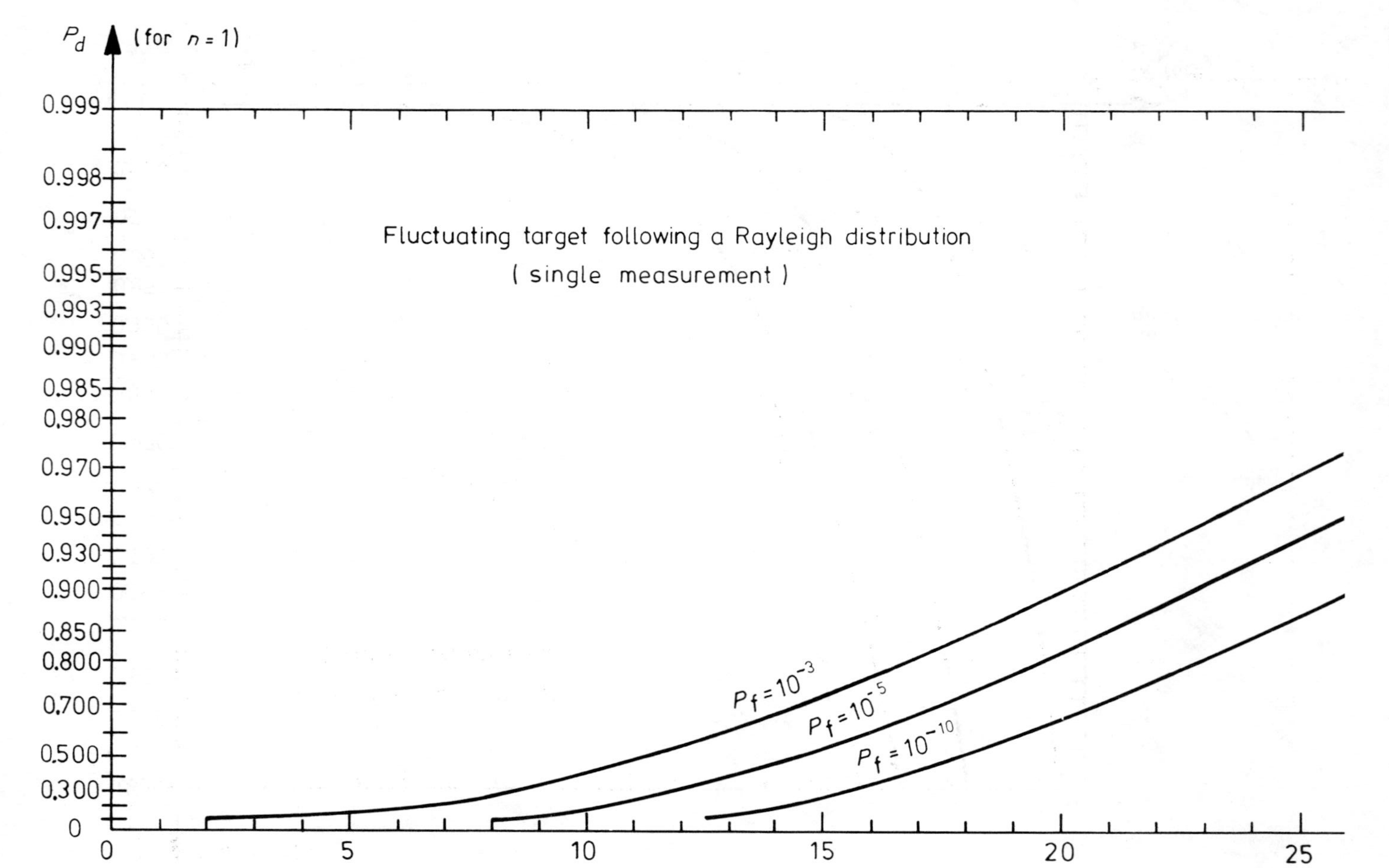

P_d (for n = 1)
0.999
0.998
0.997
0.995
0.993
0.990
0.985
0.980
0.970
0.950
0.930
0.900
0.850
0.800
0.700
0.500
0.300
0
Fluctuating target following a Rayleigh distribution
(single measurement)
P_f = 10^-3
P_f = 10^-5
P_f = 10^-10
0
5
10
15
20
25

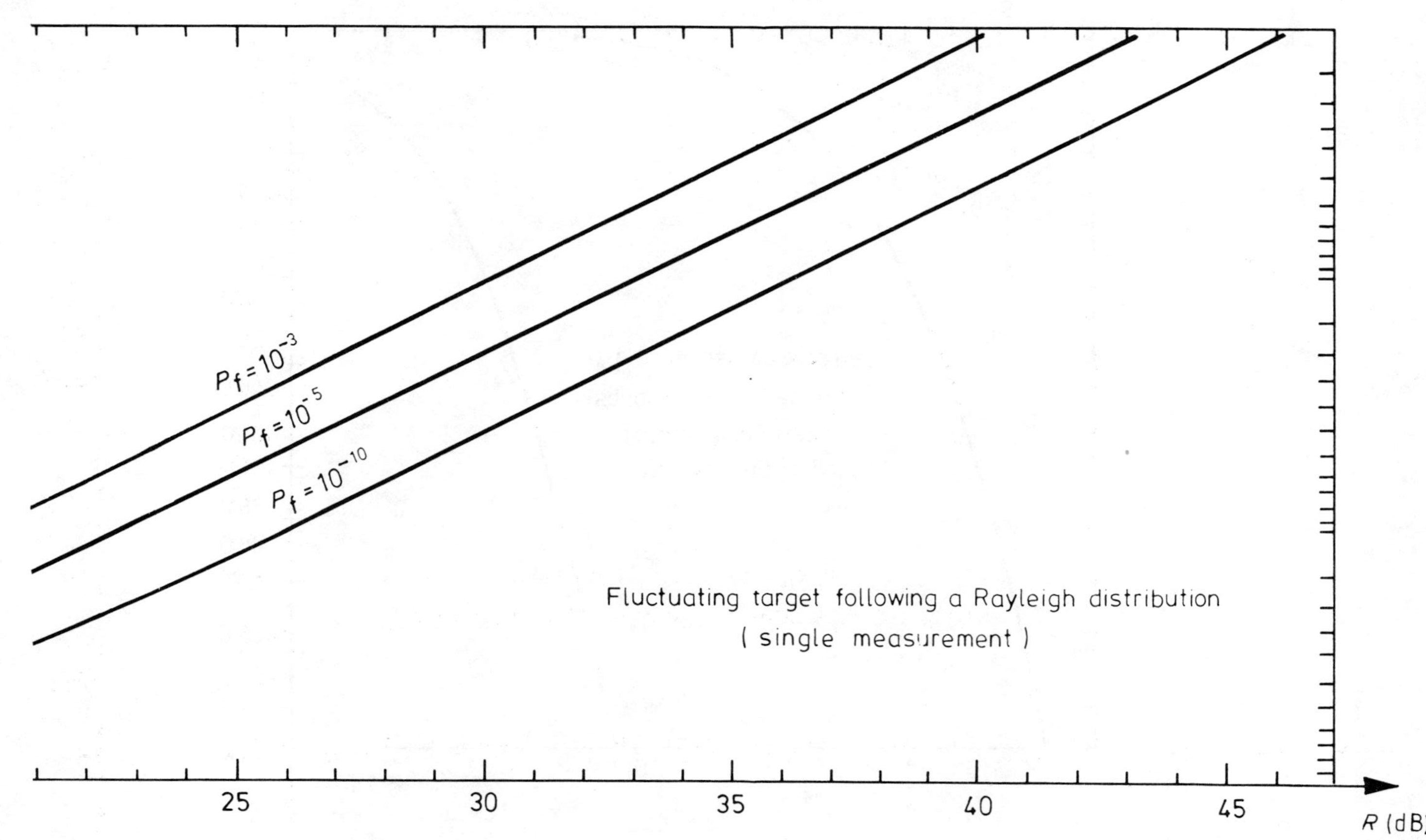

Pf = 10-3
Pf = 10-5
Pf = 10-10
Fluctuating target following a Rayleigh distribution
(single measurement)
25
30
35
40
45
R (dB)

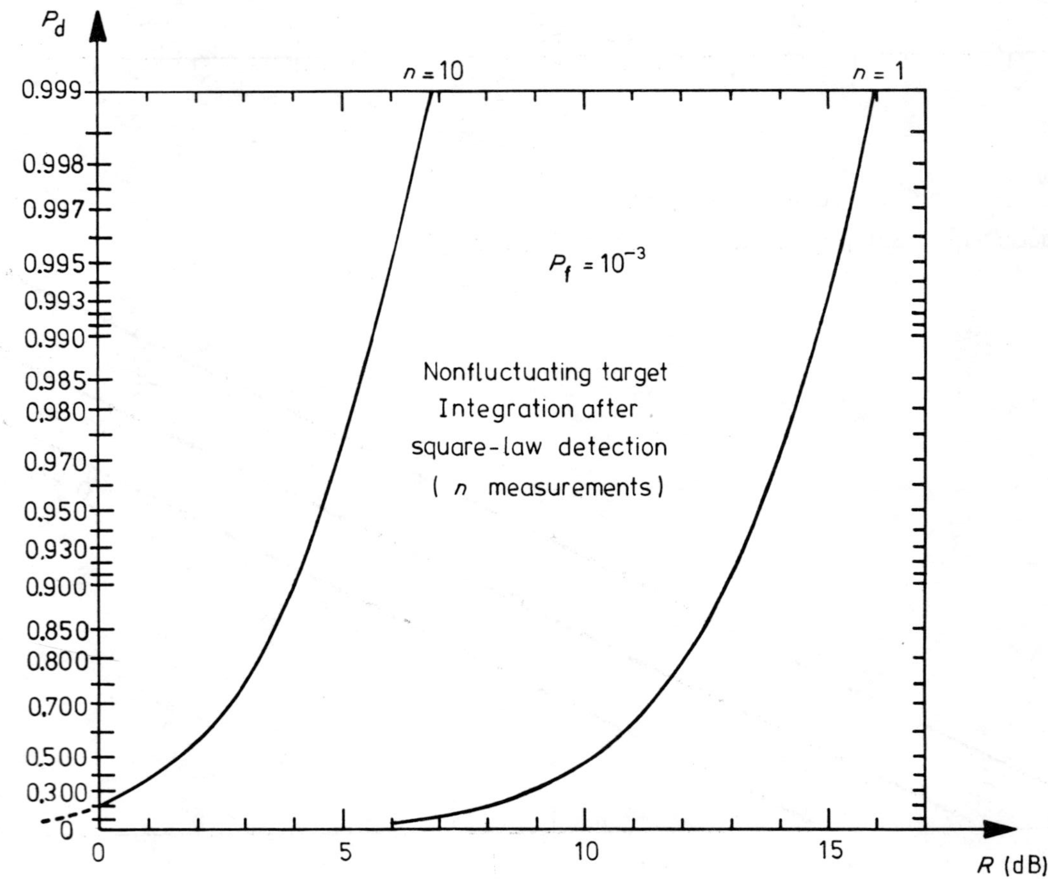

Pd
0.999
0.998
0.997
0.995
0.993
0.990
0.985
0.980
0.970
0.950
0.930
0.900
0.850
0.800
0.700
0.500
0.300
0
n = 10
n = 1
Pf = 10^-3
Nonfluctuating target
Integration after
square-law detection
(n measurements)
0
5
10
15
R (dB)